T0073049

Text Data Mining

Chengqing Zong • Rui Xia • Jiajun Zhang

Text Data Mining

Springer

Chengqing Zong
Institute of Automation
Chinese Academy of Sciences
Beijing, Beijing, China

Rui Xia
School of Computer Science & Engineering
Nanjing University of Science
and Technology
Nanjing, Jiangsu, China

Jiajun Zhang
Institute of Automation
Chinese Academy of Sciences
Beijing, Beijing, China

ISBN 978-981-16-0099-9 ISBN 978-981-16-0100-2 (eBook)
https://doi.org/10.1007/978-981-16-0100-2

Jointly published with Tsinghua University Press
The print edition is not for sale in China (Mainland). Customers from China (Mainland) please order the print book from: Tsinghua University Press.

© Tsinghua University Press 2021
This work is subject to copyright. All rights are reserved by the Publisher, whether the whole or part of the material is concerned, specifically the rights of translation, reprinting, reuse of illustrations, recitation, broadcasting, reproduction on microfilms or in any other physical way, and transmission or information storage and retrieval, electronic adaptation, computer software, or by similar or dissimilar methodology now known or hereafter developed.
The use of general descriptive names, registered names, trademarks, service marks, etc. in this publication does not imply, even in the absence of a specific statement, that such names are exempt from the relevant protective laws and regulations and therefore free for general use.
The publisher, the authors, and the editors are safe to assume that the advice and information in this book are believed to be true and accurate at the date of publication. Neither the publisher nor the authors or the editors give a warranty, expressed or implied, with respect to the material contained herein or for any errors or omissions that may have been made. The publisher remains neutral with regard to jurisdictional claims in published maps and institutional affiliations.

This Springer imprint is published by the registered company Springer Nature Singapore Pte Ltd.
The registered company address is: 152 Beach Road, #21-01/04 Gateway East, Singapore 189721, Singapore

Foreword

We are living in the Big Data era. Over 80% of real-world data are unstructured in the form of natural language text, such as books, news reports, research articles, social media messages, and webpages. Although data mining and machine learning have been popular in data analysis, most data mining methods handle only structured or semistructured data. In comparison with mining structured data, mining unstructured text data is more challenging but will also play a more essential role in turning massive data into structured knowledge. It is no wonder we have witnessed such a dramatic upsurge in the research on text mining and natural language processing and their applications in recent years.

Text mining is a confluence of natural language processing, data mining, machine learning, and statistics used to mine knowledge from unstructured text. There have already been multiple textbooks dedicated to data mining, machine learning, statistics, and natural language processing. However, we seriously lack textbooks on text mining that systematically introduce important topics and up-to-date methods for text mining. This book, "Text Data Mining," bridges this gap nicely. It is the first textbook, and a brilliant one, on text data mining that not only introduces foundational issues but also offers comprehensive and state-of-the-art coverage of the important and ongoing research themes on text mining. The in-depth treatment of a wide spectrum of text mining themes and clear introduction to the state-of-the-art deep learning methods for text mining make the book unique, timely, and authoritative. It is a great textbook for graduate students as well as a valuable handbook for practitioners working on text mining, natural language processing, data mining, and machine learning and their applications. This book is written by three pioneering researchers and highly reputed experts in the fields of natural language processing and text mining. The first author has written an authoritative and popular textbook on natural language processing that has been adopted as a standard textbook for university undergraduate and first-year graduate students in China. However, this new text mining book has completely different coverage from his NLP textbook and offers new and complementary text mining themes. Both books can be studied independently, although I would strongly encourage students working on NLP and text mining to study both.

This text mining book starts with text preprocessing, including both English and Chinese text preprocessing, and proceeds to text representation, covering the vector space model and distributed representation of words, phrases, sentences, and documents, in both statistical modeling and deep learning models. It then introduces feature selection methods, statistical learning methods, and deep neural network methods, including multilayer feed-forward neural networks, convolutional neural networks, and recurrent neural networks, for document classification. It next proceeds to text clustering, covering sample and cluster similarities, various clustering methods, and clustering evaluation. After introducing the fundamental theories and methods of text mining, the book uses five chapters to cover a wide spectrum of text mining applications, including topic modeling (which is also treated as a fundamental issue from some viewpoints but can be used independently), sentiment analysis and opinion mining, topic detection and tracking, information extraction, and automated document summarization. These themes are active research frontiers in text mining and are covered comprehensively and thoroughly, with a good balance between classical methods and recent developments, including deep learning methods.

As a data mining researcher, I have recently been deeply involved in text mining due to the need to handle the large scale of real-world data. I could not find a good text mining textbook written in English or Chinese to learn and teach. It is exciting to see that this book provides such a comprehensive and cutting-edge introduction. I believe this book will benefit data science researchers, graduate students, and those who want to include text mining in practical applications. I loved reading this book and recommend it highly to everyone who wants to learn text mining!

Jiawei Han

ACM Fellow and IEEE Fellow
Michael Aiken Chair Professor
Department of Computer Science
University of Illinois at Urbana-Champaign
Champaign, IL, USA

Preface

With the rapid development and popularization of Internet and mobile communication technologies, text data mining has attracted much attention. In particular, with the wide use of new technologies such as cloud computing, big data, and deep learning, text mining has begun playing an increasingly important role in many application fields, such as opinion mining and medical and financial data analysis, showing broad application prospects.

Although I was supervising graduate students studying text classification and automatic summarization more than ten years ago, I did not have a clear understanding of the overall concept of text data mining and only regarded the research topics as specific applications of natural language processing. Professor Jiawei Han's book *Data Mining: Concepts and Technology*, published by Elsevier, Professor Bing Liu's *Web Data Mining*, published by Springer, and other books have greatly benefited me. Every time I listen to their talks and discuss these topics with them face to face, I have benefited immensely. I was inspired to write this book for the course Text Data Mining, which I was invited to teach to graduates of the University of Chinese Academy of Sciences. At the end of 2015, I accepted the invitation and began to prepare the content design and selection of materials for the course. I had to study a large number of related papers, books, and other materials and began to seriously think of the rich connotation and extension of the term Text Data Mining. After more than a year's study, I started to compile the courseware. With teaching practice, the outline of the concept has gradually formed.

Rui Xia and Jiajun Zhang, two talented young people, helped me materialize my original writing plan. Rui Xia received his master's degree in 2007 and was admitted to the Institute of Automation, Chinese Academy of Sciences, and studied for Ph.D. degree under my supervision. He was engaged in sentiment classification and took it as the research topic of his Ph.D. dissertation. After he received his Ph.D. degree in 2011, his interests extended to opinion mining, text clustering and classification, topic modeling, event detection and tracking, and other related topics. He has published a series of influential papers in the field of sentiment analysis and opinion mining. He received the ACL 2019 outstanding paper award, and his paper on ensemble learning for sentiment classification has been cited more than

600 times. Jiajun Zhang joined our institute after he graduated from university in 2006 and studied in my group in pursuit of his Ph.D. degree. He mainly engaged in machine translation research, but he performed well in many research topics, such as multilanguage automatic summarization, information extraction, and human–computer dialogue systems. Since 2016, he has been teaching some parts of the course on Natural Language Processing in cooperation with me, such as machine translation, automatic summarization, and text classification, at the University of Chinese Academy of Sciences; this course is very popular with students. With the solid theoretical foundation of these two talents and their keen scientific insights, I am gratified that many cutting-edge technical methods and research results could be verified and practiced and included in this book.

From early 2016 to June 2019, when the Chinese version of this book was published, it took more than three years. In these three years, most holidays, weekends, and other spare times of ours were devoted to the writing of this book. It was really suffering to endure the numerous modifications or even rewriting, but we were also very happy. We started to translate the Chinese version into English in the second half of 2019. Some more recent topics, including BERT (bidirectional encoder representations from transformers), have been added to the English version.

As a cross domain of natural language processing and machine learning, text data mining faces the double challenges of the two domains and has broad application to the Internet and equipment for mobile communication. The topics and techniques presented in this book are all the technical foundations needed to develop such practical systems and have attracted much attention in recent years. It is hoped that this book will provide a comprehensive understanding for students, professors, and researchers in related areas. However, I must admit that due to the limitation of the authors' ability and breadth of knowledge, as well as the lack of time and energy, there must be some omissions or mistakes in the book. We will be very grateful if readers provide criticism, corrections, and any suggestions.

Beijing, China
20 May 2020

Chengqing Zong

Acknowledgments

During the writing process and after the completion of the first draft of the Chinese version of this book, many experts from related fields reviewed selected chapters and gave us valuable comments and suggestions. They are (in alphabetical order) Xianpei Han, Yu Hong, Shoushan Li, Kang Liu, Xiaojun Wan, Kang Xu, Chengzhi Zhang, and Xin Zhao. In addition, we also received help from several graduate students and Ph.D. candidates (also in alphabetical order): Hongjie Cai, Zixiang Ding, Huihui He, Xiao Jin, Junjie Li, Mei Li, Yuchen Liu, Cong Ma, Liqun Ma, Xiangqing Shen, Jingyuan Sun, Fanfan Wang, Leyi Wang, Qain Wang, Weikang Wang, Yining Wang, Kaizhou Xuan, Shiliang Zheng, and Long Zhou. They helped us to double check and confirm English expressions, references, and web addresses and to redraw several charts in the book, which saved us much time. We would like to express our heartfelt thanks to all of them!

We also want to sincerely thank Professor Jiawei Han for his guidance and suggestions on this book. We are honored that he was willing to write the foreword to this book despite his busy schedule. Finally, we would like to recognize Ms. Hui Xue and Qian Wang, Tsinghua University Press, and Ms. Celine Chang, and Ms. Suraj Kumar, Springer, for their great help!

Beijing, China
Nanjing, China
Beijing, China
20 May 2020

Chengqing Zong
Rui Xia
Jiajun Zhang

Contents

About the Authors

Chengqing Zong is a Professor at the National Laboratory of Pattern Recognition (NLPR), Institute of Automation, Chinese Academy of Sciences (CASIA) and an adjunct professor in the School of Artificial Intelligence at University of Chinese Academy of Sciences (UCAS). He authored the book "Statistical Natural Language Processing" (which is in Chinese and sold more than 32K copies) and has published more than 200 papers on machine translation, natural language processing, and cognitive linguistics. He served as the chair for numerous prestigious conferences, such as ACL, COLING, AAAI, and IJCAI, and has served as an associate editor for journals, such as ACM TALLIP and ACTA Automatic Sinica, and as an editorial board member for journals, including IEEE Intelligent Systems, Journal of Computer Science and Technology, and Machine Translation. He is currently the President of the Asian Federation of Natural Language Processing (AFNLP) and a member of the International Committee on Computational Linguistics (ICCL).

Rui Xia is a Professor at the School of Computer Science and Engineering, Nanjing University of Science and Technology, China. He has published more than 50 papers in high-quality journals and top-tiered conferences in the field of natural language processing and text data mining. He serves as area chair and senior program committee member for several top conferences, such as EMNLP, COLING, IJCAI, and AAAI. He received the outstanding paper award of ACL 2019, and the Distinguished Young Scholar award from the Natural Science Foundation of Jiangsu Province, China in 2020.

Jiajun Zhang is a Professor at the NLPR, CASIA and an adjunct professor in the SAIU of UCAS. He has published more than 80 conference papers and journal articles on natural language processing and text mining and received 5 best paper awards. He served as the area chair or on the senior program committees for several top conferences, such as ACL, EMNLP, COLING, AAAI, and IJCAI. He is the deputy director of China's Machine Translation Technical Committee of the Chinese Information Processing Society of China. He received the Qian Wei-Chang Science and Technology Award of Chinese Information Processing and the CIPS Hanvon Youth Innovation Award. He was supported by the Elite Scientists Sponsorship Program of China Association for Science and Technology (CAST).

Acronyms

ACE	Automatic content extraction
AMR	Abstract meaning representation
ATT	Adaptive topic tracking
AUC	Area under the ROC curve
Bagging	Bootstrap aggregating
BERT	Bidirectional encoder representations from transformer
BFGS	Broyden–Fletcher–Goldfarb–Shanno
Bi-LSTM	Bidirectional long short-term memory
BIO	Begin–inside–outside
BLEU	Bilingual evaluation understudy
BOW	Bag of words
BP	Back-propagation
BPTS	Back-propagation through structure
BPTT	Back-propagation through time
BRAE	Bilingually constrained recursive autoencoder
CBOW	Continuous bag-of-words
CFG	Context-free grammar
CNN	Convolutional neural network
CRF	Conditional random field
C&W	Collobert and Weston
CWS	Chinese word segmentation
DBSCAN	Density-based spatial clustering of applications with noise
DF	Document frequency
DL	Deep learning
DMN	Deep memory network
ELMo	Embeddings from language models
EM	Expectation maximization
EP	Expectation propagation
FAR	False alarm rate
FNN	Feed-forward neural network
GPT	Generative pretraining

GRU	Gated recurrent unit
HAC	Hierarchical agglomerative clustering
HMM	Hidden Markov model
HTD	Hierarchical topic detection
ICA	Independent component analysis
IDF	Inverse document frequency
IE	Information extraction
IG	Information gain
KBP	Knowledge base population
KDD	Knowledge discovery in databases
KKT	Karush–Kuhn–Tucker
K-L	Kullback–Leibler
L-BFGS	Limited-memory Broyden–Fletcher–Goldfarb–Shanno
LDA	Latent Dirichlet allocation
LM	Language model
LSA	Latent semantic analysis
LSI	Latent semantic indexing
LSTM	Long short-term memory
MCMC	Markov chain Monte Carlo
MDR	Missed detection rate
ME	Maximum entropy
MI	Mutual information
ML	Machine learning
MLE	Maximum likelihood estimation
MMR	Maximum marginal relevance
MUC	Message understanding conference
NB	Naïve Bayes
NCE	Noise contrastive estimation
NED	New event detection
NER	Named entity recognition
NLP	Natural language processing
NNLM	Neural network language model
PCA	Principal component analysis
PLSA	Probabilistic latent semantic analysis
PLSI	Probabilistic latent semantic indexing
PMI	Pointwise mutual information
P-R	Precision–recall
PU	Positive-unlabeled
PCFG	Probabilistic context-free grammar
PMI-IR	Pointwise mutual information—information retrieval
POS	Part of speech
PV-DBOW	Distributed bag-of-words version of the paragraph vector
PV-DM	Paragraph vector with sentence as distributed memory
Q&A	Question and answering
RAE	Recursive autoencoder

RecurNN	Recursive neural network
RG	Referent graph
RNN	Recurrent neural network
ROC	Receiver operating characteristic
ROUGE	Recall-oriented understudy for gisting evaluation
RTD	Retrospective topic detection
SCL	Structure correspondence learning
SCU	Summary content unit
SMO	Sequential minimal optimization
SO	Semantic orientation
SRL	Semantic role labeling
SS	Story segmentation
SST	Stanford sentiment treebank
STC	Suffix tree clustering
SVD	Singular value decomposition
SVM	Support vector machine
TAC	Text analysis conference
TD	Topic detection
TDT	Topic detection and tracking
TF	Term frequency
TF-IDF	Term frequency—inverted document frequency
UGC	User-generated context
UniLM	Unified pretraining language model
VBEM	Variational Bayes expectation maximization
VSM	Vector space model
WCSS	Within-cluster sum of squares
WSD	Word sense disambiguation

Chapter 1
Introduction

1.1 The Basic Concepts

Compared with generalized data mining technology, beyond analyzing various document formats (such as doc/docx files, PDF files, and HTML files), the greatest challenge in text data mining lies in the analysis and modeling of unstructured natural language text content. Two aspects need to be emphasized here: first, text content is almost always unstructured, unlike databases and data warehouses, which are structured; second, text content is described by natural language, not purely by data, and other non-text formats such as graphics and images are not considered. Of course, it is normal for a document to contain tables and figures, but the main body in such documents is text. Therefore, text data mining is de facto an integrated technology of natural language processing (NLP), pattern classification, and machine learning (ML).

The so-called mining usually has the meanings of "discovery, search, induction and refinement." Since discovery and refinement are necessary, the target results being sought are often not obvious but hidden and concealed in the text or cannot be found and summarized in a large range. The adjectives "hidden" and "concealed" noted here refer not only to computer systems but also human users. However, in either case, from the user's point of view, the hope is that the system can directly provide answers and conclusions to the questions of interest, instead of delivering numerous possible search results for the input keywords and leaving users to analyze and find the required answers themselves as in the traditional retrieval system. Roughly speaking, text mining can be classified into two types. In the first, the user's questions are very clear and specific, but they do not know the answer to the questions. For example, users want to determine what kind of relationship someone has with some organizations from many text sources. The other situation is when the user only knows the general aim but does not have specific and definite questions. For example, medical personnel may hope to determine the regularity of some diseases and the related factors from many case records. In this case, they may not be referring to a specific disease or specific factors, and the relevant data in their

© Tsinghua University Press 2021

C. Zong et al., *Text Data Mining*, https://doi.org/10.1007/978-981-16-0100-2_1

entirety need to be mined automatically by system. Certainly, there is sometimes no obvious boundary between the two types.

Text mining technology has very important applications in many fields, such as the national economy, social management, information services, and national security. The market demand is huge. For example, government departments and management can timely and accurately investigate the people's will and understand public opinions by analyzing and mining microblogs, WeChat, SMSs (short message services), and other network information for ordinary people. In the field of finance or commerce, through the in-depth excavation and analysis of extensive written material, such as news reports, financial reports, and online reviews, text mining can predict the economic situation and stock market trends for a certain period. Electronic product enterprises can acquire and evaluate their product users or market reactions at any time and capture data support for further improving product quality and providing personalized services. For national security and public security departments, text data mining technology is a useful tool for the timely discovery of social instability factors and effectively controlling the current situation. In the field of medicine and public health, many phenomena, regularities, and conclusions can be found by analyzing medical reports, cases, records, and relevant documents and materials.

Text mining, as a research field crossing multiple technologies, originated from single techniques such as text classification, text clustering, and automatic text summarization. In the 1950s, text classification and clustering emerged as an application of pattern recognition. At that time, research was mainly focused on the needs of books and on information classification, and classification and clustering are, of course, based on the topics and contents of texts. In 1958, H.P. Luhn proposed the concept of automatic summarization (Luhn 1958), which added new content to the field of text mining. In the late 1980s and early 1990s, with the rapid development and popularization of Internet technology, demand for new applications has promoted the continuous development and growth of this field. The US government has funded a series of research projects on information extraction, and in 1987, the US Defense Advanced Research Projects Agency (DARPA) initiated and organized the first Message Understanding Conference (MUC[1]) to evaluate the performance of this technology. In the subsequent 10 years, seven consecutive evaluations have made information extraction technology a research hot spot in this field. Next, a series of social media-oriented text processing technologies, such as text sentiment analysis, opinion mining, and topic detection and tracking, emerged and developed rapidly. Today, this technical field is growing rapidly not only in theory and method but also in the form of system integration and applications.

[1] https://www-nlpir.nist.gov/related_projects/muc/.

1.2 Main Tasks of Text Data Mining

As mentioned above, text mining is a domain that crosses multiple technologies involving a wide range of content. In practical applications, it is usually necessary to combine several related technologies to complete an application task, and the execution of mining technology is usually hidden behind the application system. For example, a question and answering (Q&A) system often requires several links, such as question parsing, knowledge base search, inference and filtering of candidate answers, and answer generation. In the process of constructing a knowledge base, key technologies such as text clustering, classification, named entity recognition, relationship extraction, and disambiguation are indispensable. Therefore, text mining is not a single technology system but is usually an integrated application of several technologies. The following is a brief introduction to several typical text mining technologies.

(1) Text Classification

Text classification is a specific application of pattern classification technology. Its task is to divide a given text into predefined text types. For example, according to the Chinese Library Classification (5-th Edition),[2] all books are divided into 5 categories and 22 subcategories. On the first page of www.Sina.com,[3] the content is divided into the following categories: news, finance, sports, entertainment, cars, blog, video, house and property, etc. Automatically classifying a book or an article into a certain category according to its content is a challenging task.

Chapter 5 of this book introduces text classification techniques in detail.

(2) Text Clustering

The purpose of text clustering is to divide a given text set into different categories. Generally, different results can be clustered based on different perspectives. For example, based on the text content, the text set can be clustered into news, culture and entertainment, sports or finance, and so on, while based on the author's tendency, it can be grouped into positive categories (positive views with positive and supportive attitudes) and negative categories (negative views with negative and passive attitudes).

The basic difference between text clustering and text classification is that classification predefines the number of categories and the classification process automatically classifies each given text into a certain category and labels it with a category tag. Clustering, by contrast, does not predefine the number of categories, and a given document set is divided into categories that can be distinguished from each other based on certain standards and evaluation indices. Many similarities exist between text clustering and text classification, and the adopted algorithms and

[2]https://baike.baidu.com/item/中国图书馆图书分类法/1919634?fr=aladdin.

[3]https://www.sina.com.cn/.

models have intersections, such as models of text representation, distance functions, and K-means algorithms.

Chapter 6 of this book introduces text clustering techniques in detail.

(3) Topic Model

In general, every article has a topic and several subtopics, and the topic can be expressed by a group of words that have strong correlation and that basically share the same concepts and semantics. We can consider each word as being associated with a certain topic with a certain probability, and in turn, each topic selects a certain vocabulary with a certain probability. Therefore, we can give the following simple formula:

$$p(word_i|document_j) = \sum_k p(word_i|topic_k) \times p(topic_k|document_j) \tag{1.1}$$

Thus, the probability of each word appearing in the document can be calculated.

To mine the topics and concepts hidden behind words in text, people have proposed a series of statistical models called topic models.

Chapter 7 of this book introduces the topic model in detail.

(4) Text Sentiment Analysis and Opinion Mining

Text sentiment refers to the subjective information expressed by a text's author, that is, the author's viewpoint and attitude. Therefore, the main tasks of text sentiment analysis, which is also called text orientation analysis or text opinion mining, include sentiment classification and attribute extraction. Sentiment classification can be regarded as a special type of text classification in which text is classified based on subjective information such as views and attitudes expressed in the text or judgments of its positive or negative polarity. For example, after a special event (such as the loss of communication with Malaysia Airlines MH370, UN President Ban Ki-moon's participation in China's military parade commemorating the 70th anniversary of the victory of the Anti-Fascist War or talks between Korean and North Korean leaders), there is a high number of news reports and user comments on the Internet. How can we automatically capture and understand the various views (opinions) expressed in these news reports and comments? After a company releases a new product, it needs a timely understanding of users' evaluations and opinions (tendentiousness) and data on users' age range, sex ratio, and geographical distribution from their online comments to help inform the next decisions. These are all tasks that can be completed by text sentiment analysis.

Chapter 8 of this book introduces text sentiment analysis and opinion mining techniques.

(5) Topic Detection and Tracking

Topic detection usually refers to the mining and screening of text topics from numerous news reports and comments. Those topics that most people care about, pay attention to, and track are called *hot topics*. Hot topic discovery, detection, and tracking are important technological abilities in public opinion analysis, social media computing, and personalized information services. The form of their application varies, for example, *Hot Topics Today* is a report on what is most attracting readers' attention from all the news events on that day, while *Hot Topics 2018* lists the top news items that attracted the most attention from all the news events throughout 2018 (this could also be from January 1, 2018, to a different specified date).

Chapter 9 of this book introduces techniques for topic detection and tracking.

(6) Information Extraction

Information extraction refers to the extraction of factual information such as entities, entity attributes, relationships between entities, and events from unstructured and semistructured natural language text (such as web news, academic documents, and social media), which it forms into structured data output (Sarawagi 2008). Typical information extraction tasks include named entity recognition, entity disambiguation, relationship extraction, and event extraction.

In recent years, biomedical/medical text mining has attracted extensive attention. Biomedical/medical text mining refers to the analysis, discovery, and extraction of text in the fields of biology and medicine, for example, research from the biomedical literature to identify the factors or causes related to a certain disease, analysis of a range of cases recorded by doctors to find the cause of certain diseases or the relationship between a certain disease and other diseases, and other similar uses. Compared with text mining in other fields, text mining in the biomedical/medical field faces many special problems, such as a multitude of technical terms and medical terminology in the text, including idioms and jargon used clinically, or proteins named by laboratories. In addition, text formats vary greatly based on their different source, such as medical records, laboratory tests, research papers, public health guidelines, or manuals. Unique problems faced in this field are how to express and utilize common knowledge and how to obtain a large-scale annotation corpus.

Text mining technology has also been a hot topic in the financial field in recent years. For example, from the perspective of ordinary users or regulatory authorities, the operational status and social reputation of a financial enterprise are analyzed through available materials such as financial reports, public reports, and user comments on social networks; from the perspective of an enterprise, forewarnings of possible risks may be found through the analysis of various internal reports, and credit risks can be controlled through analysis of customer data.

It should be noted that the relation in information extraction usually refers to some semantic relation between two or more concepts, and relation extraction automatically discovers and mines the semantic relation between concepts. Event extraction is commonly used to extract the elements that make up the pairs of events in a specific domain. The "event" mentioned here has a different meaning from that used in daily life. In daily life, how people describe events is consistent with their understanding of events: they refer to when, where, and what happened. The thing that happened is often a complete story, including detailed descriptions of causes, processes, and results. By contrast, in even extraction, the "event" usually refers to a specific behavior or state expressed by a certain predicate framework. For example, "John meets Mary" is an event triggered by the predicate "meet." The event understood by ordinary people is a story, while the "event" in event extraction is just an action or state.

Chapter 10 of this book introduces information extraction techniques.

(7) Automatic Text Summarization

Automatic text summarization or automatic summarization, in brief, refers to a technology that automatically generates summaries using natural language processing methods. Today, when information is excessively saturated, automatic summarization technology has very broad applications. For example, an information service department needs to automatically classify many news reports, form summaries of some (individual) event reports (report), and recommend these reports to users who may be interested. Some companies or supervisory departments want to know roughly the main content of statements (SMS, microblog, WeChat, etc.) published by some user groups. Automatic summarization technology is used in these situations.

Chapter 11 of this book introduces automatic text summarization techniques.

1.3 Existing Challenges in Text Data Mining

Study of the techniques of text mining is a challenging task. First, the theoretical system of natural language processing has not yet been fully established. At present, text analysis is to a large extent only in the "processing" stage and is far from reaching the level of deep semantic understanding achieved by human beings. In addition, natural language is the most important tool used by human beings to express emotions, feelings, and thoughts, and thus they often use euphemism, disguise, or even metaphor, irony, and other rhetoric means in text. This phenomenon is obvious, especially in Chinese texts, which presents many special difficulties for text mining. Many machine learning methods that can achieve better results in other fields, such as image segmentation and speech recognition, are often difficult to use in natural language processing. The main difficulties confronted in text mining include the following aspects.

(1) Noise or ill-formed expressions present great challenges to NLP

Natural language processing is usually the first step in text mining. The main data source for text mining processing is the Internet, but when compared with formal publications (such as all kinds of newspapers, literary works, political and academic publications, and formal news articles broadcast by national and local government television and radio stations), online text content includes large ill-formed expressions. According to a random sampling survey of Internet news texts conducted by Zong (2013), the average length of Chinese words on the Internet is approximately 1.68 Chinese characters, and the average length of sentences is

47.3 Chinese characters, which are both shorter than the word length and sentence length in the normal written text. Relatively speaking, colloquial and even ill-formed expressions are widely used in online texts. This phenomenon is common, especially in online chatting, where phrases such as "up the wall," "raining cats and dog," and so on can be found. The following example is a typical microblog message:

//@XXXX://@ YYYYY: Congratulations to the first prospective members of the Class of 2023 offered admission today under Stanford's restrictive early action program. https://stanford.io/2E7cfGF#Stanford2023

The above microblog message contains some special expressions. Existing noise and ill-formed language phenomena greatly reduce the performance of natural language processing systems. For example, a Chinese word segmentation (CWS) system trained on a corpora of normal texts such as the *People's Daily* and the *Xinhua Daily* can usually achieve an accuracy rate of more than 95%, even as high as 98%, but its performance on online text immediately drops below 90%. According to the experimental results of (Zhang 2014), using the character-based Chinese word segmentation method based on the maximum entropy (ME)classifier, when the dictionary size is increased to more than 1.75 million (including common words and online terms), the performance of word segmentation on microblog text as measured by the F_1-measure metric can only reach approximately 90%. Usually, a Chinese parser can reach approximately 87% or more on normal text, but on online text, its performance decreases by an average of 13% points (Petrov and McDonald 2012). The online texts addressed by these data are texts on the Internet and do not include the texts of dialogues and chats in microblogs, Twitter, or WeChat.

(2) Ambiguous expression and concealment of text semantics

Ambiguous expressions are common phenomena in natural language texts, for example, the word "bank" may refer to a financial bank or a river bank. The word "Apple" may refer to the fruit or to a product such as an Apple iPhone or an Apple Computer, a Mac, or Macintosh. There also exist many phenomena of syntactic ambiguity. For example, the Chinese sentence "关于(guanyu, about)鲁迅(Lu Xun, a famous Chinese writer)的(de, auxiliary word)文章(wenzhang, articles)" can be understood as "关于【鲁迅的文章】(about articles of Lu Xun)" or "【关于鲁迅】的文章 (articles about Lu Xun)." Similarly, the English sentence "I saw a boy with a telescope" may be understood as "I saw [a boy with a telescope]," meaning I saw a boy who had a telescope, or "[I saw a boy] with a telescope" meaning I saw a boy by using a telescope. The correct parsing of these ambiguous expressions has become a very challenging task in NLP. However, regrettably, there are no effective methods to address these problems, and a large number of intentionally created "special expressions/tokens" such as Chinese "words" "木有 (no)," "坑爹 (cheating)," and '奥特 (out/out-of-date)" and English words "L8er (later)," "Adorbs (adorable)," and "TL;DR (Too long, didn't read)" appear routinely in online dialogue texts.

Sometimes, to avoid directly identifying certain events or personages, the speaker will turn a sentence around deliberately, for example, asking "May I know the age of the ex-wife of X's father's son?".

Please look at the following news report:

> Mr. Smith, who had been a policeman for more than 20 years, had experienced a multitude of hardships, had numerous achievements and been praised as a hero of solitary courage. However, no one ever thought that such an steely hero, who had made addicted users frightened and filled them with trepidation, had gone on a perilous journey for a small profit and shot himself at home last night in hatred.

For most readers, it is easy to understand the incident reported by this news item without much consideration. However, if someone asks the following question to a text mining system based on this news *What kind of policeman is Mr. Smith?* and *Is he dead?* it will be difficult for any current system to give a correct answer. The news story never directly expresses what kind of policeman Mr. Smith is but uses *addicted users* to hint to readers that he is an antidrug policeman and uses "shot himself" to show that he has committed suicide. This kind of information hidden in the text can only be mined by technology with deep understanding and reasoning, which is very difficult to achieve.

(3) Difficult collection and annotation of samples

At present, the mainstream text mining methods are machine learning methods based on large-scale datasets, including the traditional statistical machine learning method and the deep learning (DL) method. These require a large-scale collection of labeled training samples, but it is generally very difficult to collect and annotate such large-scale samples. On the one hand, it is difficult to obtain much online content because of copyright or privacy issues, which prohibit publication opening and sharing. On the other hand, even when data are easy to obtain, processing these data is time-consuming and laborious because they often contain considerable noise and garbled messages, they lack a uniform format, and there is no standard criterion for data annotation. In addition, the data usually belong to a specific field, and help from experts in that specific domain is necessary for annotation. Without help from experts, it is impossible to provide high-quality annotation of the data. If the field changes, the work of data collection, processing, and annotation will have to start again, and many ill-formed language phenomena (including new online words, terms and ungrammatical expressions) vary with changing domains and over time, which greatly limits expansion of the data scale and affects the development of text mining technology.

(4) Hard to express the purpose and requirements of text mining

Text mining is unlike other theoretical problems, wherein objective functions are clearly established and then ideal answers obtained by optimizing functions and solving the extremum. In many cases, we do not know what the results of text mining will be or how to use mathematical models to describe the expected results and conditions clearly. For example, we can extract frequently used "hot" words from some text that can represent the themes and stories of these texts, but how to organize them into story outlines (summaries) expressed in fluent natural languages is not an easy task. As another example, we know that there are some regular patterns and correlations hidden in many medical cases, but we do not know what regular patterns and correlations exist and how to describe them.

(5) Unintelligent methods of semantic representation and computation model

Effectively constructing semantic computing models is a fundamental challenge that has puzzled the fields adopting NLP for a long time. Since the emergence of deep learning methods, word vector representation and various computing methods based on word vectors have played an important role in NLP. However, semantics in natural language are different from pixels in images, which can be accurately represented by coordinates and grayscales. Linguists, computational linguists, and scholars engaged in artificial intelligence research have been paying close attention to the core issues of how to define and represent the semantic meanings of words and how to achieve combination computing from lexical semantics to phrase, sentence, and, ultimately, paragraph and discourse semantics. To date, there are no convincing, widely accepted and effective models or methods for semantic computing. At present, most semantic computing methods, including many methods for word sense disambiguation, word sense induction based on topic models, and word vector combinations, are statistical probability-based computational methods. In a sense, statistical methods are "gambling methods" that choose high probability events. In many cases, the events with the highest probability will become the final selected answer. In fact, this is somewhat arbitrary, subjective, or even wrong. Since the model for computing probability is based on samples taken by hand, the actual situation (test set) may not be completely consistent with the labeled samples, which inevitably means that some small probability events become "fishes escaping from the net." Therefore, the gambling method, which is always measured by probability, can solve most of the problems that are easy to count but cannot address events that occur with small probability, are hard to find, and occur with low frequency. Those small probability events are always difficult problems to solve, that is, they are the greatest "enemy" faced in text mining and NLP.

In summary, text mining is a comprehensive application technology that integrates the challenges in various fields, such as NLP, ML, and pattern classification, and is sometimes combined with technologies to process graphics, images, videos, and so on. The theoretical system in this field has not yet been established, its prospect for application is extremely broad, and time is passing: text mining will surely become a hot spot for research and will grow rapidly with the development of related technologies.

1.4 Overview and Organization of This Book

As mentioned in Sect. 1.1, text mining belongs to the research field combining NLP, pattern classification, ML, and other related technologies. Therefore, the use and development of technical methods in this field also change with the development and transition of related technologies.

Reviewing the history of development, which covers more than half a century, text mining methods can be roughly divided into two types: knowledge engineering-based methods and statistical learning methods. Before the 1980s, text mining was mainly based on knowledge engineering, which was consistent with the historical track of rule-based NLP and the mainstream application of expert systems dominated by syntactic pattern recognition and logical reasoning. The basic idea of this method is that experts in a domain collect and design logical rules manually for the given texts based on their empirical knowledge and common sense and then the given texts are analyzed and mined through inference algorithms using the designed rules. The advantage of this method is that it makes use of experts' experience and common sense, there is a clear basis for each inference step, and there is a good explanation for the result. However, the problem is that it requires extensive human resources to deduce and summarize knowledge based on experience, and the performance of the system is constrained by the expert knowledge base (rules, dictionaries, etc.). When the system needs to be transplanted to the new fields and tasks, much of the experience-based knowledge cannot be reused, so that usually, much time is needed to rebuild a system. Since the later 1980s, and particularly after 1990, with the rapid development and broad application of statistical machine learning methods, text mining methods based on statistical machine learning obtained obvious advantages in terms of accuracy and stability and do not need to consume the same level of human resources. Especially in the era of big data on the Internet, given massive texts, manual methods are obviously not comparable to statistical learning methods in terms of speed, scale, or coverage when processing data. Therefore, statistical machine learning methods are gradually becoming the mainstream in this field. Deep learning methods, or neural network-based ML methods, which have emerged in recent years, belong to the same class of methods, which can also be referred to as data-driven methods. However, statistical learning methods also have their own defects; for example, supervised machine learning methods require many manually annotated samples, while unsupervised models usually perform poorly, and the results from the system for both supervised and unsupervised learning methods lack adequate interpretability.

In general, knowledge engineering-based methods and statistical learning methods have their own advantages and disadvantages. Therefore, in practical application, system developers often combine the two methods, using the feature engineering method in some technical modules and the statistical learning method in the others to help the system achieve the strongest performance possible through the fusion of the two methods. Considering the maturity of technology, knowledge engineering-based methods are relatively mature, and their performance ceiling is predictable. For statistical learning methods, with the continuous improvement of existing models and the continuous introduction of new models, the performance of models and algorithms is gradually improving, but there is still great room for improvement, especially in large-scale data processing. Therefore, statistical learning methods are in the ascendant. These are the reasons this book focuses on statistical learning methods.

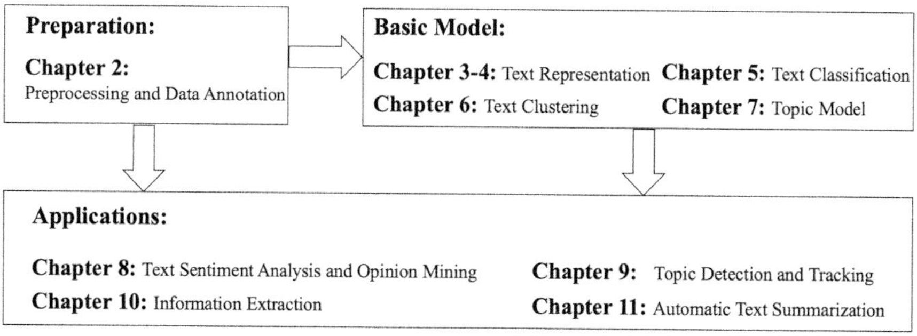

Fig. 1.1 An overview of this book

This book mainly introduces the basic methods and concepts behind text mining but does not get involved in the implementation details of specific systems, nor does it overly elaborate the task requirements and special problems of specific applications. For example, text mining technology in the biomedical and financial fields has attracted much attention in recent years, and many related technologies and resources need to be developed for these fields, such as domain knowledge bases, annotation tools, and annotation samples for domain-related data. The authors hope that the basic methods and ideas introduced in this book have a certain universality and commonality. Once readers know these fundamental methods, they can expand them and implement the system oriented to their specific task requirements.

The remaining nine chapters are organized along the following lines, as shown in Fig. 1.1.

Chapter 2 introduces the methods for data preprocessing. Data preprocessing is the preparation stage before all subsequent models and algorithms are implemented; for example, word segmentation for the Chinese, Japanese, Vietnamese, and other possible languages requires word segmentation. In most online texts, there is much noise and many ill-formed expressions. If these data are not preprocessed well, the subsequent modules will be badly affected, and it will be difficult to achieve the expected final results; indeed, it may be that the model cannot even run. The text representation described in Chaps. 3 and 4 is the basis of the models used in the subsequent chapters. If the text cannot be accurately represented, it is impossible to obtain better results using any of the mathematical models and algorithms introduced in these chapters. The text classification methods introduced in Chap. 5, the text clustering algorithms introduced in Chap. 6, and the topic models introduced in Chap. 7 are the theoretical foundations of other text mining technologies, in a sense, because classification and clustering are the two most fundamental and core problems of pattern recognition and are the two most commonly used methods in machine learning and statistical natural language processing. Most of the models and methods introduced in the following chapters can be treated as classification and clustering problems or can be solved by adopting the concepts of classification or clustering. Therefore, Chaps. 5–7 can be regarded as the theoretical foundations or basic models of the book. In addition, it should be noted that text classification, clustering, and topic models are sometimes used as the sole specific application in some tasks.

Chapters 8–11 can be regarded as an application technology for text mining. A specific task can be performed by one model of can be jointly carried out by several models and algorithms. In most practical applications, the latter method is adopted. For example, text mining tasks in the field of medicine usually involve the techniques of text automatic classification and clustering, topic modeling, information extraction, and automatic summarization, while public opinion analysis tasks for social networks may involve text classification, clustering, topic modeling, topic detection and tracking, sentiment analysis, and even automatic summarization.

With the rapid development and popularization of Internet and mobile communication technologies, requirements may emerge for new applications and new technologies to be applied to text mining. However, we believe that regardless of the application requirements and regardless of the technology behind the new name, new methods of text representation and category distance measurement, and new implementation methods and models (such as end-to-end neural network models), the basic ideas behind clustering and classification and their penetration and application in various tasks will not undergo fundamental changes. This belief is the so-called all things remain essentially the same.

1.5 Further Reading

The following chapters of this book will introduce text mining methods for different tasks and explain the objectives, solutions, and implementation methods. As the beginning of the book, this chapter mainly introduces the basic concepts and challenges of text mining. For a detailed explanation of the concept of data mining, readers can refer to the following literature: (Han et al. 2012; Cheng and Zhu 2010; Li et al. 2010b; Mao et al. 2007). Wu et al. (2008) introduced ten classical algorithms in the field of data mining. Aggarwal (2018) is a relatively comprehensive book introducing text mining technologies. By contrast, readers will find that text mining is regarded as a specific application of machine learning technology in Aggarwal's book, which focuses on discussing text information processing from the perspective of machine learning methods (especially statistical machine learning methods) without the involvement of deep learning and neural network-based methods. Moreover, only traditional statistical methods are used in various text mining tasks, such as text classification, sentiment analysis, and opinion mining, and few related works based on deep learning methods have been introduced in recent years. However, in this book, we regard text mining as the practical application of NLP technology because text is the most important mode of presentation for natural language. Since it is necessary to mine the information needed by users from text, NLP technology remains indispensable. Therefore, this book is driven by task requirements and illustrates the basic principles of text mining models and algorithms through examples and descriptions of processes from the perspective of NLP. For example, in the text representation chapter, text representation and modeling methods based on deep learning are summarized based on the granularity of words, sentences, and documents. In the text mining tasks that follow, in addition to introducing traditional classical methods, deep learning methods, which are highly recommended in recent years, are given special attention.

If Zong (2013) is treated as a foundation or teaching material for the introduction of NLP technologies, then this book is an introduction to the application of NLP technologies. The former mainly introduces the basic concepts, theories, tools, and methods of NLP, while this book focuses on the implementation methods and classical models of NLP application systems.

Other books elaborate on specific technologies of text mining and have good reference value. For example, Liu (2011, 2012, 2015) introduce the concepts and technologies related to web data mining, sentiment analysis, and opinion mining in detail; Marcu (2000) and Inderjeet (2001) provide detailed descriptions of automatic summarization technology, especially the introduction of early summarization technology. Relevant recommendations will be introduced in the "Further Reading" section in each following chapter.

In addition, it should be noted that the authors of this book consider the readers as already having a foundation in pattern recognition and machine learning by default. Therefore, many basic theories and methods are not introduced in detail, their detailed derivation is omitted, and they are only cited as tools. If readers want to know the detailed derivation process, we recommend that they read the following books: (Li 2019; Yu 2017; Zhang 2016; Zhou 2016), etc.

Exercises

1.1 Please summarize the difference between KDD and text data mining.

1.2 Please read an article on natural language processing and learn how to compute the metrics of precision, recall, and the F_1-measure to evaluate the performance of a natural language processing tool.

1.3 Please make a comparison table to show the strengths and weaknesses of rule-based methods and statistical methods.

1.4 Please give some examples for the use of text data mining techniques in real life. Specifically, explain the inputs (source text) and the outputs (what results should we expect to see?).

Chapter 2
Data Annotation and Preprocessing

2.1 Data Acquisition

Data acquisition sources and methods are different for different text mining tasks. Considering the sources of data, there are usually two situations. The first is open domain data. For example, when building a system for mining public opinion from social media, the data naturally come from all available public social networks, including mobile terminals. Although the subject of the mined text may be limited to one or a few specific topics, the data source is open. The situation is closed domain data. For example, the data processed by text mining tasks oriented toward the financial field are proprietary data from banks and other financial industries; similarly, the texts processed by tasks oriented toward hospitals exist in a private network from the internal institutions of the hospital, and they cannot be obtained by public users. Of course, the so-called open domain and closed domain are not absolute, and when implementing a system in practice, it is often not sufficient to solely rely on the data in a specific domain because they will mainly contain professional domain knowledge, while much of the data associated with common sense exists in public texts. Therefore, closed data need to be supplemented with data obtained from public websites (including Wikipedia, Baidu Encyclopedia, etc.), textbooks, and professional literature. Relatively speaking, the data from public networks (especially social networks) contain more noise and ill-formed expressions, so it takes more time to clean and preprocess them.

The following is an example of how movie reviews are obtained to illustrate the general method of data acquisition.

Before acquiring data, one must first know which websites generally contain the required data. The website *IMDb*[1] provides users with comments on movies, and there are many links to movies on the web page, as shown in Fig. 2.1.

[1] https://www.imdb.com/.

© Tsinghua University Press 2021

C. Zong et al., *Text Data Mining*, https://doi.org/10.1007/978-981-16-0100-2_2

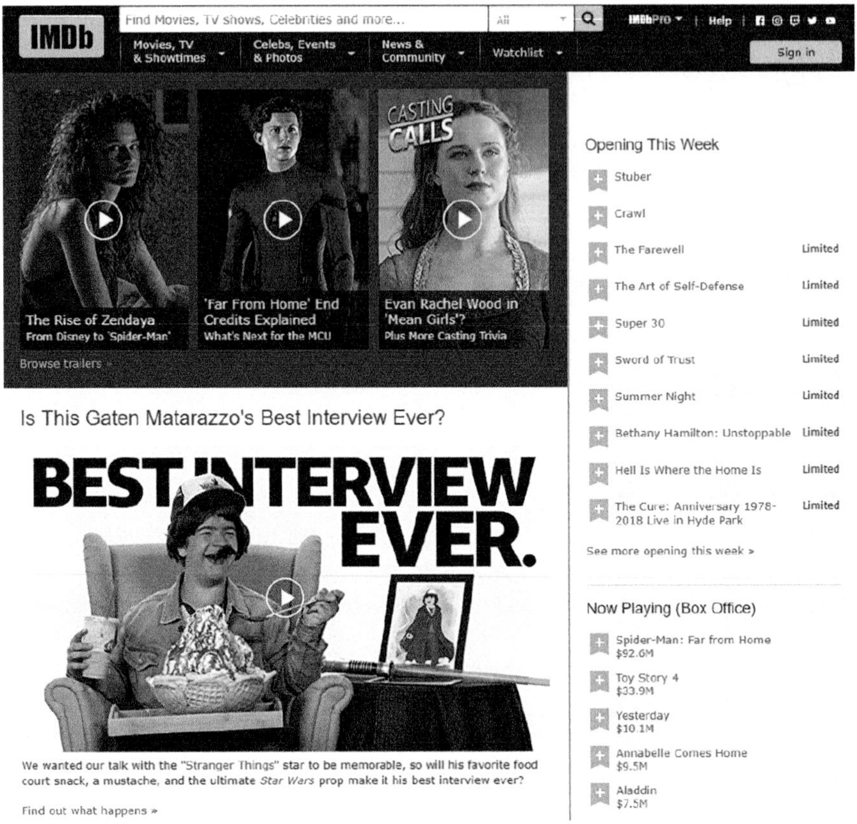

Fig. 2.1 The IMDb homepage

Taking "Mission: Impossible - Fallout" as an example, it can be seen that there are many comments on this film. As shown in Fig. 2.2, a total of 269,543 people gave comments, and the average score given is 7.7 (see the right top corner of Fig. 2.2).

A comment with its score is provided at the bottom of the main page belonging to the "Mission: Impossible - Fallout" film, as shown in Fig. 2.3, but this is not comprehensive. Click "See all 1,597 user reviews" in the bottom row of Fig. 2.3 to view the link to all comments in a comments page. At the end of the comments page, there is a "Load More" button. The user can click the button to obtain extra comments and download the data connected with a link by using Python's urllib2 library.

When using the Python programming language to crawl data from a website, the user must first check and then abide by the robot protocol of the website, which defines what website data can be crawled and what cannot be crawled. Figure 2.4 shows the robot protocol content of IMDb. The "Disallow" in the protocol limits the content that cannot be crawled (much of the search-related content cannot be

Fig. 2.2 The main page for "Mission: Impossible - Fallout"[2]

crawled). There is no restriction on the crawling of movie reviews, so it is legal and valid to crawl these contents, but the process must conform to the crawling time interval. This means that in the process of crawling, the speed should be reduced as much as possible. In fact, crawling process reflects access to the web server. If crawling makes too frequent requests, it will affect the operation of the web server. In addition, it is better to crawl a website when network traffic is low (e.g., at night) to avoid interfering with the normal operations of the website.

The data downloaded from the web page usually have a good structure. The `beautiful Soup` toolkit for Python can be employed to extract the content and obtain the links for the next page. When parsing a web page, the row delimiter ('\r', '\n') of the web page should be deleted, and there may be special symbols in the downloaded data such as " " and "<", which represent a space and the

[2]https://www.imdb.com/title/tt4912910/.

User Reviews

★★★★★★★★★★★ **Best action movie you'll see in years**
29 September 2019 | by zaloc – See all my reviews

The good: Wow. The music, the stunts, the actors, the cities and landscapes. Everything put together amazingly well. And an ending scene that you will never forget. Cruise and McQuarrie did a stunning work here.

The bad: Im no expert, but Cavill's acting seem from time to time, poor. Something that you don't notice in the fighting scenes. Also, some people may find the plot quite complicated, and it is. You really need to pay attention. I recommend watching MI5, and, if you have time, all the others. There is reference to all of them along the movie.

See this movie on a big screen with good sound!!! Enjoy the full experience

42 of 63 people found this review helpful. Was this review helpful to you? Yes   No | Report this

Review this title | See all 1,597 user reviews »

Fig. 2.3 The review page for "Mission: Impossible - Fallout"

less-than sign, respectively; these can be replaced if they are not necessary. Table 2.1 shows the corresponding meaning of some special symbols used in web pages.

After obtaining the comment content, it is necessary to clean the data and remove any noise words or text that is too short (which is usually meaningless). The processing procedure is specifically given as follows:

(1) **Noise processing:** There are likely to be English words or letters in downloaded Chinese text or symbols from other languages. This requires identification of the language type. The `Langdect` toolkit in Python can be used to help identify and delete those data that are not needed. In addition, the crawled microblog data may contain advertisement links, "@," and so on, which requires special handling. These links can be deleted directly, and the symbol "@" is usually followed by a user name, which can be determined by using simple methods such as rule-based or template-based approaches and then deleted.

(2) **Conversion of traditional Chinese characters:** There may be some traditional Chinese characters in the downloaded simplified Chinese text, which need to be converted into simplified characters. The conversion process can be performed with the help of the open source toolkit OpenCC[3] or other tools.

(3) **Remove comments that are too short:** For English comments, the number of words in the text can be directly counted by numerating spaces. However, for Chinese, Japanese, or other language text, it is necessary to segment the characters into words first before counting the number of words. If the length of a piece of text is shorter than a certain threshold (e.g., 5), it is usually removed.

---

[3] https://opencc.byvoid.com/.

```
# robts.txt for https://www.imdb.com properties
User-agent: *
Disallow:  /OnThisDay
Disallow:  /ads/
Disallow:  /ap/
Disallow:  /mymovies/
Disallow:  /r/
Disallow:  /register
Disallow:  /registration/
Disallow:  /search/name-text
Disallow:  /search/title-text
Disallow:  /find
Disallow:  /find$
Disallow:  /find/
Disallow:  /tvschedule
Disallow:  /updates
Disallow:  /watch/_ajax/option
Disallow:  /_json/video/mon
Disallow:  /_json/getAdsForMediaViewer/
Disallow:  /list/ls*/_ajax
Disallow:  /*/*/rg*/mediaviewer/rm*/tr
Disallow:  /*/rg*/mediaviewer/rm*/tr
Disallow:  /*/mediaviewer/*/tr
Disallow:  /title/tt*/mediaviewer/rm*/tr
Disallow:  /name/nm*/mediaviewer/rm*/tr
Disallow:  /gallery/rg*/mediaviewer/rm*/tr
Disallow:  /tr/
```

Fig. 2.4 The robot protocol of IMDb[4]

(4) The mappings of labels: Websites usually provide the labels for their categories, and the number of categories is potentially different from that of the predefined classifier employed, so it is necessary to map the labels or categories from one type to another. For example, the evaluation score in downloaded data uses a 5-point system, while the sentiment classifier can only distinguish the sentiment into two categories, positive and negative, so the samples with scores of 4 and 5 can be taken as positive samples, those with scores of 1 and 2 can be treated as negative samples, and the "neutral" samples with scores of 3 can be deleted. Certainly, if a classifier is trained with three categories, i.e., positive, neural, and negative, you would annotate those samples with a score of 3 as neutral and preserve the middle category.

[4]https://www.imdb.com/robots.txt.

Table 2.1 Corresponding table of special symbols used in web data[5]

Displayed result	Description	Entity name	Entity number
	Space		
<	Less than	&lt;	&#60;
>	Greater than	&gt;	&#62;
&	Ampersand	&amp;	&#38;
"	Quotation mark	&quot;	&#34;
'	Apostrophe	' (do not support IE)	&#39;
¢	Cent	&cent;	&#162;
£	Pound	&pound;	&#163;
¥	Yen	&yen;	&#165;
	Euro	&euro;	&#8384;
§	Section	&sect;	&#167;
©	Copyright	&copy;	&#169;
®	Registered trademark	&reg;	&#174;
™	Trademark	&trade;	&#8482;
×	Times sign	&times;	&#215;
÷	Division sign	&divide;	&#247;

The methods for acquiring open domain data for other tasks are very similar, but the annotation methods are different; for example, for automatic text summarization or information extraction, the annotation work is much more complicated than simply annotating categories.

2.2 Data Preprocessing

After data acquisition, it is usually necessary to further process the data. The main tasks include:

(1) Tokenization: This refers to a process of segmenting a given text into lexical units. Latin and all inflectional languages (e.g., English) naturally use spaces as word separators, so only a space or punctuation is required to realize lexicalization, but there are no word separation marks in written Chinese and some other agglutinative languages (e.g., Japanese, Korean, Vietnamese), so word segmentation is required first. This issue is mentioned above.

(2) Removing stop words: Stop words mainly refer to functional words, including auxiliary words, prepositions, conjunctions, modal words, and other high-frequency words that appear in various documents with little text information, such as *the, is, at, which, on,* and so on in English or 的(de), 了(le) and 是(shi)

---

[5]https://www.w3school.com.cn/html/html_entities.asp.

in Chinese. Although "是(be)" is not a functional word, it has no substantive meaning for the distinction of text because of its high frequency of occurrence, so it is usually treated as a stop word and removed. To reduce the storage space needed by the text mining system and improve its operating efficiency, stop words are automatically filtered out during the phase of representing text. In the process of implementation, a list of stop words is usually established, and all words in the list are directly deleted before features are extracted.

(3) **Word form normalization**: In the text mining task for Western languages, the different forms of a word need to be merged, i.e., word form normalization, to improve the efficiency of text processing and alleviate the problem of data sparsity caused by discrete feature representation. The process of word form normalization includes two concepts. One is lemmatization, which is the restoration of arbitrarily deformed words into original forms (capable of expressing complete semantics), such as the restoration of *cats* into *cat* or *did* into *do*. Another is stemming, which is the process of removing affixes to obtain roots (not necessarily capable of expressing complete semantics), such as *fisher* to *fish* and *effective* to *effect*.

The process of word form normalization is usually realized by rules or regular expressions. The Porter stemming algorithm is a widely used stemming algorithm for English that adopts a rule-based implementation method (Porter 1980). The algorithm mainly includes the following four steps: (a) dividing letters into vowels and consonants; (b) utilizing rules to process words with suffixes of `-s`, `-ing`, and `-ed`; (c) designing special rules to address complicated suffixes (e.g.,`-ational`, etc.); and (d) fine-tuning the processing results by rules. The basic process of the algorithm is presented in Fig. 2.5.

In the Porter stemming algorithm, only a portion of the main rewriting rules are given from Step 2 to Step 4, and the rest are not introduced individually, as this is simply an example to illustrate the basic ideas behind it.

The implementation code for the algorithm can be obtained from the following web page:

https://tartarus.org/martin/PorterStemmer/

In addition, the NLTK toolkit in Python also provides calling functions for the algorithm.

It should be noted that there is no uniform standard for stemming results, and different stemming algorithms for words in the same language may have different results. In addition to the Porter algorithm, the Lovins stemmer (Lovins 1968) and the Paice stemmer (Paice 1990) are also commonly used for English word stemming.

Algorithm The Porter Stemming Algorithm

Input: An English word;
Output: The stem or original type of input word;
Algorithm:
Step 1: Distinguishing vowels and consonants by using the following rules:
(1) Letters `a`, `e`, `i`, `o`, `u` are vowels;
(2) The letter `y` has the following three cases:
(a) If `y` is the beginning of a word, it is judged as a consonant. e.g., `y` is a consonant in the word `young`;
(b) If the previous letter of `y` is a vowel, `y` is judged as a consonant. e.g., `y` is a consonant in the word `boy`;
(c) If the previous letter of `y` is a consonant, `y` is judged as a vowel. e.g., `y` is a vowel in the word `fly`.
(3) All other letters except `a`, `e`, `i`, `o`, `u`, `y` are consonants.
Step 2: Processing words with `-s`, `-ing` and `-ed` suffixes by using the following rules:
(1) Words ending with `-s` are treated as follows:
(a) If the word ends with `-sses`, then restore it to `-ss`. e.g., the word `caresses` should be restored to `caress`;
(b) If the word ends with `-ies`, then delete `-es`. e.g., `cries` becomes `cri`;
(c) If the word ends with `-s` and one of all letters before `s` is a vowel at least, consider the following two cases:
(i) if the vowel is adjacent the last `s`, the word will not change. e.g., the word `gas` is the original type and does not need to change;
(ii) Otherwise, delete the last letter `s`. e.g., `gaps` restore to `gap`.
(2) If the word ends with `-ing` and the previous part of the word contains a vowel letter except for `ing`, delete `ing`. e.g., the word `doing` restore to `do`.
Step 3: Use the following rules to process words with other suffixes.
(1) If the word ends with `-y` and the previous part of `-y` contains vowel letters, `-y` is changed to `i`. e.g., the word `happy` is rewritten as `happi`.
(2) If the word ends with `-ational` and the previous section of `-ational` contains vowel letters, `-ational` is rewritten as `-ate`, for example, the word `relational` is rewritten as `relate`.
Step 4: Fine-tuning by the following rules:
For the words ending with `e`, if the number of consonants is greater than 1 except for the first letter and the last letter, the last letter `e` is removed, for example, `relate` is changed to `relat`.

Fig. 2.5 The Porter stemming algorithm

2.3 Data Annotation

Data annotation is the foundation of supervised machine learning methods. In general, if the scale of annotated data is larger, the quality is higher, and if the coverage is broader, the performance of the trained model will be better. For different text mining tasks, the standards and specifications for data annotation are different, as is the complexity. For example, only category labels need to be annotated on each document for text classification tasks, while for some complex tasks, much more information needs to be marked, e.g., the boundary and type of each "entity" in the records should be marked for the analysis of electronic medical records. The "entity" mentioned here is not just the named entity (person name, place name, organization name, time, number, etc.), as there are also many specialized terms in the medical field, such as disease names, the presence of certain symptoms, the absence certain symptoms, the frequency with which some symptoms occur, the factors of deterioration, irrelevant factors, and the degree. See the following two examples:[6]

(1) *Mr. Shinaberry is a 73-year-old gentleman who returned to [Surgluthe Leon Calcner Healthcare]$_{Hosp}$ to the emergency room on [9/9/02]$_{Time}$ with [crescendo spontaneous angina]$_{Sym}$ and [shortness of breath]$_{Sym}$. He is [three-and-one-half months]$_{Dur}$ after a presentation with [subacute left circumflex thrombosis]$_{Dis}$,[ischemic mitral regurgitation]$_{Dis}$, [pulmonary edema]$_{Dis}$ and a small [nontransmural myocardial infarction]$_{Dis}$. [Dilatation of the left circumflex]$_{Treat}$ resulted in extensive dissection but with eventual achievement of a very good [angiographic and clinical result]$_{TR}$ after [placement of multiple stents]$_{Treat}$, and his course was that of gradual recovery and uneventful return home.*
(2) *Mr. Brunckhorst is a 70-year-old man who recently had been experiencing an increase in frequency of [chest pain]$_{Sym}$ with exertion. He was administered an [exercise tolerance test]$_{Test}$ that was predictive of [ischemia]$_{TR}$ and revealed an [ejection fraction of approximately 50%]$_{TR}$. As a result of his [positive exercise tolerance test]$_{TR}$, he was referred for [cardiac catheterization]$_{Treat}$ on March 1998, which revealed three [vessel coronary artery disease]$_{Dis}$. At this time, he was referred to the [cardiac surgery service]$_{Treat}$ for revascularization.*

In the examples, the label $Time$ indicates time, Sym indicates the presence of such symptoms, $Hosp$ indicates the name of the hospital, $Test$ indicates laboratory tests, TR indicates the results of the laboratory tests, Dis indicates the name of the disease, $Treat$ indicates the method of treatment, and Dur indicates the duration.

In the task of analyzing electronic medical records, usually, more than 20 different labels are defined. When annotating, it is often necessary to develop an annotation tool that can not only annotate the boundaries and types of all "entities" but also the relationships between them. In example (1) above, an annotation tool gives the relation graph shown in Fig. 2.6.

Of course, this type of relation graph is convenient and intuitive for annotators and domain experts to check and annotate. In fact, all specific marks are stored in the system. It is difficult to complete this kind of annotation task, which requires guidance based on professional knowledge, without the involvement of experts in the field.

For research on multimodal automatic summarization methods, we annotated a dataset including text, images, audio, and video. Different from synchronous multimodal data (e.g., movies), the dataset consists of asynchronous multimodal data, i.e., the pictures and sentences in the text or video and the sentences do not have

[6]https://portal.dbmi.hms.harvard.edu/projects/n2c2-nlp/.

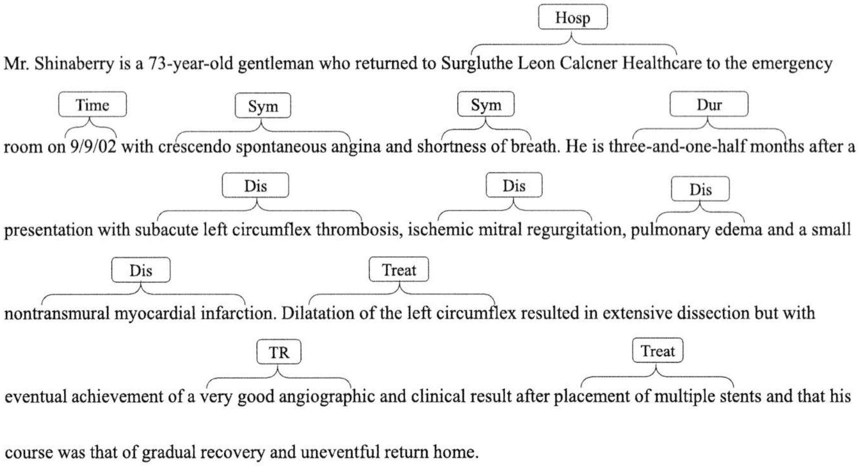

Fig. 2.6 An example of medical record annotation

a one-to-one correspondence. The dataset is centered on topics from Chinese and English news. There are multiple news documents with pictures for the same topic. For each topic, word-limited Chinese and English text summaries are presented.

During the data collection process, we chose 25 news topics in Chinese and English over the past 5 years, such as the Ebola virus in Africa, protesting against the deployment of the "Sade" antimissile system, or Li Na winning the Australian Open Tennis Championship. For each topic, we collected 20 news documents and 5 to 10 videos for the same period, making sure that the collected news texts were not significantly different in length. Generally, the length of each news item did not exceed 1,000 Chinese characters (or English words), and each video was within 2 min in length. The main reason for these restrictions is that overly long text or video will seriously increase the difficulty of manual annotation, which may lead to a too great divergence in the results annotated by different people.

During the annotation, the annotation policies given by the Document Understanding Conference and Text Analysis Conference were used for reference, and ten graduate students were invited to annotate the corpus. They were asked to read the news documents first, watch the videos on the same topic, and then write a summary independently. The policies for writing the summary are as follows: (1) the summary should retain the most important information from the news documents and videos; (2) there should be little to no redundant information in the summary; (3) the summary should have good readability; and (4) the summary should be within the length limitation (the Chinese summary does not exceed 500 Chinese characters, and the English abstract does not exceed 300 English words).

In the end, for each topic, three summaries independently written by different annotators were selected as reference answers.

At present, most of summaries generated by the existing automatic summarization systems are text without any other modal information, such as images.

Considering that a multimodal summary can enhance the user experience, we have also presented the summary data by text and picture. The annotation of this dataset involves two tasks: the writing of text summaries and the selection of pictures. The requirements for text summaries are the same as the methods described previously. To select the picture, two graduate students were invited to independently pick out the three most important pictures for each topic, and then we asked the third annotator to select three pictures as the final reference based on the results from the first two annotators. The basic policies for selecting pictures are that the pictures should be closely related to the news topic and they should be closely related to the content of the text summary.

The abovementioned corpora for summarization studies have been released on the following website: http://www.nlpr.ia.ac.cn/cip/dataset.htm/. Readers who are interested in multimodal summarization can download it from this website.

In summary, data annotation is a time-consuming and laborious task that often requires considerable manpower and financial support, so data sharing is particularly important. The methods introduced in this section are just examples, and more detailed specifications, standards, and instructions are required in data annotation. For many complex annotation tasks, developing convenient and easy-to-use annotation tools is a basic requirement for annotating large-scale data.

2.4 Basic Tools of NLP

As mentioned earlier, text mining involves many techniques from NLP, pattern classification, and machine learning and is one of technology with a clear application goal in across domains. Regardless of technologies applied for data preprocessing and annotation as described earlier or for the realization of data mining methods as will be described later, many basic techniques and tools are required, such as word segmenters, syntactic parsers, part-of-speech taggers, and chunkers. Some NLP methods are briefly introduced in the following.

2.4.1 Tokenization and POS Tagging

The purpose of tokenization is to separate text into a sequence of "words," which are usually called "tokens." The tokens include a string of successive alphanumeric characters, numbers, hyphens, and apostrophes. For example, "that's" will be separated into two tokens, that, 's; "rule-based" will be divided into three tokens, rule, -, based; and "TL-WR700N" will be divided into three tokens, TL, -, WR700N. The NLTK toolkit provides a tokenization package.[7]

[7]https://www.nltk.org/api/nltk.tokenize.html.

As we know, a word is usually expressed in different forms in the documents because of grammatical reasons, such as take, takes, taken, took, and taking. And also, many words can derive different expressions with the same meaning, such as token, tokenize, and tokenization. In practice, especially when using statistical methods, it is often necessary to detect the words sharing the same stem and meaning but in different forms and consider them the same when performing semantic understanding tasks. So, stemming and lemmatization are usually necessitated to reduce inflectional forms and sometimes derivationally related forms of a word to a common base form.[8]

For many Asian languages, such as Chinese, Japanese, Korean, and Vietnamese, the tokenization is usually expressed as word segmentation because their words are not separated by white spaces. The following takes the Chinese as example to briefly introduce the methods to segment Chinese words.

The Chinese word segmentation (CWS) is usually the first step in Chinese text processing, as noted above. There has been much research on CWS methods. From the early dictionary-based segmentation methods (such as the maximum matching method and shortest path segmentation method), to the statistical segmentation method based on *n*-gram, to the character-based CWS method later, dozens of segmentation methods have been proposed. Among them, the character-based CWS method is a landmark and an innovative method. The basic idea is that there are only four possible positions for any unit in a sentence, including Chinese characters, punctuation, digits, and any letters (collectively referred to as a "character"): the first position of the word (marked as B), the last position of the word (marked as E), the middle position of the word (marked as M), or a single character word (marked as S). B, E, M, and S are thus called the word position symbols. B and E always appear in pairs. Please see the following examples:

Chinese sentence: 约翰在北京见到了玛丽。(John met Mary in Beijing)
Segmentation result: 约翰(John)/ 在(in)/ 北京(Beijing)/ 见到了(met)/ 玛丽(Mary)。
The segmentation results can be represented by position symbols: 约/B 翰/E 在/S 北/B 京/E 见/B 到/M 了/E 玛/B 丽/E 。/S

In this way, the task of CWS becomes the task of sequence labeling, and the classifier can be trained with large-scale labeled samples to carry out the task of labeling every unit in the text as a unique word position symbol. In practice, people also try to fuse or integrate several methods, such as the combination of the *n*-gram-based generative method and the character-based discriminative method (Wang et al. 2012) or the combination of the character-based method and the deep learning method, to establish a word segmenter with better performance.

Part-of-speech tagging refers to automatically tagging each word in a sentence with a part-of-speech category. For example, the sentence "天空是蔚蓝的(the sky is blue.)" is annotated as "天空/NN 是/NV 蔚蓝/AA 的/Au x 。/PU" after word

[8]https://nlp.stanford.edu/IR-book/html/htmledition/stemming-and-lemmatization-1.html.

segmentation and part-of-speech tagging. The symbol NN represents noun, VV represents verb, AA represents adjective, Aux represents structural auxiliary, and PU represents punctuation. Part-of-speech tagging is the premise and foundation of syntactic parsing; it is also an important feature of text representation and is of great help to named entity recognition, relation extraction, and text sentiment analysis.

Part-of-speech tagging is a typical problem of sequence tagging. For Chinese text, this task is closely related to automatic word segmentation. Therefore, these two tasks are integrated in many CWS toolkits and are even achieved by a single model, such as the early CWS method based on the hidden Markov model (HMM).

At present, some CWS and part-of-speech tagging tools can be found in the following websites:

https://github.com/FudanNLP/fnlp
http://www.nlpr.ia.ac.cn/cip/software.htm
https://nlp.stanford.edu/software/tagger.shtml

In recent years, in deep learning methods or neural network-based methods, the text can be dealt with in character level or in sub-word level, but not in word level. So the tokenization and POS tagging can also be skipped.

2.4.2 Syntactic Parser

Syntactic parsing includes the tasks of constituent, or phrase structure, parsing, and dependency parsing. The purpose of phrase structure parsing is to automatically analyze the phrase structure relation in a sentence and to output the syntactic structure tree of the parsing sentence. The purpose of dependency parsing is to automatically analyze the relation of semantic dependency between words in a sentence. For example, Fig. 2.7 is a phrase structure tree of the sentence "The policemen have arrived at the scene and are carefully investigating the cause of the accident." The node symbols VV, NN, ADVP, NP, VP, and PU in Fig. 2.7 are part-of-speech symbols and phrase markers, respectively. IP is the root node symbol of the sentence. Figure 2.8 is the dependency tree corresponding to this sentence.

The arrow in Fig. 2.8 indicates the dependency (or domination) relation. The starting end of the arrow is the dominant word, and the pointing end of the arrow is the dominant word. The symbols on the directed arcs indicate the type of dependency relation. SBJ indicates a subject relation, i.e., the word at the end of the arrow is the subject of the word at the start of the arrow. OBJ indicates the object relation, that is, the word at the end of the arrow is the object of the word at the start of the arrow. VMOD indicates the verb modification relation, that is, the word at the end of the arrow modifies the verb at the beginning of the arrow. NMOD is a noun modification relation, that is, the word at the end of the arrow modifies the noun at the beginning of the arrow. ROOT denotes the root node of the clause. PU denotes the punctuation mark of the clause.

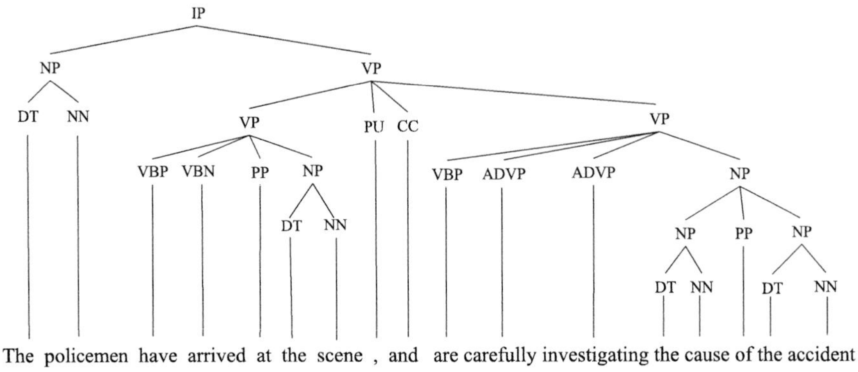

Fig. 2.7 An example of a phrase structure parsing tree

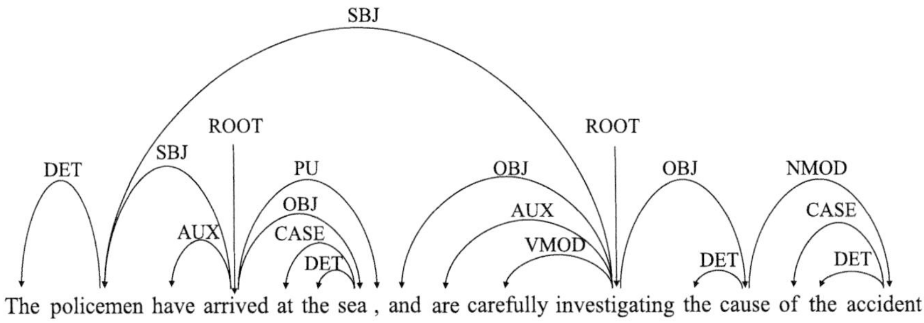

Fig. 2.8 An example of a dependency parsing tree

The phrase structure tree of a sentence can be converted into a dependency tree one by one. The basic idea of the conversion can be described as follows: first, determine the core predicate of the sentence as the only root node of the sentence, and then define the extraction rule of the central word. Next, the central word of each phrase is extracted, and the noncenter word is dominated by the central word.

In NLP, the phrase structure analyzer is usually called the syntactic parser, and the dependency analyzer is called the dependency parser.

The following web pages provide some parsers:

Berkeley Parser: https://github.com/nikitakit/self-attentive-parser
Charniak Parser: https://www.cs.brown.edu/people/ec/#software
http://www.nlpr.ia.ac.cn/cip/software.htm

The syntactic parser is usually employed to parse a complete sentence, and ultimately, we hope to obtain a full parsing tree for the sentence, so it is also called full parsing. In practice, sometimes it is not necessary to obtain a complete syntactic parsing tree but only to identify the basic noun phrase (base NP) or basic verb phrase (base VP) included in the sentence. For example, the sentence *Foreign-funded enterprises also play an important role in China's economy* contains the base

NPs *foreign-funded enterprises, China's economy*, and *important role* and contains the base VP *play*. The parsing technique for identifying a specific type of phrase in a sentence is usually called shallow parsing. At present, the shallow parsing method in use is more similar to the character-based word segmentation method. The tagging unit can be either the word or the character. The word or character position tag can adopt four tagging systems using B, E, M, and S and can also adopt three tagging systems using B, I, and O. For example, NP-B denotes the first word (character) of a base NP, NP-I denotes that the word (character) is inside the base NP, and NP-O denotes that the word (character) does not belong to the NP. The classifier model is similar to the methods used in CWS and named entity recognition. Readers can refer to Chap. 9 for a detailed introduction of named entity recognition methods.

2.4.3 N-gram Language Model

N-gram is a traditional language model (LM) that plays a very important role in NLP. The basic idea is as follows: for a character string (phrase, sentence, or fragment) $s = w_1 w_2 \cdots w_l$ composed of l (l is a natural number, $l \geq 2$) basic statistical units, its probability can be calculated by the following formula:

$$\begin{aligned} p(s) &= p(w_1)p(w_2|w_1)p(w_3|w_1 w_2)\cdots p(w_l|w_1\cdots w_{l-1}) \\ &= \prod_{i=1}^{l} p(w_i|w_1\cdots w_{i-1}) \end{aligned} \tag{2.1}$$

The *basic statistical units* mentioned here may be characters, words, punctuation, digits, or any other symbols constituting a sentence, or even phrases, part-of-speech tags, etc., which are collectively referred to "words" for convenience of expression. In Eq. (2.1), it means that the probability of generating the i-th ($1 \leq i \leq l$) word is determined by the previously (the "previously" usually refers to the left in the written order of the words) generated $i-1$ words $w_1 w_2 \cdots w_{i-1}$. With increasing sentence length, the historical number of conditional probabilities increases exponentially. To simplify the complexity of the calculation, it is assumed that the probability of the current word is only related to the previous $n-1$ (n is an integer, $1 \leq n \leq l$) words. Thus, Eq. (2.1) becomes

$$p(s) = \prod_{i-1}^{l} p(w_i|w_1\cdots w_{i-1}) \approx \prod_{i=1}^{l} p(w_i|w_{i-1}) \tag{2.2}$$

When $n = 1$, the probability of word w_i appearing at the i-th position is independent of the previous words, and the sentence is a sequence of independent words. This calculation model is usually called a one-gram model, which is recorded as a unigram, unigram, or monogram. Each word is a unigram. When $n = 2$, the probability of word w_i appearing at the i-th position is only related to the previous word w_{i-1}. This calculation model is called the two-gram model. Two adjacent co-occurrence words are called two-grams, usually signed as bigrams or bi-grams. For example, for the sentence *We helped her yesterday*, the following sequence of words, *We helped*, *helped her*, and *her yesterday*, are all bigrams. In this case, the sentence is regarded as a chain composed of bigrams, called a first-order Markov chain. By that analogy, when $n = 3$, the probability of the word w_i appearing at the i-th position is only related to the previous word w_{i-1} and word w_{i-2} $(i \geq 2)$. This calculation model is called a three-gram model. The sequence of three adjacent co-occurrence words is called three grams, usually signed as trigrams or tri-grams. Sequences composed of trigrams can be regarded as second-order Markov chains.

When calculating the n-gram model, a key problem is smoothing the data to avoid the problems caused by zero probability events (n-gram). For this reason, researchers have proposed several data smoothing methods, such as additive smoothing, discounting methods, and deleted interpolation methods. At the same time, to eliminate the negative influence of training samples from different fields, topics, and types on the model's performance, researchers have also proposed methods for language model adaptation, which will not be described in detail here. Readers can refer to (Chen and Goodman 1999) and (Zong 2013) if interested.

The neural network language model (NNLM) has played an important role in NLP in recent years. For details about this model, please refer to Chap. 3 in this book.

2.5 Further Reading

In addition to the NLP technologies introduced above, word sense disambiguation (WSD), semantic role labeling (SRL), and text entailment are also helpful for text data mining, but their performance has not reached a high level (e.g., the accuracy of semantic role labeling for normal text is only 80%). Relevant technical methods have been described in many NLP publications. The readers can refer to (Manning and Schütze 1999; Jurafsky and Martin 2008; Zong 2013) if necessary.

Exercises

2.1 Please collect some articles from newspapers and some text from social network sites, such as Twitter, QQ, or other microblog sites. Compare the different language expressions and summarize your observations.

2.2 Collect some sentences, parse them using a constituent parser and a dependency parser, and then compare the different results, i.e., the syntactic parsing tree and the dependency of words in the sentences.

2.3 Collect some text from social network sites in which there are some noise and stop words. Please make a list of the stop words and create a standard or policy to determine what words are noise. Then, implement a program to remove the stop words and noise words from the given text. Aim to deliver higher generalization for the algorithm of the program.

2.4 Collect some corpora, tokenize it by implementing an existing tool or a program, and extract all n-grams in the corpora ($n \in N$ and $1 < n \leq 4$).

2.5 Collect some medical instructions or some chapters from text books for medical students and annotate all named entities, other terms, and their relations in the collected instructions or text corpora.

2.6 Use a Chinese word segmenter to perform Chinese word segmentation for different styles of Chinese corpora, such as from public newspapers, specific technical domains, and social network sites. Analyze the segmentation results, and evaluate the correct rate of the Chinese word segmenter. How do the results compare to the segmentation results for the same corpora when using a different segmenter?

Chapter 3
Text Representation

3.1 Vector Space Model

3.1.1 Basic Concepts

The vector space model (VSM) was proposed by G. Salton et al. in the field of information retrieval in the late 1960s and is the simplest text representation method (Salton et al. 1975). It was first used in the SMART information retrieval system and gradually became the most commonly used text representation model in text mining. Before introducing VSM in detail, we first provide some basic concepts.

- **Text**: Text is a sequence of characters with certain granularities, such as phrases, sentences, paragraphs, or a whole document. For the convenience of description, we use the document to represent a piece of text in the following. Note that the vector space model is applicable to not only documents but also text at other granularities, such as a sentence, a phrase, or even a word.
- **Term**: This is the smallest inseparable language unit in VSM, and it can denote characters, words, phrases, etc. In VSM, a piece of text is regarded as a collection of terms, expressed as $(t_1, t_2, \ldots, t_n)$, where t_i denotes the i-th term.
- **Term weight**: For text containing n terms, each term t is assigned a weight w according to certain principles, indicating that term's importance and relevance in the text. In this way, a text can be represented by a collection of terms with their corresponding weights: $(t_1 : w_1, t_2 : w_2, \ldots, t_n : w_n)$, abbreviated to $(w_1, w_2, \ldots, w_n)$.

The vector space model assumes that a document conforms to the following two requirements: (1) each term t_i is unique (i.e., there is no duplication); (2) the terms have no order. We can regard $t_1, t_2, \ldots, t_n$ as an n-dimensional orthogonal coordinate system, and a text can be represented as an n-dimensional vector: $(w_1, w_2, \ldots, w_n)$. Normally, we denote $\boldsymbol{d} = (w_1, w_2, \ldots, w_n)$ as the representation

© Tsinghua University Press 2021
C. Zong et al., *Text Data Mining*, https://doi.org/10.1007/978-981-16-0100-2_3

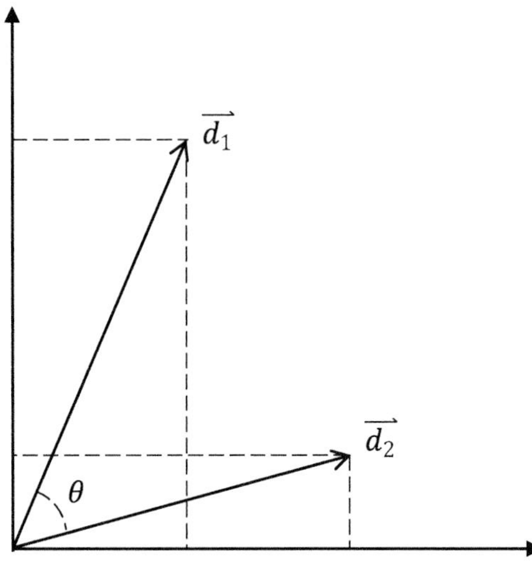

Fig. 3.1 Vector space model

of text in the vector space model. As shown in Fig. 3.1, texts $\boldsymbol{d}_1$ and $\boldsymbol{d}_2$ are, respectively, represented by two n-dimensional vectors in the vector space.

There are two problems that need to be solved in VSM construction: how to design the set of terms and how to calculate the term weights.

3.1.2 Vector Space Construction

Before text representation based on VSM, it is usually first necessary to preprocess the text according to the techniques described in Chap. 2, such as tokenization, the removal of stop words and lexical normalization. Then, we need to convert the text into a sequence of tokens.

The vector space model needs a collection of terms $(t_1, t_2, \ldots, t_n)$. If words are used as terms, the collection can be regarded as a vocabulary. The vocabulary can be generated from the corpus or imported from an external lexicon. The terms can be viewed as a bag of words; thus, the vector space model is also called the bag-of-words (BOW) model.

After construction of terms, the vector space is fixed. Last, a piece of text is represented as a vector in the vector space through term weight calculation. Some common term weighting methods are listed as follows:

- Boolean (BOOL) weight: This method indicates whether a feature term appears in the current document. If it is in the document, the weight is 1; otherwise, it is 0. The Boolean weight t_i in document $\boldsymbol{d}$ is denoted as

$$\mathrm{BOOL}_i = \begin{cases} 1 & \text{if } t_i \text{ appears in document } \boldsymbol{d} \\ 0 & \text{otherwise} \end{cases} \tag{3.1}$$

- Term frequency (TF): This weight represents the frequency of a term in the current document. TF assumes that frequent terms contain more information than infrequent ones, and thus the more frequently terms appear in a document, the more important they are. TF can be expressed as follows:

$$\mathrm{tf}_i = N(t_i, \boldsymbol{d}) \tag{3.2}$$

For a few high-frequency words, e.g., some stop words, the absolute frequency will be much higher than the average, and this will affect text representation. To lower such impact, we can use the logarithmic term frequency instead:

$$f_i = \log(\mathrm{tf}_i + 1) \tag{3.3}$$

- Inverse Document Frequency (IDF): IDF is a global statistical feature that reflects the importance of terms throughout the corpus. Document frequency (DF) denotes the number of documents that contain the specific term in the corpus. The higher the DF of a term is, the lower the amount of effective information it contains. On the basis of DF, IDF is defined as follows:

$$\mathrm{idf}_i = \log \frac{N}{\mathrm{df}_i} \tag{3.4}$$

where df_i denotes the DF of feature t_i and N is the total number of documents in the corpus. The IDF of a rare term is high, whereas the IDF of a frequent term is low.
- Term Frequency-Inverted Document Frequency (TF-IDF): This method is defined as the product of TF and IDF:

$$\mathrm{tf\_idf}_i = \mathrm{tf}_i \cdot \mathrm{idf}_i \tag{3.5}$$

TF-IDF assumes that the most discriminative features are those that appear frequently in the current document and rarely in other documents.

In Fig. 3.2, we use words as terms and TF as term weights to build a vector space model to represent the following text: “Artificial intelligence is a branch of computer science, which attempts to produce an intelligent machine that can respond in a similar way to human intelligence.”

The vocabulary includes the following words: “education,” “intelligence,” “human,” “sports,” “football,” “games,” “AI,” “science,” “text,” “artificial,” “computer,” etc. The weight of each word is its frequency in the text.

Fig. 3.2 The feature weight based on feature frequency

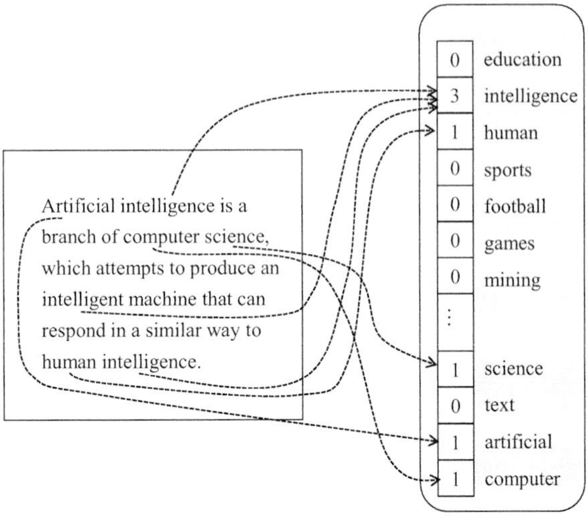

3.1.3 Text Length Normalization

Documents have different lengths, and the length has an effect on the text representation. For an extreme example, if we duplicate the content of a document twice and use the TF weights to represent the document, each weight in the new vector will be doubled, although the expanded text does not increase the amount of information.

Therefore, to reduce the influence of length on text representation, it is necessary to normalize the feature vectors; this is also called text length normalization or length normalization for short. The common length normalization methods for text $\boldsymbol{d} = (w_1, w_2, \ldots, w_n)$ include:

a. L_1 Norm normalization

$$\boldsymbol{d}_1 = \frac{\boldsymbol{d}}{||\boldsymbol{d}||_1} = \frac{\boldsymbol{d}}{\sum_i w_i} \tag{3.6}$$

The normalized vectors are on the hyperplane $w_1 + w_2 + \cdots + w_n = 1$ in the vector space.

b. L_2 Norm normalization

$$\boldsymbol{d}_2 = \frac{\boldsymbol{d}}{||\boldsymbol{d}||_2} = \frac{\boldsymbol{d}}{\sqrt{\sum_i w_i^2}} \tag{3.7}$$

The normalized vectors are on the spherical surface $w_1^2 + w_2^2 + \cdots + w_n^2 = 1$ in the vector space. Note that the L_1 norm and L_2 norm can be generalized to the L_p norm.

c. Maximum word frequency normalization

$$\boldsymbol{d}_{max} = \frac{\boldsymbol{d}}{||\boldsymbol{d}||_\infty} = \frac{\boldsymbol{d}}{\max_i\{w_i\}} \tag{3.8}$$

It should be noted that, unlike nondimensional scale normalization, which is commonly used in machine learning and data mining tasks, text representation normalization is a process to remove the effect of text length.

3.1.4 Feature Engineering

The vector space model assumes that the coordinates in space are orthogonal, e.g., the terms constituting the document are independent of each other, regardless of their positions. Such a hypothesis de facto neglects word order, syntax, and the semantic information of the original document. For example, it is obviously unreasonable that the two texts "John is quicker than Mary" and "Mary is quicker than John," which express exactly opposite semantics, have the same text representation in VSM.

Therefore, according to task requirements, terms can be defined as keywords, chunks, and phrases, along with their positions, part-of-speech tags, syntactic structures, and semantic information. In text mining tasks, the process of manually defining such features is called "feature engineering."

We list some commonly used linguistic features as follows:

(1) *n*-gram features

The basic VSM usually takes words as terms, which neglects word order information. n-gram features take the contiguous sequence of n items as the basic unit and thereby capture part of the word order information. Take the sentence "I strongly recommend this movie" as an example. Its unigram, bigram, and trigram features are shown in Table 3.1.

Of these, the unigram is simply the word feature. n-gram features have been widely used in text classification, text clustering, and other tasks. However, as n

Table 3.1 An example of n-gram features

	I strongly recommend this movie
Unigram	I, strongly, recommend, this, movie
Bigram	I strongly, strongly recommend, recommend this, this movie
Trigram	I strongly recommend, strongly recommend this, recommend this movie

increases, the dimension of the feature space will grow dramatically, the feature vector will become sparser, and the statistical quality will be diminished, while the computational cost is increased. Furthermore, it is difficult to capture a long-distance relationship between words; for this kind of relational information, we must resort to more in-depth language processing techniques.

(2) Syntactic features

Syntactic analysis is the process of analyzing a sentence into its constituents based on grammar rules; it results in a parse tree showing their syntactic relation to each other and that may also contain semantic and other information. Dependency parsing is an important branch of syntactic analysis that describes language structure through the dependency relationship between words (Zong 2013). As a structured text representation, the dependency tree takes words as nodes and expresses the dominant and dominated relationship of words by the directional relationship between nodes. The dependency tree of the sentence "I strongly recommend this movie" is shown in Fig. 3.3.

A simple method of extracting dependency relations is to extract interdependent word pairs as terms, such as "recommend-movie" in the above example. In this way, the long-distance dependency of "recommend" and "movie" can be captured.

(3) Lexicon features

Polysemy and synonymy are common phenomena in natural language. It is important for natural language processing to identify whether two words express the same meaning and to identify the specific meaning of polysemous words in documents. External lexicons (e.g., WordNet in English, HowNet in Chinese) can help here, as we can use the semantic concepts defined in these lexicons as substitutes or supplements to words. This approach can alleviate the issues of ambiguity and diversity in natural language and improve our ability to represent the text.

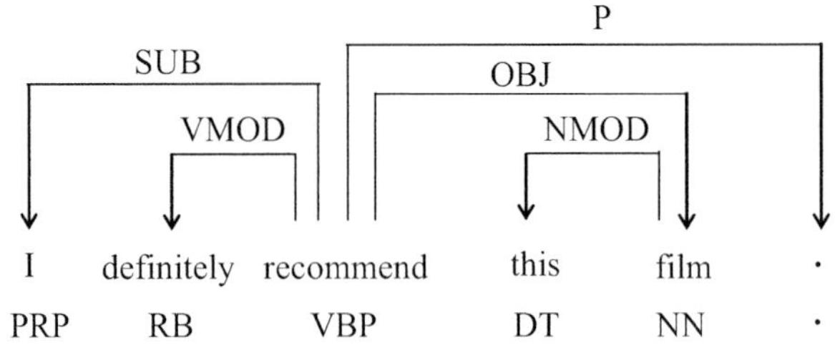

Fig. 3.3 An example of a dependency tree

3.1.5 *Other Text Representation Methods*

In addition to the traditional vector space model representing texts with high-dimensional sparse vectors, there is also a class of distributed text representation methods. Unlike the vector space model, distributed representation achieves a low-dimensional dense text representation using topic modeling or representation learning. Typical methods include concept representation and deep representation.

(1) Concept Representation

The traditional vector space model is an explicit text representation method that cannot capture the implicit semantic relationships in the text. Topic models, such as latent semantic analysis (LSA), probabilistic latent semantic analysis (PLSA), and latent Dirichlet allocation (LDA), can better capture polysemy and synonymy and mine implicit topics (also called concepts) in texts. Topic models also provide a concept representation method that transforms the high-dimensional sparse vectors in the traditional vector space model into low-dimensional dense vectors to alleviate the curse of dimensionality. We will introduce the topic model in Chap. 7.

(2) Representation Learning

The goal of deep learning for text representation is to learn low-dimensional dense vectors of text at different granularities through machine learning. In recent years, with the improvement of computing power, distributed text representation methods based on artificial neural networks have achieved great success in natural language processing, and a series of these methods have emerged. Compared with the traditional vector space model, the vector dimension of distributed representations is much lower, which can effectively alleviate the data sparsity problem and improve computational efficiency. The learning method can fully capture the semantic information and other deep information of the text in the process of constructing the text representation, and thereby both avoid the complex feature engineering required by the traditional vector space model and achieve efficient performance in many text mining tasks. In later chapters, we will introduce deep text representation and its applications in different text mining tasks.

It is also worth noting that the goal of text representation is to construct a good representation suitable for specific natural language processing tasks. For different tasks, the text representation will have a different emphasis. For example, for the sentiment analysis task, it is necessary to embody more emotional attributes in the vector space construction and representation learning process. For topic detection and tracking tasks, more event description information must be embedded. Therefore, text representation is often related to tasks, and there is essentially no general and ideal text representation for all types of tasks. When evaluating text representation methods, it is also necessary to combine the characteristics of different tasks.

The bag-of-words model is the most popular text representation method in text data mining tasks such as text classification and sentiment analysis. As mentioned earlier, the bag-of-words model regards each text as a collection of words, the size of which is determined by the vocabulary that appears in all documents. Each element in the collection indicates whether a particular word appears in the current text, or it represents the statistical weight of that particular word in the current text. It can be seen that Boolean and statistical weights are based on the string matching of the words. Therefore, discrete symbol representation is the basis of the bag-of-words model. The discrete symbol representation of a word is equivalent to a one-hot representation. That is, each word is represented by a Boolean vector whose dimension is the size of the vocabulary, where the corresponding position of the current word is 1, and all the rest are 0s. For example, if there are 50,000 different words in the training samples of the text classification task, then the size of the vocabulary is 50,000. We can number all words according to the order in which they appear in the training samples. For example, if the word "text" appears first and the word "mining" occurs last, the indices of "text" and "mining" are 1 and 50,000, respectively. Each word has a unique number that corresponds to a 50,000-dimension vector. For example, "text" corresponds to $[1, 0, 0, \ldots, 0]$, namely, all the other 49,999 positions are 0 except that the first position is 1.

There are two potential problems in this kind of representation: first, the discrete symbol matching method is prone to generate sparse data. Second, any two words in the one-hot representation method are independent of each other; that is, this method cannot capture the semantic similarity between words. In recent years, research on learning distributed text representation in low-dimensional continuous semantic vector space has generated much interest. This approach surpasses the traditional bag-of-words model and achieves state-of-the-art performance in many text mining tasks, such as text classification, sentiment analysis, and information extraction. In the remainder of this chapter, we will introduce the learning methods of distributed representations for words, phrases, sentences, and documents.

3.2 Distributed Representation of Words

The word is the smallest linguistic unit with independent meaning, and it is also the basic unit of phrases, sentences, and documents. Traditional one-hot representation methods cannot describe the grammatical and semantic information of words. Thus, research began to focus on how to encode grammatical and semantic information in word representations. Harris and Firth proposed and clarified the distributed hypothesis of words in 1954 and 1957: the semantics of a word are determined by its context. That is, words with similar contexts have similar meanings (Harris 1954; Firth 1957). If we capture all the context information of a word, we obtain the semantics of this word, and therefore, the richer the context is, the better the distributed representation describing the semantic information of the words will be. Since the 1990s, with the development of statistical learning methods and the rapid growth of text data, approaches to learning the distributed representations of words have attracted increasing attention. Generally, the core concept behind distributed representation is the use of a low-dimensional real-valued vector to represent a word so that words with similar semantics are close in the vector space. This section introduces several methods that learn distributed representations of words.

The distributed hypothesis indicates that the quality of word representation largely depends on the modeling of context information. In vector space models, the most commonly used context is a collection of words in a fixed window, but richer contexts such as n-grams are difficult to use. For example, if an n-gram is used as the context, the number of n-gram will increase exponentially as n grows, inevitably resulting in a data sparsity problem and the curse of dimensionality. Popular neural network models are composed of a series of simple operations, such as linear transformation and nonlinear activation, which in theory can simulate arbitrary functions. Therefore, complex contexts can be modeled through simple neural networks, enabling the distributed representations of words to capture more syntactic and semantic information.

The training data in the neural network model are formalized as a collection of sentences $D = \{w_{i_1}^{m_i}\}_{i=1}^{M}$, where m_i represents the number of words contained in the ith sentence and $w_{i_1}^{m_i}$ represents the word sequence in the sentence $w_{i_1}, w_{i_2}, \ldots, w_{m_i}$. The vocabulary can be obtained by enumerating the words appearing in the training data D after text preprocessing. Assuming that each word is mapped into a d-dimensional distributed vector (commonly referred to as word embedding), then vocabulary V corresponds to a word embedding matrix, i.e., $\boldsymbol{L} \in \mathbb{R}^{|V| \times d}$. The goal of the neural network model is to optimize the word embedding matrix $\boldsymbol{L}$ and learn accurate representations for each word. Next, we introduce several commonly used neural network models for word representation learning.

3.2.1 *Neural Network Language Model*

Word embedding was initially employed in learning neural network language model, which is used to calculate the occurrence probability of a piece of text and measure its fluency. Given a sentence $w_1, w_2, \ldots, w_m$ consisting of m words, its occurrence possibility can be calculated by the chain rule:

$$\begin{aligned} p(w_1 w_2 \cdots w_m) = p(w_1)p(w_2|w_1) \cdots p(w_i|w_1, \cdots, w_{i-1}) \\ \cdots p(w_m|w_1, \cdots, w_{m-1}) \end{aligned} \tag{3.9}$$

Traditional language models commonly use the maximum likelihood estimation method to calculate the conditional probability $p(w_i|w_1, \cdots, w_{i-1})$:

$$p(w_i|w_1, \cdots, w_{i-1}) = \frac{\text{count}(w_1, \cdots, w_i)}{\text{count}(w_1, \cdots, w_{i-1})} \tag{3.10}$$

The larger i is, the less likely the phrase $w_1, \ldots, w_i$ is to appear, and the less accurate the maximum likelihood estimation will be. Therefore, the typical solution is to apply the $(n-1)$-order Markov chain (the n-gram language model). Suppose that the probability of the current word only depends on the preceding $(n-1)$ words:

$$p(w_i|w_1, \cdots, w_{i-1}) \approx p(w_i|w_{i-n+1}, \cdots, w_{i-1}) \tag{3.11}$$

When $n = 1$, the model is a unigram model where the words are independent of each other. $n = 2$ denotes a bigram model where the probability of the current word relies on the previous word. $n = 3$, $n = 4$, and $n = 5$ are the most widely used n-gram language models (see Sect. 2.4.3 for more details about n-gram language model). This approximation method makes it possible to calculate the language model probability of any word sequence. However, probability estimation methods based on matching discrete symbols, such as words and phrases, still face serious data sparsity problems and cannot capture semantic similarity between words. For example, the semantics of two bigrams "very boring" and "very uninteresting" are similar, and the probabilities of $p(\text{boring}|\text{very})$ and $p(\text{uninteresting}|\text{very})$ should be very close. However, in practice, the frequency of the two bigrams in the corpus may vary greatly, resulting in a large difference between these two probabilities.

Bengio et al. proposed a language model based on a feed-forward neural network (FNN) (Bengio et al. 2003). The basic approach maps each word into a low-dimensional real-valued vector (word embedding) and calculates the probability $p(w_i|w_{i-n+1}, \cdots, w_{i-1})$ of the n-gram language model in the continuous vector space. Figure 3.4a shows a three-layer feed-forward neural network language model. The $(n-1)$ words from the historical information are mapped into word embeddings and then concatenated to obtain $\boldsymbol{h}_0$.

$$\boldsymbol{h}_0 = [\boldsymbol{e}(w_{i-n+1}); \cdots ; \boldsymbol{e}(w_{i-1})] \tag{3.12}$$

where $\boldsymbol{e}(w_{i-1}) \in \mathbb{R}^d$ denotes the d-dimensional word embedding corresponding to the word w_{i-1}, which can be obtained by retrieving the word embedding matrix $\boldsymbol{L} \in \mathbb{R}^{|V|\times d}$.[1] $\boldsymbol{h}_0$ is then fed into the linear and nonlinear hidden layers to learn an abstract representation of the $(n-1)$ words.

$$\boldsymbol{h}_1 = f(\boldsymbol{U}^1 \times \boldsymbol{h}_0 + \boldsymbol{b}^1) \tag{3.13}$$

$$\boldsymbol{h}_2 = f(\boldsymbol{U}^2 \times \boldsymbol{h}_1 + \boldsymbol{b}^2) \tag{3.14}$$

where the nonlinear activation function can be $f(\cdot) = \tanh(\cdot)$. Finally, the probability distribution of each word in V can be calculated by the softmax function:

$$p(w_i|w_{i-n+1}, \cdots, w_{i-1}) = \frac{\exp\{\boldsymbol{h}_2 \cdot \boldsymbol{e}(w_i)\}}{\sum_{k=1}^{|V|} \exp\{\boldsymbol{h}_2 \cdot \boldsymbol{e}(w_k)\}} \tag{3.15}$$

[1]The word embeddings are usually randomly initialized and updated during training.

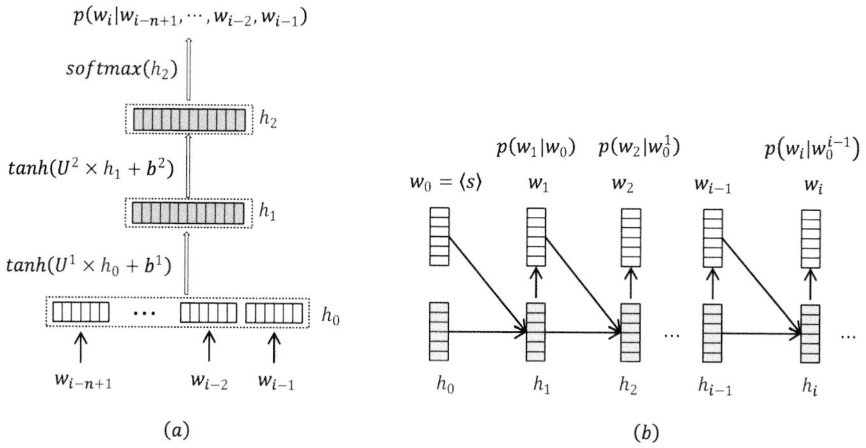

Fig. 3.4 Illustration of neural network language models: (**a**) feed-forward neural network language model, (**b**) recurrent neural network language model

In the above formulas, the weighting matrices $\boldsymbol{U}^1$, $\boldsymbol{U}^2$, $\boldsymbol{b}^1$, $\boldsymbol{b}^2$ and word embedding matrix $\boldsymbol{L}$ are all trainable neural network parameters θ. The training process optimizes the parameter θ to maximize the log-likelihood of the complete set of training data

$$\theta^* = \underset{\theta}{\operatorname{argmax}} \sum_{m=1}^{M} \log p(w_{i_1}^{m_i}) \tag{3.16}$$

where M is the size of the training data. After training the language model, the optimized word embedding matrix $\boldsymbol{L}^*$ is obtained, which contains the distributed vector representations of all words in the vocabulary V. Note that the logarithms in this book are based on 2 if not denoted otherwise.

Since an FNN can only model the context of a fixed window and cannot capture long-distance context dependency, Mikolov et al. proposed using a recurrent neural network (RNN) to directly model probability $p(w_i|w_1, \cdots, w_{i-1})$ (Mikolov et al. 2010), aiming at utilizing all historical information $w_1, \cdots, w_{i-1}$ to predict the probability of the current word w_i. The key point of recurrent neural networks is that they calculate the hidden layer representation $\boldsymbol{h}_i$ at each time step:

$$\boldsymbol{h}_i = f(\boldsymbol{W} \times \boldsymbol{e}(w_{i-1}) + \boldsymbol{U} \times \boldsymbol{h}_{i-1} + \boldsymbol{b}) \tag{3.17}$$

The hidden layer representation $\boldsymbol{h}_{i-1}$of time $i-1$ contains historical information from time step 0 to $(i-1)$. (The historical information of time 0 is usually set to empty, i.e., $\boldsymbol{h}_0 = 0$.) $f(\cdot)$ is a nonlinear activation function, which can be $f(\cdot) = \tanh(\cdot)$. Based on the i-th hidden layer representation $\boldsymbol{h}_i$, the probability of the next word $p(w_i|w_1, \ldots, w_{i-1})$ can be calculated directly by the softmax function,

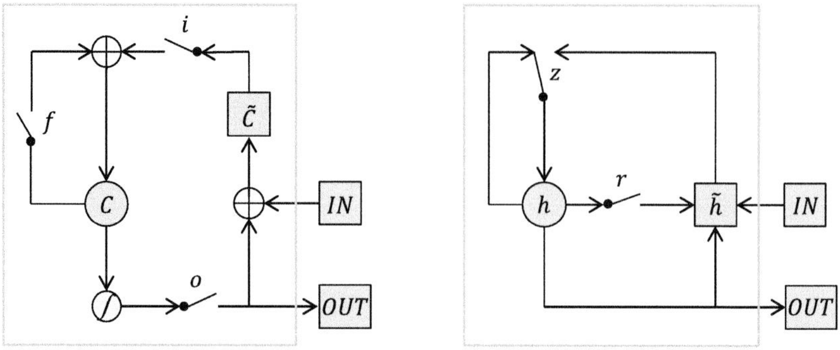

(a) (b)

Fig. 3.5 Illustration of neural units in LSTM and GRU networks: (**a**) LSTM unit, (**b**) GRU unit

as shown in Fig. 3.4b. The optimization method for neural network parameters and word embedding matrices is similar to a feed-forward neural network, which maximizes the log-likelihood of training data.

To further investigate the information passing between hidden layers ($\boldsymbol{h}_{i-1}$) and ($\boldsymbol{h}_i$) and effectively encode long-distance historical information, $f(\cdot)$ can be implemented by long short-term memory (LSTM) (Hochreiter and Schmidhuber 1997) (Fig. 3.5a) or gated recurrent unit (GRU) (Cho et al. 2014) (Fig. 3.5b) networks. For both LSTM and GRU, the input includes the hidden layer representation at the previous step $\boldsymbol{h}_{i-1}$ and the output of the previous step w_{i-1}, and the output is the hidden layer representation of the current step $\boldsymbol{h}_i$.

As shown in Fig. 3.5a, the LSTM is controlled by three gates and one memory cell. The calculation process is as follows:

$$\boldsymbol{i}_i = \sigma(\boldsymbol{W}_i \times \boldsymbol{e}(w_{i-1}) + \boldsymbol{U}_i \times \boldsymbol{h}_{i-1} + \boldsymbol{b}_i) \tag{3.18}$$

$$\boldsymbol{f}_i = \sigma(\boldsymbol{W}_f \times \boldsymbol{e}(w_{i-1}) + \boldsymbol{U}_f \times \boldsymbol{h}_{i-1} + \boldsymbol{b}_f) \tag{3.19}$$

$$\boldsymbol{o}_i = \sigma(\boldsymbol{W}_o \times \boldsymbol{e}(w_{i-1}) + \boldsymbol{U}_o \times \boldsymbol{h}_{i-1} + \boldsymbol{b}_o) \tag{3.20}$$

$$\tilde{\boldsymbol{c}}_i = \tanh(\boldsymbol{W}_c \times \boldsymbol{e}(w_{i-1}) + \boldsymbol{U}_c \times \boldsymbol{h}_{i-1} + \boldsymbol{b}_c) \tag{3.21}$$

$$\boldsymbol{c}_i = \boldsymbol{f}_i \odot \boldsymbol{c}_{i-1} + \boldsymbol{i}_i \odot \tilde{\boldsymbol{c}}_i \tag{3.22}$$

$$\boldsymbol{h}_i = \boldsymbol{o}_i \odot \tanh(\boldsymbol{c}_i) \tag{3.23}$$

where $\sigma(x) = \frac{1}{1+e^{(-x)}}$, $\boldsymbol{i}_i$, $\boldsymbol{f}_i$, and $\boldsymbol{o}_i$ denote the input gate, forget gate and output gate, respectively, and $\boldsymbol{c}_i$ denotes the memory cell. The LSTM is expected to selectively encode historical and current information through the three gates.

As shown in Fig. 3.5b, a GRU is a simplified version of an LSTM that omits the memory cell.

$$\boldsymbol{r}_i = \sigma(\boldsymbol{W}_r \times \boldsymbol{e}(w_{i-1}) + \boldsymbol{U}_r \times \boldsymbol{h}_{i-1} + \boldsymbol{b}_r) \tag{3.24}$$

$$z_i = \sigma(\boldsymbol{W}_z \times \boldsymbol{e}(w_{i-1}) + \boldsymbol{U}_z \times \boldsymbol{h}_{i-1} + \boldsymbol{b}_z) \tag{3.25}$$

$$\tilde{\boldsymbol{h}}_i = \tanh(\boldsymbol{W} \times \boldsymbol{e}(w_{i-1}) + \boldsymbol{U} \times (\boldsymbol{r}_i \odot \boldsymbol{h}_{i-1}) + \boldsymbol{b}) \tag{3.26}$$

$$\boldsymbol{h}_i = z_i \odot \tilde{\boldsymbol{h}}_i + (1 - z_i) \odot \boldsymbol{h}_{i-1} \tag{3.27}$$

where $\boldsymbol{r}_i$ and z_i are the reset gate and the update gate, respectively. LSTMs and GRUs can effectively capture long-distance semantic dependencies, and thus they have better performance in many text mining tasks, such as text summarization and information extraction (Nallapati et al. 2016; See et al. 2017).

3.2.2 C&W Model

In a neural network language model, word embedding is not the goal but only the by-product. Collobert and Weston (2008) proposed a model that directly aims at learning and optimizing word embeddings. The model is named after the first letter of these two researchers' names and thus called the C&W model.

The goal of a neural network language model is to accurately estimate the conditional probability $p(w_i|w_1, \ldots, w_{i-1})$. Therefore, it is necessary to calculate the probability distribution of the whole vocabulary by using the matrix operation from the hidden layer to the output layer using the softmax function at every time step. The computational complexity is $O(|\boldsymbol{h}| \times |V|)$, where $|\boldsymbol{h}|$ is the number of neurons in the highest hidden layer (usually hundreds or one thousand) and $|V|$ is the size of the vocabulary (usually tens of thousands to hundreds of thousands). This matrix operation greatly reduces the efficiency of model training. Collobert and Weston argued that it is not necessary to learn a language model if the goal is only to learn word embeddings. Instead, the model and objective function can be designed directly from the perspective of the distributed hypothesis: given an n-gram ($n = 2C+1$) $(w_i, C) = w_{i-C} \cdots w_{i-1} w_i w_{i+1} \cdots w_{i+C}$ in the training corpus (here, C is the window size), randomly replace the central word w_i with other words w_i' in the vocabulary and obtain a new n-gram $(w_i, C) = w_{i-C} \cdots w_{i-1} w_i' w_{i+1} \cdots w_{i+C}$, where (w_i, C) is no doubt more reasonable than (w_i', C). When scoring each n-gram, then the score of (w_i, C) must be higher than (w_i', C):

$$s(w_i, C) > s(w_i', C) \tag{3.28}$$

As shown in Fig. 3.6, a simple feed-forward neural network model only needs to calculate the score of the n-gram, which distinguishes the real n-gram input from

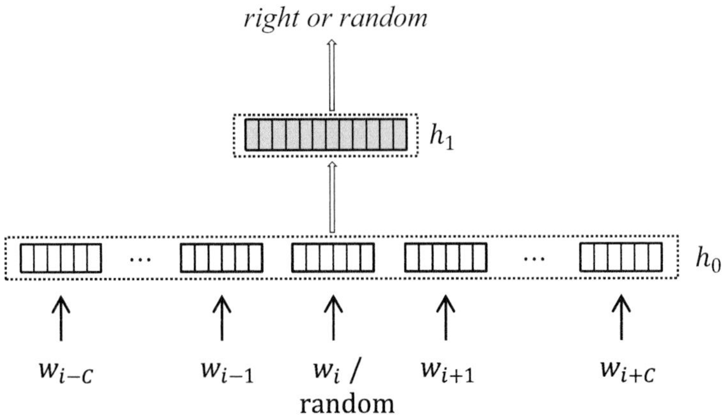

Fig. 3.6 Illustration of the C&W model

randomly generated text. We call n-gram (w_i, C) in the training data a positive sample and randomly generated n-gram (w_i', C) a negative sample.

To calculate $s(w_i, C)$, we first acquire the corresponding word embeddings from the word embedding matrix $\boldsymbol{L}$; these are then concatenated, and we obtain the representation of the first layer $\boldsymbol{h}_0$:

$$\boldsymbol{h}_0 = [\boldsymbol{e}(w_{i-C}); \cdots; \boldsymbol{e}(w_{i-1}); \boldsymbol{e}(w_i); \boldsymbol{e}(w_{i+1}); \cdots; \boldsymbol{e}(w_{i+C})] \tag{3.29}$$

$\boldsymbol{h}_0$ is passed through the hidden layer, resulting in $\boldsymbol{h}_1$:

$$\boldsymbol{h}_1 = f(\boldsymbol{W}_0 \times \boldsymbol{h}_0 + \boldsymbol{b}_0) \tag{3.30}$$

where $f(\cdot)$ is a nonlinear activation function. After linear transformation, the score of n-gram (w_i, C) becomes

$$s(w_i, C) = \boldsymbol{W}_1 \times \boldsymbol{h}_1 + \boldsymbol{b}_1 \tag{3.31}$$

where $\boldsymbol{W}_1 \in \mathbb{R}^{(1\times|\boldsymbol{h}_1|)}$, $\boldsymbol{b}_1 \in \mathbb{R}$. It can be seen that the matrix operation of the C&W model between the hidden layer and the output layer is very simple, reducing the computational complexity from $O(|\boldsymbol{h}|\times|V|)$ to $O(|\boldsymbol{h}|)$ and improving the efficiency of learning word vector representations.

In the optimization process, the C&W model expects that the score of each positive sample will be larger than that of the corresponding negative sample by a constant margin:

$$s(w_i, C) > s(w_i', C) + 1 \tag{3.32}$$

For the entire training corpus, the C&W model needs to traverse every n-gram in the corpus and minimize the following functions:

$$\sum_{(w_i,C)\in D} \sum_{(w_i',C)\in N_{w_i}} \max(0, 1 + s(w_i', C) - s(w_i, C)) \tag{3.33}$$

in which N_{w_i} is the negative sample set for the positive sample (w_i, C). In practice, we can choose one or several negative samples for each positive sample.

3.2.3 CBOW and Skip-Gram Model

The hidden layer is an indispensable component of both the neural network language model and the C&W model, and the matrix operation from the input layer to the hidden layer is also a key time-consuming step. To further simplify the neural network and learn word embeddings more efficiently, Mikolov et al. proposed two kinds of neural network models without hidden layers: the continuous bag-of-words (CBOW) model and the skip-gram model (Mikolov et al. 2013b).

(1) CBOW Model

As shown in Fig. 3.7, the concept behind the CBOW model is similar to that of the C&W model: input contextual words and predict the central word. However, unlike the C&W model, CBOW still takes the probability of target words as the optimization goal, and it simplifies the network structure by focusing on two aspects. First, the input layer is no longer a concatenation of the corresponding contextual word embeddings, but the average of these word embeddings, ignoring the word order information; second, it omits the hidden layer, instead connecting the input layer and the output layer directly and calculating the probability of the central word by logistic regression.

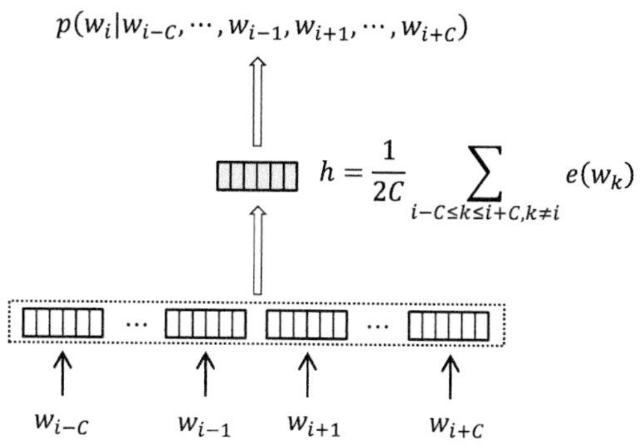

Fig. 3.7 Illustration of the CBOW model

Formally, given any n-gram ($n = 2C + 1$) $(w_i, C) = w_{i-C} \cdots w_{i-1} w_i w_{i+1} \cdots w_{i+C}$ in the training corpus as input, the average word embedding of contextual words can be calculated as follows:

$$\boldsymbol{h} = \frac{1}{2C} \sum_{i-C \leq k \leq i+C, k \neq i} \boldsymbol{e}(w_k) \tag{3.34}$$

$\boldsymbol{h}$ is directly taken as the semantic representation of the context to predict the probability of the middle word w_i

$$p(w_i | C_{w_i}) = \frac{\exp\{\boldsymbol{h} \cdot \boldsymbol{e}(w_i)\}}{\sum_{k=1}^{|V|} \exp\{\boldsymbol{h} \cdot \boldsymbol{e}(w_k)\}} \tag{3.35}$$

where C_{w_i} denotes word contexts for the word w_i in a C-sized window.

In the CBOW model, word embedding matrix $\boldsymbol{L}$ is the only parameter in the neural network. For the whole training corpus, the CBOW model optimizes $\boldsymbol{L}$ to maximize the log-likelihood of all words:

$$\boldsymbol{L}^* = \underset{\boldsymbol{L}}{\operatorname{argmax}} \sum_{w_i \in V} \log p(w_i | C_{w_i}) \tag{3.36}$$

(2) Skip-Gram Model

Unlike the CBOW model, the skip-gram model has the opposite process, that is, it aims to predict all the contextual words given only the central word. Figure 3.8 shows the basic idea of the skip-gram model.

Given any n-gram in the training corpus $(w_i, C) = w_{i-C} \cdots w_{i-1} w_i w_{i+1} \cdots w_{i+C}$, the skip-gram model predicts the probability of every word w_c in the context $C_{w_i} = w_{i-C} \cdots w_{i-1} w_{i+1} \cdots w_{i+C}$ by using the word embedding $\boldsymbol{e}(w_i)$ of the central word w_i:

$$p(w_c | w_i) = \frac{\exp\{\boldsymbol{e}(w_i) \cdot \boldsymbol{e}(w_c)\}}{\sum_{k=1}^{|V|} \exp\{\boldsymbol{e}(w_i) \cdot \boldsymbol{e}(w_k)\}} \tag{3.37}$$

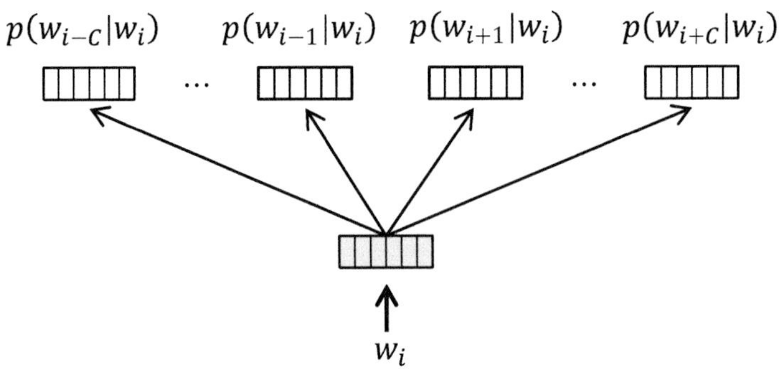

Fig. 3.8 Illustration of the skip-gram model

The objective function of the skip-gram model is similar to that of the CBOW model. It optimizes the word embedding matrix $\boldsymbol{L}$ to maximize the log-likelihood of all contextual words for each n-gram in the training data:

$$\boldsymbol{L}^* = \underset{\boldsymbol{L}}{\operatorname{argmax}} \sum_{w_i \in V} \sum_{w_c \in C_{w_i}} \log p(w_c|w_i) \tag{3.38}$$

3.2.4 Noise Contrastive Estimation and Negative Sampling

Although CBOW and skip-gram greatly simplify the structure of the neural network, it is still necessary to calculate the probability distribution of all the words in vocabulary V by using the softmax function. To speed up the training efficiency, inspired by the C&W model and noise contrastive estimation (NCE) method, Mikolov et al. (2013a) proposed negative sampling (NEG) technology.

Taking the skip-gram model as an example, each word w_c in the context $C_{w_i} = w_{i-C} \cdots w_{i-1} w_{i+1} \cdots w_{i+C}$ is predicted by the central word w_i. Negative sampling and noise contrastive estimation methods select K negative samples $w'_1, w'_2, \cdots, w'_K$ from a probability distribution $p_n(w)$ for each positive sample to maximize the likelihood of positive samples while minimizing the likelihood of negative samples.

For a positive sample w_c and K negative samples $w'_1, w'_2, \cdots, w'_K$, the noise contrastive estimation method first normalizes the probability of $K+1$ samples:

$$\begin{aligned} p(l=1, w|w_i) &= p(l=1) \times p(w|l=1, w_i) \\ &= \frac{1}{K+1} p_\theta(w|w_i) \end{aligned} \tag{3.39}$$

$$\begin{aligned} p(l=0, w|w_i) &= p(l=0) \times p(w|l=0, w_i) \\ &= \frac{K}{K+1} p_n(w) \end{aligned} \tag{3.40}$$

$$\begin{aligned} p(l=1|w, w_i) &= \frac{p(l=1, w|w_i)}{p(l=0, w|w_i) + p(l=1, w|w_i)} \\ &= \frac{p_\theta(w|w_i)}{p_\theta(w|w_i) + K p_n(w)} \end{aligned} \tag{3.41}$$

$$\begin{aligned} p(l=0|w, w_i) &= \frac{p(l=0, w|w_i)}{p(l=0, w|w_i) + p(l=1, w|w_i)} \\ &= \frac{K p_n(w)}{p_\theta(w|w_i) + K p_n(w)} \end{aligned} \tag{3.42}$$

where w denotes a sample. $l = 1$ indicates that it is from the positive samples and follows the output probability distribution of the neural network model $p_\theta(w|w_i)$.[2] $l = 0$ indicates that the sample is from the negative samples and obeys the probability distribution of noisy samples $p_n(w)$. The objective function of noise contrastive estimation is

$$J(\theta) = \log p(l = 1|w_c, w_i) + \sum_{k=1}^{K} \log p(l = 0|w_k, w_i) \tag{3.43}$$

The objective function of negative sampling is the same as that of noise contrastive estimation. The difference is that the negative sampling method does not normalize the probability of the samples but directly uses the output of the neural network language model:

$$p(l = 1|w_c, w_i) = \frac{1}{1 + \mathrm{e}^{-\boldsymbol{e}(w_i)\cdot \boldsymbol{e}(w_c)}} \tag{3.44}$$

Then, the objective function can be simplified as follows:

$$\begin{aligned} J(\theta) &= \log p(l = 1|w_c, w_i) + \sum_{k=1}^{K} \log p(l = 0|w_k, w_i) \\ &= \log p(l = 1|w_c, w_i) + \sum_{k=1}^{K} \log(1 - p(l = 1|w_k, w_i)) \\ &= \log \frac{1}{1 + \mathrm{e}^{-\boldsymbol{e}(w_i)\cdot \boldsymbol{e}(w_c)}} + \sum_{k=1}^{K} \log \left(1 - \frac{1}{1 + \mathrm{e}^{-\boldsymbol{e}(w_k)\cdot \boldsymbol{e}(w_c)}}\right) \\ &= \log \frac{1}{1 + \mathrm{e}^{-\boldsymbol{e}(w_i)\cdot \boldsymbol{e}(w_c)}} + \sum_{k=1}^{K} \log \left(\frac{1}{1 + \mathrm{e}^{\boldsymbol{e}(w_k)\cdot \boldsymbol{e}(w_c)}}\right) \\ &= \log \sigma(\boldsymbol{e}(w_i) \cdot \boldsymbol{e}(w_c)) + \sum_{k=1}^{K} \log \sigma(-\boldsymbol{e}(w_k) \cdot \boldsymbol{e}(w_c)) \end{aligned} \tag{3.45}$$

Mikolov et al. found that the model can obtain decent performance when the number of negative samples is $K = 5$. In other words, the negative sampling method can greatly lower the complexity of probability estimation, remarkably improving the learning efficiency of word embeddings.

---

[2] $p_\theta(w|w_i) = \frac{\exp\{\boldsymbol{h}\cdot \boldsymbol{e}(w)\}}{\sum_{k=1}^{|V|} \exp\{\boldsymbol{h}\cdot \boldsymbol{e}(w_k)\}} = \frac{\exp\{\boldsymbol{h}\cdot \boldsymbol{e}(w)\}}{z(w)}$, and $z(w)$ is usually set as a constant 1.0 in NCE.

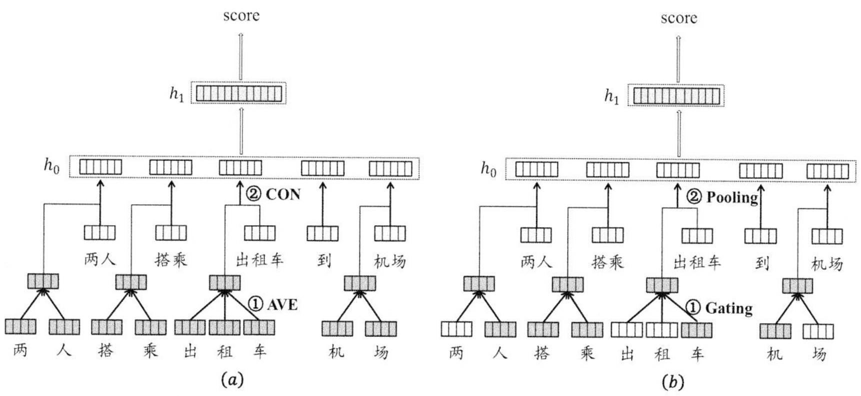

Fig. 3.9 Distributed representation based on the hybrid character-word method

3.2.5 Distributed Representation Based on the Hybrid Character-Word Method

Learning word representations based on distributed hypotheses requires sufficient context information to capture word semantics. That is, a word should have sufficiently high occurrence frequency. However, according to Zipf's Law, most words appear rarely in the corpus. For these words, it is impossible to obtain a high-quality word embedding.

Although words are the smallest semantic unit that can be used independently, they are not the smallest linguistic unit; for example, English words are composed of letters, and Chinese words are composed of characters. Taking Chinese words as an example, researchers found that 93% of Chinese words satisfy or partially satisfy the characteristics of semantic composition,[3] which means that these words are semantically transparent. If a word is semantically transparent, it indicates that the semantics of this word can be composed from its internal Chinese characters. As shown in Fig. 3.9, the semantics of the word 出租车 (chuzuche, taxi) can be obtained by the composition of the semantics of the three Chinese characters 出 (chu, out), 租 (zu, rent), and 车 (che, car). Compared to the size of the word vocabulary, the Chinese character set is limited: according to national standard GB2312, there are fewer than 7000 commonly used Chinese characters. In addition, the frequency of Chinese characters in the corpus is relatively high, leading to high-quality character embeddings under the distributed hypothesis. Therefore, if we can exploit the semantic representation of Chinese characters and design a reasonable semantic composition function, then we can greatly enhance the representation ability of Chinese words (especially low-frequency words). Based on this idea, increasing attention is being given to learning the distributed representation based

[3] 30% satisfy and 70% partially satisfy the semantic composition property.

on the hybrid character-word method (Chen et al. 2015a; Xu et al. 2016; Wang et al. 2017a).

There are many kinds of methods that can be applied to learn the distributed representation based on hybrid character words, with two main differences between them: how to design a reasonable semantic composition function and how to integrate the compositional semantics of Chinese characters with the atomic semantics of Chinese words. We will next take the C&W model as an example to introduce two methods based on hybrid character-word mechanisms.

The goal of these methods is still to distinguish real n-grams from noisy random n-grams, and thus the core task is still to calculate the score of an n-gram. Figure 3.9a is a simple and direct hybrid character-word method. Suppose a Chinese word $w_i = c_1c_2 \cdots c_l$ consists of l characters (e.g., 出租车 (chuzuche, taxi) consists of three characters). This method first learns the semantic vector composition representation $\boldsymbol{x}(c_1c_2 \cdots c_l)$ of the Chinese character string $c_1c_2 \cdots c_l$ and the atomic vector representation $\boldsymbol{e}(w_i)$ of the Chinese word w_i. Assuming that each Chinese character makes an equal contribution, then $\boldsymbol{x}(c_1c_2 \cdots c_l)$ can be obtained by averaging the character embeddings

$$\boldsymbol{x}(c_1c_2 \cdots c_l) = \frac{1}{l}\sum_{k=1}^{l} \boldsymbol{e}(c_k) \tag{3.46}$$

where $\boldsymbol{e}(c_k)$ denotes the vector representation of the character c_k. To obtain the final word embedding, the method concatenates the compositional representation of characters and the embedding of atomic Chinese words directly:

$$\boldsymbol{X}_i = [\boldsymbol{x}(c_1c_2 \cdots c_l); \boldsymbol{e}(w_i)] \tag{3.47}$$

$\boldsymbol{h}_0$, $\boldsymbol{h}_1$, and the final score is calculated in the same manner as that of the C&W model.

It is obvious that the above method does not consider the different contributions of the internal Chinese characters on the compositional semantics, nor does it consider the different contributions of compositional semantics and atomic semantics on the final word embedding. For example, in the Chinese word 出租车 (taxi), the character 车 (car) is the most important, while 租 (rent) and 出 (out) only play a modifying role with a relatively small contribution. Clearly, different Chinese characters should not be equally treated. Furthermore, some words are semantically transparent, so greater consideration should be given to compositional semantics, while others are nontransparent (such as 苗条 (miaotiao, slim)), requiring greater reliance on the atomic semantics of the word. Figure 3.9b shows a hybrid character-word method that takes into account both of the above factors. First, the compositional semantics of the characters are obtained through a gate mechanism

$$\boldsymbol{x}(c_1c_2\cdots c_l) = \sum_{k=1}^{l} \boldsymbol{v}_k \odot \boldsymbol{e}(c_k) \tag{3.48}$$

where $\boldsymbol{v}_k \in \mathbb{R}^d$ (d is the embedding size of $\boldsymbol{e}(c_k)$) denotes the controlling gate that controls the contribution of character c_k to the word vector $\boldsymbol{x}(c_1c_2\cdots c_l)$. The gate can be calculated in the following way:

$$\boldsymbol{v}_k = \tanh(\boldsymbol{W} \times [\boldsymbol{x}(c_k); \boldsymbol{e}(w_i)]) \tag{3.49}$$

in which $\boldsymbol{W} \in \mathbb{R}^{d\times 2d}$. The compositional semantics and atomic semantics are integrated through the max-pooling method:

$$\boldsymbol{x}_j^* = \max(\boldsymbol{x}(c_1c_2\cdots c_l)_j, \boldsymbol{e}(w_i)_j) \tag{3.50}$$

This means that the j-th element of the final vector is the larger one of the compositional representation and the atomic representation at the index j. Through the pooling mechanism, the final semantics of a word depend more on the properties of the word (transparent or nontransparent). Extensive experiments demonstrate that word embeddings considering inner characters are much better.

3.3 Distributed Representation of Phrases

In statistical natural language processing, the phrase generally refers to a continuous word sequence and not only to noun phrases, verb phrases, or prepositional phrases in the syntactic view. There are two main types of methods for learning distributed representations of phrases. The first treats the phrase as an indivisible semantic unit and learns the phrase representation based on the distributed hypothesis. The second considers phrasal semantics as being composed of internal words and aims to learn the composition mechanism among words.

Compared to words, phrases are much more infrequent, and the quality of phrasal vector representation based on distributed hypotheses cannot be guaranteed. Mikolov et al. consider only some common English phrases (such as "New York Times" and "United Nations") as inseparable semantic units and treat them as words (such as "New_York_Times" and "United_Nations"), then using CBOW or skip-gram to learn the corresponding distributed representations. It is easy to see that this method cannot be applied to the majority of phrases.

3.3.1 Distributed Representation Based on the Bag-of-Words Model

For phrases, representation learning based on compositional semantics is a more natural and reasonable method. The central problem is how to compose the semantics of words into the semantics of phrases. Given a phrase $ph = w_1 w_2 \cdots w_i$ consisting of i words, the simplest method of semantic composition is to use the bag-of-words model (Collobert et al. 2011), which averages the word embeddings or draws the maximum of each dimension in the word embeddings:

$$\boldsymbol{e}(ph) = \frac{1}{i}\sum_{k=1}^{i} \boldsymbol{e}(w_k) \tag{3.51}$$

$$\boldsymbol{e}(ph) = \max_{k=1}^{d}(\boldsymbol{e}(w_1)_k, \boldsymbol{e}(w_2)_k, \cdots, \boldsymbol{e}(w_i)_k) \tag{3.52}$$

Obviously, this method does not consider the contributions of different words in the phrase, nor does it model the order of words. The former problem can be solved by adding weights to each word embedding

$$\boldsymbol{e}(ph) = \frac{1}{i}\sum_{k=1}^{i} v_k \times \boldsymbol{e}(w_k) \tag{3.53}$$

where v_k can be the word frequency or TF-IDF of w_k. We can apply a gate mechanism to control the contribution of different words, as done in the hybrid character-word method.

3.3.2 Distributed Representation Based on Autoencoder

As mentioned earlier, there is another problem in the phrase representation learning method based on the bag-of-words model: it cannot capture the word order information of the phrase. In many cases, different word orders mean completely different semantics. For example, two phrases, "cat eats fish" and "fish eats cat," share the same words but have opposite meanings. Therefore, the distributed representations of phrases require effectively modeling the word order. In this section, we will introduce a typical method, namely, the recursive autoencoder (RAE) (Socher et al. 2011b).

As the name implies, the recursive autoencoder merges the vector representations of two subnodes from bottom to top in a recursive way until the phrase vector representation is obtained. Figure 3.10 shows an example where a recursive

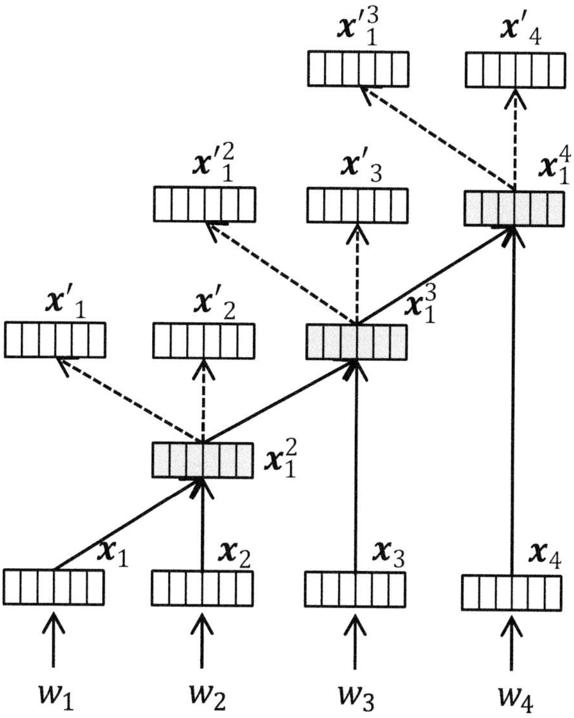

Fig. 3.10 Illustration of the recursive autoencoder

autoencoder is applied to a binary tree. Each node in the tree uses the same standard autoencoder.

The goal of a standard autoencoder is to learn a compact and abstract vector representation for a given input. For example, given the representations of the first two input words $\boldsymbol{x}_1$ and $\boldsymbol{x}_2$ in Fig. 3.10, the standard autoencoder learns an abstract representation $\boldsymbol{x}_1^2$ in the following way:

$$\boldsymbol{x}_1^2 = f(\boldsymbol{W}^{(1)} \times [\boldsymbol{x}_1; \boldsymbol{x}_2] + \boldsymbol{b}^{(1)}) \tag{3.54}$$

where $\boldsymbol{W}^{(1)} \in \mathbb{R}^{d\times 2d}$, $\boldsymbol{b}^{(1)} \in \mathbb{R}^d$, $f(\cdot) = \tanh(\cdot)$. The input includes two d-dimensional vectors $\boldsymbol{x}_1$ and $\boldsymbol{x}_2$, and the output is a d-dimensional vector $\boldsymbol{x}_1^2$, which is expected to be a compressed abstract representation of $\boldsymbol{x}_1$ and $\boldsymbol{x}_2$. To guarantee the quality of $\boldsymbol{x}_1^2$, the input should be reconstructed from the output $\boldsymbol{x}_1^2$

$$[\boldsymbol{x}'_1; \boldsymbol{x}'_2] = f(\boldsymbol{W}^{(2)} \times \boldsymbol{x}_1^2 + \boldsymbol{b}^{(2)}) \tag{3.55}$$

where $W^{(2)} \in \mathbb{R}^{2d \times d}$, $b^{(1)} \in \mathbb{R}^{2d}$, $f(\cdot) = \tanh(\cdot)$. The standard autoencoder requires the error between the input $[x_1; x_2]$ and the reconstructed input $[x_1'; x_2']$ to be as small as possible:

$$E_{rec}([x_1; x_2]) = \frac{1}{2}||[x_1; x_2] - [x_1'; x_2']||^2 \tag{3.56}$$

With x_1^2 and x_3 as input, the same autoencoder can obtain the representation x_1^3 of the phrase w_1^3. Then, with x_1^3 and x_4 as input, we can obtain the representation x_1^4 of the whole phrase.

As an unsupervised method, the recursive autoencoder takes the sum of the phrase reconstruction errors as the objective function.

$$E_\theta(ph_i) = \underset{bt \in A(ph_i)}{\text{argmin}} \sum_{nd \in bt} E_{rec}(nd) \tag{3.57}$$

where $A(ph_i)$ denotes all possible binary trees corresponding to the phrase ph_i, nd is any arbitrary node on the particular binary tree b_t, and $E_{rec}(nd)$ denotes the reconstruction error of the node nd.

To test the quality of the vector representation of the complete phrase, we can evaluate whether phrases with similar semantics will be clustered in the semantic vector space. Suppose the phrase training set is $S(ph)$; for an unknown phrase ph^*, we use the cosine distance between the phrase vectors to measure the semantic similarity between any two phrases. Then, we search a phrase list $List(ph^*)$ that is similar to ph^* from $S(ph)$ and verify whether $List(ph^*)$ and ph^* are truly semantically similar. The first column in Table 3.2 gives four test phrases with different lengths in English. The second column shows a list of similar candidate phrases found by an unsupervised recursive autoencoder RAE in the vector space.

Table 3.2 A comparison between RAE and BRAE in the semantic representation of a phrase

Input phrase	RAE	BRAE
Military force	Core force	Military power
	Main force	Military strength
	Labor force	Armed forces
At a meeting	To a meeting	At the meeting
	At a rate	During the meeting
	A meeting	At the conference
Do not agree	One can accept	Do not favor
	I can understand	Will not compromise
	Do not want	Not to approve
Each people in this nation	Each country regards	Every citizen in this country
	Each country has its	At the people in the country
	Each other, and	People all over the country

RAE can capture the structural information of the phrase to some extent, such as "military force" and "labor force", "do not agree" and "do not want". However, it still lacks the ability to encode the semantics of the phrase.

Ideally, if some phrases exist with correct semantic representation as supervised information, then the recursive autoencoder can learn phrase representation in a supervised way. However, the correct semantic representation does not exist in reality. To make the vector representation describe enough semantic information, Zhang et al. proposed a novel framework, the bilingually constrained recursive autoencoder (BRAE) (Zhang et al. 2014). The assumption is that two phrases for which one would be translated as the other have the same semantics, so they should share the same vector representation. Based on this premise, the phrase vector representation of both languages can be trained simultaneously with a co-training method. To this end, two recursive autoencoders are first used to learn the initial representations of language X and language Y in an unsupervised manner. Then, these two recursive autoencoders are optimized by minimizing the semantic distance between the translation pair (ph_x, ph_y) in languages X and Y. Figure 3.11 shows the basic architecture of this method.

The objective function of this method consists of two parts: the reconstruction error of the recursive autoencoder and the semantic error between the translation pair

$$E(ph_x, ph_y; \theta) = \alpha E_{rec}(ph_x, ph_y; \theta) + (1 - \alpha) E_{sem}(ph_x, ph_y; \theta) \tag{3.58}$$

where $E_{rec}(ph_x, ph_y; \theta)$ denotes the reconstruction error of these two phrases ph_x, ph_y, $E_{sem}(ph_x, ph_y; \theta)$ denotes the semantic error between the two phrases,

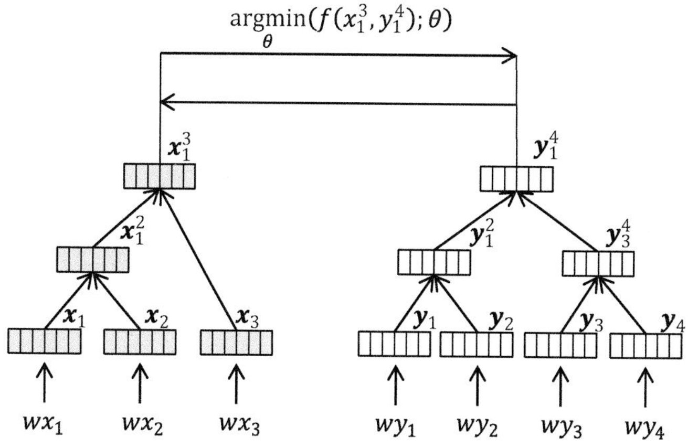

Fig. 3.11 Illustration of the bilingual constrained recursive autoencoder

and α is a weight that balances the reconstruction error and the semantic error. Specifically, the reconstruction error $E_{rec}(ph_x, ph_y; \theta)$ includes two parts:

$$E_{rec}(ph_x, ph_y; \theta) = E_{rec}(ph_x; \theta) + E_{rec}(ph_y; \theta) \tag{3.59}$$

The method for calculating the phrase reconstruction error is the same as that used by the unsupervised recursive autoencoder. $E_{sem}(ph_x, ph_y; \theta)$ contains semantic errors of two directions:

$$E_{sem}(ph_x, ph_y; \theta) = E_{sem}(ph_x|ph_y; \theta) + E_{sem}(ph_y|ph_x; \theta) \tag{3.60}$$

$$E_{sem}(ph_x|ph_y; \theta) = \frac{1}{2}||\boldsymbol{x}(ph_x) - f(\mathbf{W}_x^l \boldsymbol{y}(ph_y) + \boldsymbol{b}_x^l)||^2 \tag{3.61}$$

$$E_{sem}(ph_y|ph_x; \theta) = \frac{1}{2}||\boldsymbol{y}(ph_y) - f(\mathbf{W}_y^l \boldsymbol{y}(ph_x) + \boldsymbol{b}_y^l)||^2 \tag{3.62}$$

For a phrase set $PH_{xy} = (ph_x^{(i)}, ph_y^{(i)})_{i=1}^N$ with N translation pairs, the method attempts to minimize the error on the whole set:

$$J_{BRAE}(PH_{xy}; \theta) = \frac{1}{N} \sum_{(ph_x, ph_y) \in PH_{xy}} E(ph_x, ph_y; \theta) + \frac{\lambda}{2}||\theta||^2 \tag{3.63}$$

The second item indicates the parameter regularization term. In addition to minimizing the semantic distance between translation phrases, it can also maximize the semantic distance between nontranslation phrases

$$E_{sem}^*(ph_x, ph_y; \theta) = \max\{0, E_{sem}(ph_x, ph_y; \theta) - E_{sem}(ph_x, ph_y'; \theta) + 1\} \tag{3.64}$$

where (ph_x, ph_y) is a translation pair and (ph_x, ph_y') is a nontranslation pair that is randomly sampled. Through the co-training mechanism, we will ultimately obtain a phrase representation model for two languages.

The performance of the BRAE model is shown in the third column in Table 3.2. Compared with the unsupervised RAE, BRAE can encode the semantic information of phrases. For example, for the input phrase "do not agree," BRAE can find phrases having similar semantics but quite different words: "will not compromise" and "not to approve." This demonstrates that bilingually constrained recursive autoencoder BRAE can learn more accurate phrase embeddings.

3.4 Distributed Representation of Sentences

Since words and phrases are often not the direct processing objects in text mining tasks, learning the representation of words and phrases mainly adopts general (or task-independent) distributed representation methods. Relatively speaking, sentences are the direct processing object in many text mining tasks, such as

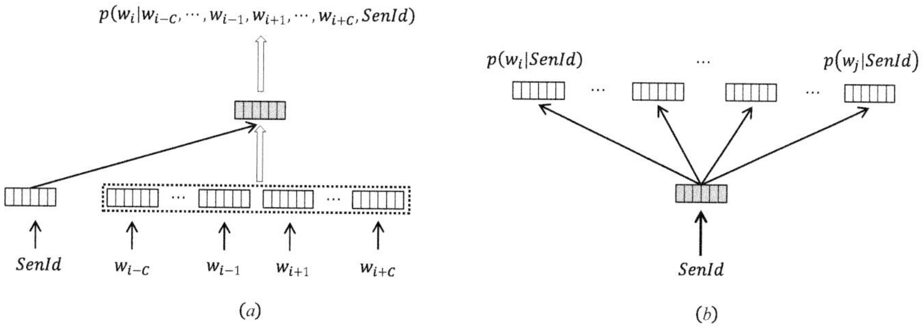

Fig. 3.12 Illustration of the PV-DM and PV-DBOW models. (**a**) PV-DM model. (**b**) PV-DBOW model

sentence-oriented text classification, sentiment analysis, and entailment inference. Therefore, it is crucial to learn the distributed representation of sentences. There are two main types of methods: general and task-dependent.

3.4.1 General Sentence Representation

The basis of the general sentence representation is very close to the unsupervised method. It designs a simple sentence representation model based on neural networks and optimizes the network parameters on large-scale monolingual training data $D = \{w_{i_1}^{m_i}\}_{i=1}^{M}$. We will introduce three classical general sentence representation methods in this section.

(1) PV-DM and PV-DBOW

Le and Mikolov extended the CBOW and skip-gram models used in word representation learning to learn both word and sentence embeddings (Le and Mikolov 2014). For M sentences in dataset D, each sentence D_i corresponds to an index i, which can uniquely represent the sentence. Assuming that the dimension of the sentence vector is P, then the vectors for sentences in the training set correspond to a matrix $\boldsymbol{PV} \in \mathbb{R}^{M\times P}$. The i-th sentence vector corresponds to the i-th row in $\boldsymbol{PV}$.

Extending the CBOW model, we can build a sentence representation model PV-DM.[4] As shown in Fig. 3.12a, PV-DM regards the sentence as the memory unit to capture the global information for any internal word. For an n-gram $(w_i, C) = w_{i-C}\cdots w_{i-1}w_iw_{i+1}\cdots w_{i+C}$ and its sentence index $SenId$, taking

[4]Paragraph Vector with sentence as Distributed Memory.

$C_{w_i} = w_{i-C} \cdots w_{i-1} w_{i+1} \cdots w_{i+C}$ as input, we calculate the average of the sentence vector and the word vectors in their contexts (or use vector concatenation)

$$\boldsymbol{h} = \frac{1}{2C+1}\big(\boldsymbol{e}(SenID) + \sum_{i-C \le k \le i+C, k \ne i} \boldsymbol{e}(w_k)\big) \tag{3.65}$$

where $\boldsymbol{e}(SenId)$ denotes the sentence vector corresponding to the $SenId$-th row in $\boldsymbol{PV}$. The calculation method, objective function, and training process for the probability of the central word $p(w_i|w_{i-C}, \cdots, w_{i-1}, w_{i+1}, \cdots, w_{i+C}, SenId)$ are all consistent with those of the CBOW model.

Extending the skip-gram model, we can build a sentence representation model PV-DBOW.[5] As shown in Fig. 3.12b, the model takes sentences as input and the randomly sampled words in the sentence as output, requiring that the model be able to predict any word in the sentence. Its objective function and training process are the same as those of the skip-gram model.

The PV-DM and PV-DBOW models are simple and effective, but they can only learn vector representations for the sentences appearing in the training set. If we want to obtain a vector representation for a sentence that has never been seen, we need to put it into the training set and retrain the model. Therefore, the generalization ability of this model is highly limited.

(2) Distributed Representation Based on Bag-of-Words Model

General sentence representation methods based on semantic composition has become increasingly popular in recent research. One of these methods represents sentences based on the bag-of-words model, treating the semantics of a sentence as a simple composition of internal word semantics. The simplest method is to use the average of the word embeddings

$$\boldsymbol{e}(s) = \frac{1}{n}\sum_{k=1}^{n} \boldsymbol{e}(w_k) \tag{3.66}$$

where $\boldsymbol{e}(w_k)$ denotes the word embedding corresponding to the k-th word w_k, which can be obtained by word embedding learning methods, such as CBOW or skip-gram. n denotes the length of the sentence, and $\boldsymbol{e}(s)$ is the sentence vector representation. It is worth noting that different words should make different contributions to the semantics of the sentence. For example, in the sentence "the Belt and Road forum will be held in Beijing tomorrow," "the Belt and Road" are obviously the most important words. Therefore, when composing the semantics of words, one key problem is how to assign appropriate weights to each word

$$\boldsymbol{e}(s) = \frac{1}{n}\sum_{k=1}^{n} v_k \times \boldsymbol{e}(w_k) \tag{3.67}$$

[5]Distributed Bag-of-Words version of the Paragraph Vector.

where v_k denotes the weight of the word w_k. v_k can be approximately estimated by TF-DF or self-information in information theory. Wang et al. (2017b) proposed a weight calculation method based on self-information (SI for short). They calculate v_k as follows:

$$v_k = \frac{\exp(\mathrm{SI}_k)}{\sum_{i=1}^{n} \exp(\mathrm{SI}_i)} \tag{3.68}$$

where $\mathrm{SI}_k = -\log(p(w_k|w_1 w_2 \cdots w_{k-1}))$ denotes the self-information of the word w_k and can be estimated by a language model. The larger the self-information of the word w_k is, the more information it carries, so it should be given greater weight in sentence representation. Although this kind of sentence representation method based on the bag-of-words model is very simple, it demonstrates high competitiveness in natural language processing tasks such as similar sentence discrimination and text entailment.

(3) Skip-Thought Model

The skip-thought method is also based on semantic composition (Kiros et al. 2015). It is similar to the PV-DBOW model, and the basic idea is also derived from the skip-gram model. However, unlike PV-DBOW, which uses the sentence to predict its internal words, the skip-thought model uses the current sentence D_k to predict the previous sentence D_{k-1} and the next sentence D_{k+1}. The model assumes that the meanings of the sentences $D_{k-1} D_k D_{k+1}$, which appear continuously in the text, are close to each other. Therefore, the previous sentence and the next sentence can be generated based on the semantics of the current sentence D_k.

Figure 3.13 gives an overview of the skip-thought model. The model has two key modules: one is responsible for encoding the current sentence D_k, and the other decodes D_{k-1} and D_{k+1} from the semantic representation of D_k. The encoder uses a recurrent neural network in which each hidden state employs a gated recurrent unit (GRU). The encoding process is consistent with the recurrent neural network language model. As shown on the left side in Fig. 3.13, after obtaining the hidden representation $\boldsymbol{h}_i^k$ of each position in the current sentence, the hidden representation $\boldsymbol{h}_n^k$ of the last position will be employed as the semantic representation of the whole sentence.

The decoder is similar to the GRU-based neural network language model, the only difference being that the input at each time step includes the hidden representation of the previous time step $\boldsymbol{h}_{j-1}$ and the output word w_{j-1} as well as the hidden representation $\boldsymbol{h}_n^k$ of the sentence D_k. The computation process of the GRU unit is as follows (taking the prediction of the previous sentence as an example):

$$\boldsymbol{r}^j = \sigma\left(\boldsymbol{W}_r^{k-1} \times \boldsymbol{e}\left(w_{j-1}^{k-1}\right) + \boldsymbol{U}_r^{k-1} \times \boldsymbol{h}_{j-1}^{k-1} + \boldsymbol{C}_r^{k-1} \times \boldsymbol{h}_n^k + \boldsymbol{b}_r^{k-1}\right) \tag{3.69}$$

$$\boldsymbol{z}^j = \sigma\left(\boldsymbol{W}_z^{k-1} \times \boldsymbol{e}\left(w_{j-1}^{k-1}\right) + \boldsymbol{U}_z^{k-1} \times \boldsymbol{h}_{j-1}^{k-1} + \boldsymbol{C}_z^{k-1} \times \boldsymbol{h}_n^k + \boldsymbol{b}_z^{k-1}\right) \tag{3.70}$$

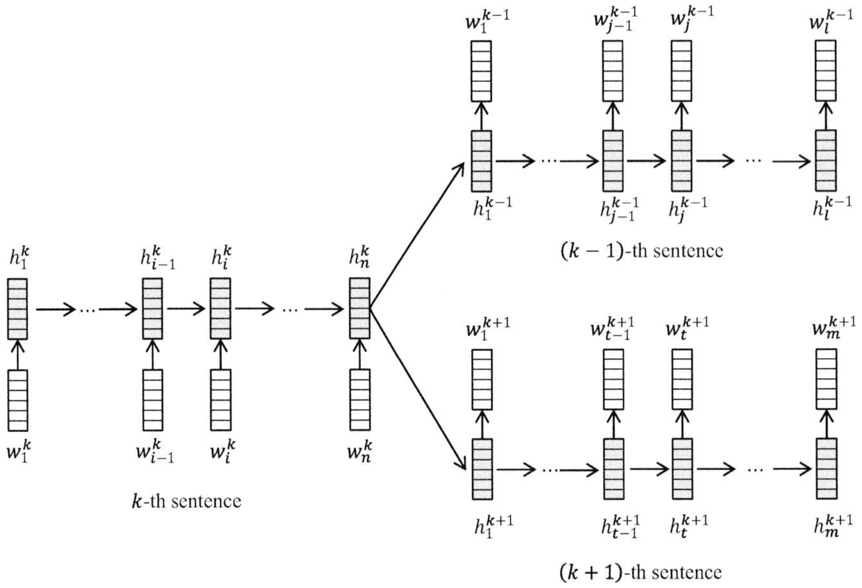

Fig. 3.13 The skip-thought model

$$\tilde{\boldsymbol{h}}_j = \tanh\left(\boldsymbol{W} \times \boldsymbol{e}\left(w_{j-1}^{k-1}\right) + \boldsymbol{U} \times \left(\boldsymbol{r}^j \odot \boldsymbol{h}_{j-1}^{k-1}\right) + \boldsymbol{C}^{k-1} \times \boldsymbol{h}_n^k + \boldsymbol{b}\right) \tag{3.71}$$

$$\boldsymbol{h}_j^{k-1} = z^j \odot \tilde{\boldsymbol{h}}_j + \left(1 - z^j\right) \odot \boldsymbol{h}_{j-1}^{k-1} \tag{3.72}$$

Given $\boldsymbol{h}_j^{k-1}$, the generated word sequence $w_1^{k-1} w_2^{k-1} \cdots w_{j-1}^{k-1}$ and the hidden representation $\boldsymbol{h}_n^k$ of the sentence D_k, the probability of the next word w_j^{k-1} can be calculated as follows:

$$p\left(w_j^{k-1} | w_{<j}^{k-1}, \boldsymbol{h}_n^k\right) \propto \exp\left(\boldsymbol{e}\left(w_j^{k-1}\right), \boldsymbol{h}_j^{k-1}\right) \tag{3.73}$$

The generation process for the next sentence D_{k+1} is similar.

The objective function of the skip-thought model is the summation of the likelihood of the previous and next sentences

$$\sum_{k=1}^{M} \left\{ \sum_{j=1}^{l} p\left(w_j^{k-1} | w_{<j}^{k-1}, \boldsymbol{h}_n^k\right) + \sum_{i=1}^{m} p\left(w_i^{k+1} | w_{<i}^{k+1}, \boldsymbol{h}_n^k\right) \right\} \tag{3.74}$$

where M is the number of sentences in the training set and l and m are the lengths of the previous and next sentences, respectively.

The skip-thought model combines the concepts of semantic composition and distributed hypotheses. If the training set is from continuous text, the skip-thought model can obtain high-quality sentence vector representations.

3.4.2 Task-Oriented Sentence Representation

Task-oriented sentence representations are optimized to maximize the performance of specific text processing tasks. For example, in the sentence-level sentiment analysis task, the vector representation of sentences ultimately predicts their sentiment polarity. That is, the final sentence representations will be sensitive to specific tasks. In this section, we introduce two task-oriented methods for sentence representation learning: a recursive neural network (RecurNN) (Socher et al. 2013) and a convolutional neural network (CNN) (Kim 2014).

(1) Sentence Representation Based on a Recursive Neural Network

A recursive neural network is a deep learning model suitable for tree structures. Given the vector representations of the child nodes, the recursive neural network recursively learns the vector representation of the parent node in a bottom-up manner until it covers the whole sentence. Given a sentence, its tree structure (usually a binary tree) can be obtained by using a syntactic parser. Figure 3.14 shows a sentence with its binary tree, and each leaf node corresponds to a d-dimensional vector for an input word. The recursive neural network merges the word embeddings

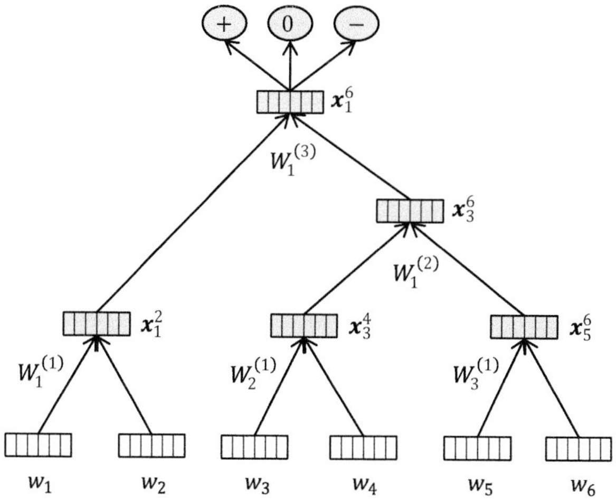

Fig. 3.14 Sentence representation model based on a recursive neural network

of the leaf nodes along the tree structure to obtain the vector representations $\boldsymbol{x}_1^2$, $\boldsymbol{x}_3^4$, $\boldsymbol{x}_5^6$ of phrases w_1^2, w_3^4, w_5^6, respectively.

$$\boldsymbol{x}_1^2 = f\left(\boldsymbol{W}_1^{(1)} \times [\boldsymbol{x}_1; \boldsymbol{x}_2] + \boldsymbol{b}_1^{(1)}\right) \tag{3.75}$$

$$\boldsymbol{x}_3^4 = f\left(\boldsymbol{W}_2^{(1)} \times [\boldsymbol{x}_3; \boldsymbol{x}_4] + \boldsymbol{b}_2^{(1)}\right) \tag{3.76}$$

$$\boldsymbol{x}_5^6 = f\left(\boldsymbol{W}_3^{(1)} \times [\boldsymbol{x}_5; \boldsymbol{x}_6] + \boldsymbol{b}_3^{(1)}\right) \tag{3.77}$$

Then, taking child nodes $\boldsymbol{x}_3^4$ and $\boldsymbol{x}_5^6$ as input, we can obtain the vector representation $\boldsymbol{x}_3^6$ corresponding to phrase w_3^6:

$$\boldsymbol{x}_3^6 = f\left(\boldsymbol{W}_1^{(2)} \times [\boldsymbol{x}_3^4; \boldsymbol{x}_5^6] + \boldsymbol{b}_1^{(2)}\right) \tag{3.78}$$

Finally, taking child nodes $\boldsymbol{x}_1^2$ and $\boldsymbol{x}_3^6$ as input, the vector representation $\boldsymbol{x}_1^6$ of the whole sentence can be obtained

$$\boldsymbol{x}_1^6 = f\left(\boldsymbol{W}_1^{(3)} \times [\boldsymbol{x}_1^2; \boldsymbol{x}_3^6] + \boldsymbol{b}_1^{(3)}\right) \tag{3.79}$$

with the parameter matrices $\boldsymbol{W}_1^{(1)}, \boldsymbol{W}_2^{(1)}, \boldsymbol{W}_3^{(1)}, \boldsymbol{W}_1^{(2)}, \boldsymbol{W}_1^{(3)} \in \mathbb{R}^{d\times 2d}$ and the biases $\boldsymbol{b}_1^{(1)}, \boldsymbol{b}_2^{(1)}, \boldsymbol{b}_3^{(1)}, \boldsymbol{b}_1^{(2)}, \boldsymbol{b}_1^{(3)} \in \mathbb{R}^d$. If the task is to predict the sentiment polarity (positive, negative, or neutral), the probability distribution of sentiment polarities can be calculated by taking $\boldsymbol{x}_1^6$ as the sentence representation through the softmax function

$$\boldsymbol{t} = \text{softmax}(\boldsymbol{W} \times \boldsymbol{x}_1^6 + \boldsymbol{b}) \tag{3.80}$$

in which $\boldsymbol{W} \in \mathbb{R}^{3\times d}$, $\boldsymbol{b} \in \mathbb{R}^3$, and 3 correspond to the number of polarities (1 for positive, -1 for negative, and 0 for neutral). Given training data $D = (D_i, L_i)_{i=1}^n$ consisting of n pairs of "sentence, sentiment polarity," the recursive neural network minimizes the cross-entropy to optimize the network parameters θ (including parameter matrices, biases, and word embeddings)

$$\theta^* = \underset{\theta}{\text{argmin}}\left\{-\sum_{i=1}^{n}\sum_{l}\delta_{L_i}(l)\log p\left(D_i, l\right)\right\} \tag{3.81}$$

where $L_i \in \{-1, 0, 1\}$ is the true target label. If $l = L_i$, then $\delta_{L_i}(l) = 1$; otherwise, $\delta_{L_i}(l) = 0$; $p(D_i, l)$ denotes the probability of sentiment polarity l in $\boldsymbol{t}$.

From Fig. 3.14, it can be found that the recursive neural network is very similar to the recursive autoencoder. There are three main differences. First, the recursive neural network takes a specific binary tree as input, while the recursive autoencoder needs to search for an optimal binary tree. Second, the recursive neural network does not need to calculate the reconstruction error at each node. Third, the recursive neural network can use either the same parameters at different nodes or different parameters according to the type of child nodes. For example, the parameter matrices $\boldsymbol{W}_1^{(1)}, \boldsymbol{W}_2^{(1)}, \boldsymbol{W}_3^{(1)}, \boldsymbol{W}_1^{(2)}, \boldsymbol{W}_1^{(3)}$ and biases $\boldsymbol{b}_1^{(1)}, \boldsymbol{b}_2^{(1)}, \boldsymbol{b}_3^{(1)}, \boldsymbol{b}_1^{(2)}, \boldsymbol{b}_1^{(3)}$ can either be the same or different.

(2) Sentence Representation Based on a Convolutional Neural Network

Recurrent neural networks are based on tree structures, which are suitable for tasks that depend on word order and hierarchical structures, such as sentiment analysis and syntactic parsing. For the task of sentence topic classification, some key information in the sentence plays a conclusive role in topic prediction. Therefore, a convolutional neural network becomes a classical model for performing this task. As shown in Fig. 3.15, for a sentence, a convolutional neural network takes the embeddings of each word as input, sequentially summarizes the local information of the window-sized context by convolution, extracts the important global information by pooling, and then passes through other network layers (dropout layer, linear and nonlinear layer, etc.) to obtain a fixed-sized vector representation that is utilized to describe the semantic information of the whole sentence.

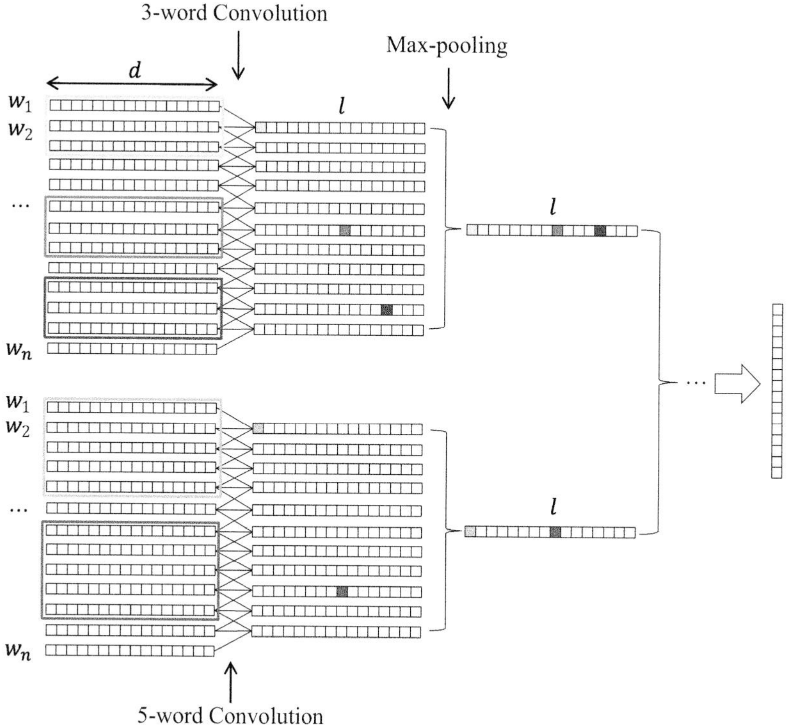

Fig. 3.15 Sentence representation model based on a convolutional neural network

Formally, given a sentence consisting of n words $w_1 w_2 \cdots w_n$, the words are first mapped into a list of word embeddings $\boldsymbol{X} = [\boldsymbol{x}_1, \boldsymbol{x}_2, \ldots, \boldsymbol{x}_n]$ by using a pretrained or randomly initialized word embedding matrix $\boldsymbol{L} \in \mathbb{R}^{|V| \times d}$. For a window $\boldsymbol{x}_{i:i+h-1}$ with a length h, the convolution layer applies the convolution operator[6] F_t ($1 \leq t \leq T$, where T denotes the number of convolution operators) to obtain a local feature y_i^t

$$y_i^t = F_t(\boldsymbol{W} \times \boldsymbol{x}_{i:i+h-1} + b) \tag{3.82}$$

where $F_t(\cdot)$ denotes the nonlinear activation function, $\boldsymbol{W} \in \mathbb{R}^{1 \times hd}$, $b \in \mathbb{R}$, $y_i^t \in \mathbb{R}$. The convolution operator F_t traverses the whole sentence from $\boldsymbol{x}_{1:h-1}$ to $\boldsymbol{x}_{n-h+1:n}$ and obtains a list of features $\boldsymbol{y}^t = [y_1^t, y_2^t, \ldots, y_{n-h+1}^t]$. We can see that $\boldsymbol{y}^t \in \mathbb{R}^{n-h+1}$ is a variable-length vector whose dimension depends on the sentence length n. The sentences vary in length (from several words to hundreds), and accordingly, the dimension of $\boldsymbol{y}^t$ differs for different sentences.

To convert the variable-length $\boldsymbol{y}^t$ into a fixed-length output, pooling is an indispensable operation, and maximum pooling is the most popular mechanism (Collobert et al. 2011; Kim 2014). It is believed that $\hat{y}^t = \max(\boldsymbol{y}^t)$ represents the most important feature obtained by the convolution operator in the whole sentence. T convolution operators result in a T-dimensional vector $\boldsymbol{y} = [\hat{y}^1, \hat{y}^2, \ldots, \hat{y}^T]$.

The window size h is an empirical value. To make the model robust, the convolutional neural network generally utilizes windows with different sizes of h. For example, $h = 3$ and $h = 5$ are used in Fig. 3.15, where each window corresponds to a T-dimensional vector $\boldsymbol{y} = [\hat{y}^1, \hat{y}^2, \ldots, \hat{y}^T]$. These vectors obtained by different windows can be concatenated into a fixed-sized vector, which is then fed into other network layers, such as for feed-forward neural networks. For the task of sentence topic classification, cross-entropy minimization is the objective and can be adopted to optimize the network parameters, similar to the sentiment analysis task.

3.5 Distributed Representation of Documents

The document is usually the direct processing object in many natural language processing tasks, such as text classification, sentiment analysis, text summarization, and discourse parsing. The key for these tasks is to deeply understand the document, and the premise of document understanding is to represent the document. The distributed representation of documents can capture global semantic information efficiently, so it has become an important research direction. The central issue is how to learn document representation from the representations of its internal words,

[6] This is also called filter, and it performs information filtering for a window-sized context.

phrases, and sentences. This section will introduce two kinds of methods: general purpose and task-oriented purpose.

3.5.1 General Distributed Representation of Documents

(1) Document Representation Based on the Bag-of-Words Model

In the general distributed representation of documents, a document can be regarded as a special form of sentence, that is, the concatenation of all sentences. Therefore, we can learn distributed document representation using the methods employed by sentence representation learning. For instance, the bag-of-words model based on compositional semantics can easily obtain the distributed representation of document $D = (D_i)_{i=1}^{M}$ ($D_i = s_i$ denotes the i-th sentence) from words

$$\boldsymbol{e}(D) = \frac{1}{|D|} \sum_{k=1}^{|D|} v_k \times \boldsymbol{e}(w_k) \tag{3.83}$$

in which v_k is the weight of word w_k and $|D|$ is the number of different words in document D. The average of word vector $v_k = \frac{1}{|D|}$ or the weighted average of word vector $v_k = \text{tf\_idf}(w_k)$ can be used. This method is simple and efficient, but it considers neither the order of words in a sentence nor the relationship between sentences in a document.

(2) Document Representation Based on the Hierarchical Autoencoder

To solve the problem in the bag-of-words model, Li et al. (2015) proposed a hierarchical autoencoder model. Its underlying concept is that a document representation $\boldsymbol{e}(D)$ for M-sentence document $D = (D_i)_{i=1}^{M}$ is good enough as long as we can reconstruct the original document D from the representation $\boldsymbol{e}(D)$.

The hierarchical autoencoder model is divided into two modules: one is an encoder model to learn the document representation $\boldsymbol{e}(D)$ from D, and the other is a reconstruction model that reconstructs the original document D from the representation $\boldsymbol{e}(D)$. In the encoder model, long short-term memory (LSTM) is first used to obtain the representation $\boldsymbol{e}(s_i)$ of each sentence. These sentence representations are then used as input to the second LSTM to model the sentence sequences in the document, ultimately resulting in the document representation $\boldsymbol{e}(D)$ (where $\boldsymbol{e}(s_i)$ and $\boldsymbol{e}(D)$ are the LSTM hidden representations corresponding to the end mark of the sentence and of the document, respectively)

$$\boldsymbol{e}(s_i) = \boldsymbol{h}^s_{\text{end}_s}(\text{enc}) \tag{3.84}$$

$$\boldsymbol{h}^s_t(\text{enc}) = \text{LSTM}\left(\boldsymbol{e}(w_t), \boldsymbol{h}^s_{t-1}(\text{enc})\right) \tag{3.85}$$

$$\boldsymbol{e}(D) = \boldsymbol{h}^D_{\text{end}_D}(\text{enc}) \tag{3.86}$$

$$\boldsymbol{h}^D_t(\text{enc}) = \text{LSTM}\left(\boldsymbol{e}(s_t), \boldsymbol{h}^D_{t-1}(\text{enc})\right) \tag{3.87}$$

where enc denotes the encoder LSTM.

The reconstruction (decoder) model aims to reconstruct document D from its distributed representation $\boldsymbol{e}(D)$, and it also employs the hierarchical LSTM: it first reconstructs the sentence hidden representation $\boldsymbol{h}^s_t(\text{dec})$ (dec denotes decoder LSTM) and then reconstructs all the words in the sentence s_t

$$\boldsymbol{h}^D_t(\text{dec}) = \text{LSTM}\left(\boldsymbol{e}'(s_{t-1}), \boldsymbol{h}^D_{t-1}(\text{dec}), \boldsymbol{c}^D_t\right) \tag{3.88}$$

$$\boldsymbol{h}^s_t(\text{dec}) = \text{LSTM}\left(\boldsymbol{e}(w_{t-1}), \boldsymbol{h}^s_{t-1}(\text{dec})\right) \tag{3.89}$$

where $\boldsymbol{h}^D_0(\text{dec}) = \boldsymbol{e}(D)$, $\boldsymbol{e}'(s_{t-1})$ is the hidden representation corresponding to the end mark of the previous sentence s_{t-1} and $\boldsymbol{c}^D_t$ is the context representation of the encoder model, which can be calculated by an attention mechanism

$$\boldsymbol{c}^D_t = \sum_{k=1}^{M} a_k \boldsymbol{h}^D_k(\text{enc}) \tag{3.90}$$

$$a_k = \frac{\exp(v_k)}{\sum_{k'} \exp(v_{k'})} \tag{3.91}$$

$$v_k = \boldsymbol{v}^{\text{T}} \times f\left(\boldsymbol{W}_1 \times \boldsymbol{h}^D_{t-1}(\text{dec}) + \boldsymbol{W}_2 \boldsymbol{h}^D_k(\text{enc})\right) \tag{3.92}$$

in which a_k is the weight of each sentence in the encoder model, $\boldsymbol{W}_1, \boldsymbol{W}_2 \in \mathbb{R}^{d \times d}$. $\boldsymbol{h}^s_0(\text{ dec}) = \boldsymbol{e}'(s_0)$ is the hidden representation of the reconstructed sentence. Based on $\boldsymbol{h}^s_t(\text{dec})$, the probability of the reconstruction word w_t can be computed as follows:

$$p\left(w_t|\cdot\right) = \text{softmax}\left(\boldsymbol{e}(w_t), \boldsymbol{h}^s_t(\text{dec})\right) \tag{3.93}$$

The objective function of this neural network is to maximize the likelihood of the original document, that is, the reconstructed word at each time should be the same as that in the corresponding position of the original document.

In Fig. 3.16, the document contains two sentences. The first LSTM layer is used to encode two sentences and obtain the representations $\boldsymbol{e}(s_1)$ and $\boldsymbol{e}(s_2)$ (the hidden representation corresponding to the sentence end mark). Then, the second LSTM layer is used to encode sentence sequences $\boldsymbol{e}(s_1)$ and $\boldsymbol{e}(s_2)$ and obtain the document representation $\boldsymbol{e}(D)$. Taking document representation $\boldsymbol{e}(D)$ as input, we can calculate the context of the encoder representations $\boldsymbol{e}(s_1)$ and $\boldsymbol{e}(s_2)$ by the attention mechanism. Then, the hidden representation $\boldsymbol{h}^D_t(\text{dec})$ of each sentence is reconstructed, and each word is generated to reconstruct sentences. After training,

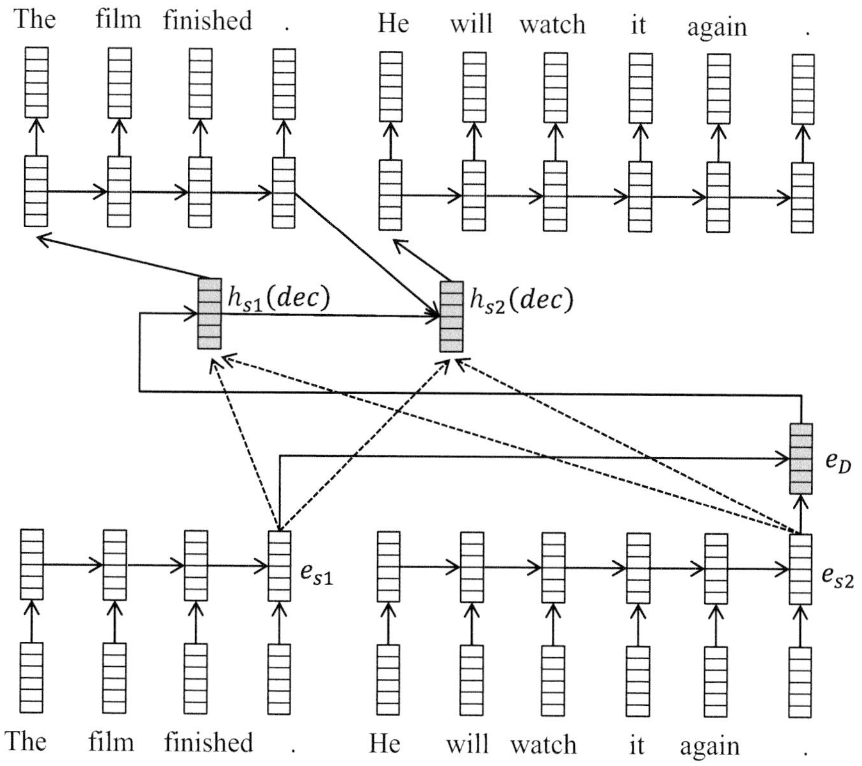

Fig. 3.16 Sentence representation model based on an autoencoder

the hierarchical autoencoder model can obtain the distributed representation $\boldsymbol{e}(D)$ for any document.

3.5.2 Task-Oriented Distributed Representation of Documents

The task-oriented distributed representation of documents, which has the final goal of optimizing the performance of downstream tasks, has been widely applied in tasks such as text classification and sentiment analysis. This section will introduce a task-oriented document representation method proposed by Tang et al. (2015b).

In this method, documents are regarded as composed of sentences, and sentences are regarded as composed of words. Therefore, semantic composition from words to sentences and from sentences to documents is the key concept in document representation. Assume that document $D = (D_i)_{i=1}^{M}$ is composed of M sentences and the ith sentence $D_i = s_i = w_{i,1} \cdots w_{i,n}$ is composed of n words. Then, a learning model based on document representation can be divided into three

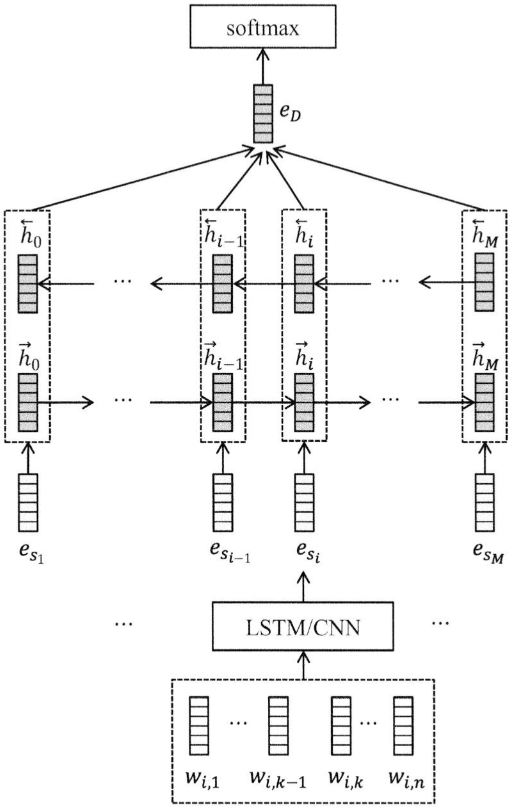

Fig. 3.17 Document representation model based on hierarchical autoencoder

layers: the bottom layer of sentence representation and the middle and top layers of document representation, as shown in Fig. 3.17.

The layer of sentence representation learns the semantic composition from the words in sequence $w_{i,1} \cdots w_{i,n}$ to sentence s_i. The former sections in this chapter have introduced several sentence representation models, such as recurrent neural networks, recursive neural networks, and convolutional neural networks. Among them, recurrent neural networks and convolutional neural networks are widely used. Both of these networks can be applied to obtain the distributed sentence representation:

$$\boldsymbol{e}_{s_i} = \text{LSTM}(w_{i,1} \cdots w_{i,n}) \tag{3.94}$$

$$\boldsymbol{e}_{s_i} = \text{CNN}(w_{i,1} \cdots w_{i,n}) \tag{3.95}$$

In practice, we can compare these two network architectures and choose the one with better performance.

The layer of document representation is used to learn the semantic composition from sentences in sequence $s_1 \dots s_M$ to document D. One popular model for this process is bidirectional LSTM networks. Taking distributed sentence representation $\boldsymbol{e}_{s_1} \cdots \boldsymbol{e}_{s_M}$ as input, bidirectional LSTM learns the forward hidden states $\overrightarrow{\boldsymbol{h}_i}$ and backward states $\overleftarrow{\boldsymbol{h}_i}$ of each sentence s_i:

$$\overrightarrow{\boldsymbol{h}_i} = \text{LSTM}\left(\boldsymbol{e}_{s_i}, \overrightarrow{\boldsymbol{h}}_{i-1}\right) \tag{3.96}$$

$$\overleftarrow{\boldsymbol{h}_i} = \text{LSTM}\left(\boldsymbol{e}_{s_i}, \overleftarrow{\boldsymbol{h}}_{i+1}\right) \tag{3.97}$$

The bidirectional hidden representations are concatenated into a single representation $\boldsymbol{h}_i = [\overrightarrow{\boldsymbol{h}_i}, \overleftarrow{\boldsymbol{h}_i}]$ for the sentence s_i. Based on the hidden representation of each sentence, the document representation can be obtained by the average or attention mechanism

$$\boldsymbol{e}_D = \sum_{i=1}^{M} v_i \boldsymbol{h}_i \tag{3.98}$$

where $v_i = \frac{1}{M}$ or v_i is the weight learned by an attention mechanism.

Given the distributed document representation $\boldsymbol{e}_D$, the classification layer first applies a feed-forward network to convert $\boldsymbol{e}_D$ into vector $\boldsymbol{x} = [x_1, \dots, x_C]$ whose dimension is the category number C. Then, the softmax function is used to convert vector $\boldsymbol{x}$ into probability distribution $\boldsymbol{p} = [p_1, \dots, p_C]$:

$$\boldsymbol{x} = f(\boldsymbol{W} \times \boldsymbol{e}_D + \boldsymbol{b}) \tag{3.99}$$

$$p_k = \frac{\exp(x_k)}{\sum_{k'=1}^{C} \exp(x_{k'})} \tag{3.100}$$

In the text classification or sentiment analysis task, large-scale labeled training data $T = \{(D, L)\}$ exist, where D is the document and L is the correct category corresponding to the document. The objective function aims to minimize the cross-entropy over the training data:

$$Loss = -\sum_{D \in T} \sum_{k=1}^{C} L_k(D) \log(p_k(D)) \tag{3.101}$$

If document D belongs to category k, then $L_k(D) = 1$; otherwise, $L_k(D) = 0$. After training, the sentence layer and the document layer can learn the distributed representation for any document.

3.6 Further Reading

Since words are the basic language unit that composes phrases, sentences, and documents, word representation learning is the basis and the most critical research direction. The research frontier of representation learning for words mainly focuses on the following four directions: (1) how to fully exploit information on the internal structure of words (Xu et al. 2016; Bojanowski et al. 2017; Pinter et al. 2017); (2) how to more effectively use contextual information (Ling et al. 2015; Hu et al. 2016; Li et al. 2017a) and other sources of external knowledge such as dictionaries and knowledge graphs (Wang et al. 2014; Tissier et al. 2017); (3) how to better interpret word representations (Arora et al. 2016; Wang et al. 2018); and (4) how to effectively evaluate the quality of word representations (Yaghoobzadeh and Schütze 2016). Lai et al. (2016) summarized the mainstream methods for word representation and offered proposals on how to learn better word representations.

The learning representations of phrases, sentences, and documents mostly focuses on the mechanism of semantic composition. For example, Yu and Dredze (2015) proposed a semantic composition function model for feature fusion to learn the distributed representation of phrases. Wang and Zong (2017) investigated the advantages and disadvantages of different composition mechanisms in the representation learning of phrases. Hashimoto and Tsuruoka (2016) studied whether the semantics of a phrase can be obtained from the semantics of its internal words. Learning sentence representations pays more attention to semantic composition (Gan et al. 2017; Wieting and Gimpel 2017) and the utilization of linguistic knowledge (Wang et al. 2016b). There are usually two methods for document representation: one based on compositional semantics and the other based on topic models. Making full use of their advantages and delivering better document representations have become hot research topics (Li et al. 2016b).

Exercises

3.1 Please compare the vector space model and the distributed representation model and summarize the advantages and disadvantages of each.

3.2 Please analyze noisy contrastive estimation and negative sampling and comment on the advantages and disadvantages of the two methods.

3.3 Please design an algorithm to detect whether a Chinese word is semantically transparent or not.

3.4 Please analyze the unsupervised recursive autoencoder to determine why it cannot learn the semantic representations of phrases.

3.5 For learning distributed sentence representations, a recurrent neural network can capture the word order information, and a convolutional neural network can summarize the key information for a window-sized context. Please design a model that could combine the merits of both models.

3.6 Please comment on whether it is reasonable to represent a whole document with a single distributed vector. If it is not reasonable, please design a new method to represent the documents.

Chapter 4
Text Representation with Pretraining and Fine-Tuning

4.1 ELMo: Embeddings from Language Models

As introduced in the previous section, a word can be well represented by its context. Thus, the quality of word representations (embeddings) depends at least on two factors. The first is whether the contexts are sufficiently rich: do we have abundant text data containing diverse contexts for each word? The second is whether the context is well captured and exploited. In other words, the word representations will remain unsatisfactory if the model cannot effectively utilize and represent all the contexts of a word. When word embeddings are applied to downstream tasks, another important issue arises that must also be addressed: are word embeddings context dependent? For example, a large language model based on a recurrent neural network (RNN) produces word embeddings as a byproduct. The usage will be context independent if we directly apply the pretrained word embeddings into the downstream tasks, but it will be context dependent if we first employ the pretrained RNN to obtain the dynamic representations according to the test sentence and then apply them to downstream tasks.

Generally, ELMo,[1] proposed by Peters et al. (2018), is the first successful model to attempt to solve all the above problems, and it achieves remarkable performance improvements in several downstream text-processing tasks. ELMo employs the pretraining framework. In the pretraining stage, a bidirectional LSTM-based language model is trained on the 1B Word Benchmark set (including approximately 30 million sentences).[2] In the specific applications, the pretrained bidirectional LSMT first performs on the test sentences, and then task- and context-dependent word embeddings are calculated according to dynamic hidden representations in the neural model. Last, the context-dependent word embeddings are fine-tuned in task-dependent models to perform specific text-processing tasks.

[1] Codes and models can be found at https://allennlp.org/elmo.

[2] https://github.com/ciprian-chelba/1-billion-word-language-modeling-benchmark.

© Tsinghua University Press 2021

C. Zong et al., *Text Data Mining*, https://doi.org/10.1007/978-981-16-0100-2_4

4.1.1 Pretraining Bidirectional LSTM Language Models

ELMo employs the bidirectional LSTM-based language model for pretraining. Given a sentence $(SOSx_1 \cdots x_{j-1}x_j \cdots x_n EOS)$ (SOS and EOS are special symbols indicating the start and end of the sentence), a forward language model computes the probability of x_j conditioned on its left contexts $p(x_j|SOS, x_1, \cdots, x_{j-1})$, while a backward language model calculates the probability of x_j conditioned on its right contexts $p(x_j|x_{j+1}, \cdots, x_n, EOS)$. Intuitively, both of the bidirectional contexts can be captured.

As illustrated in Fig. 4.1, the bottom layer first projects each symbolic token into a distributed representation using CNN over characters. Then, both forward and backward LSTMs are employed to learn two language models utilizing L layers. To calculate $p(x_j|SOS, x_1, \cdots, x_{j-1})$, the forward language model passes the token embedding $\boldsymbol{x}_{j-1}$ through L forward LSTM layers, resulting in the top representation $\overrightarrow{\boldsymbol{h}}^L_{j-1}$. Then, a softmax function is adopted to compute the probability of x_j:

$$p(x_j|SOS, x_1, \cdots, x_{j-1}) = \text{softmax}(\overrightarrow{\boldsymbol{h}}^L_{j-1}, x_j) = \frac{\overrightarrow{\boldsymbol{h}}^L_{j-1} \cdot \boldsymbol{x}_j}{\sum_{x'} \overrightarrow{\boldsymbol{h}}^L_{j-1} \cdot \boldsymbol{x}'} \tag{4.1}$$

Similarly, the backward language model employs L backward LSTM layers to obtain $\overleftarrow{\boldsymbol{h}}^L_{j+1}$ and compute $p(x_j|x_{j+1}, \cdots, x_n, EOS)$. The network parameters of bidirectional LSMTs are optimized to maximize the following log likelihood of

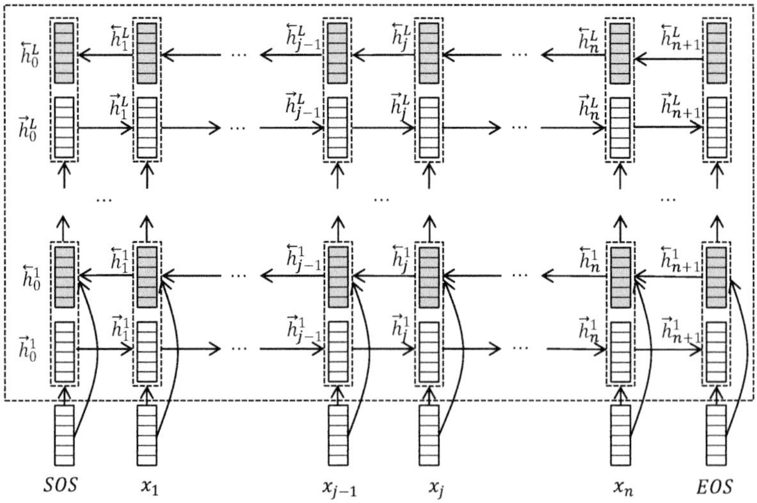

Fig. 4.1 The architecture of ELMo

forward and backward language models over T training sentences (30 million used in the original ELMo work):

$$\sum_{t=1}^{T}\sum_{j=0}^{n+1}\Big(\log p(x_j^{(t)}|SOS, x_1^{(t)}, \cdots, x_{j-1}^{(t)}; \theta) + \log p(x_j^{(t)}|x_{j+1}^{(t)}, \cdots, x_n^{(t)}, EOS; \theta)\Big) \tag{4.2}$$

4.1.2 Contextualized ELMo Embeddings for Downstream Tasks

After pretraining, bidirectional LSTMs are available. Instead of directly applying the learned fixed word embeddings $\boldsymbol{x}_j$ (static embedding) into downstream tasks, ELMo embeddings are dynamic depending on specific contexts in the test sentences. Specifically, each test sentence in a downstream task is input into the pretrained bidirectional LSTMs, resulting in $(2L+1)$-layer representations, including one input layer and L hidden layers of forward and backward LSTMs ($L = 2$ in the original ELMo work). All the representations of x_j can be rewritten as follows:

$$\boldsymbol{R}_j = \{\boldsymbol{x}_j, (\overrightarrow{\boldsymbol{h}}_j^l, \overleftarrow{\boldsymbol{h}}_j^l)|l = 1, \cdots, L\} = \{\boldsymbol{h}_j^l|l = 1, \cdots, L\} \tag{4.3}$$

where $\boldsymbol{h}_j^0 = \boldsymbol{x}_j$ denotes the input layer representation and $\boldsymbol{h}_j^l = [\overrightarrow{\boldsymbol{h}}_j^l; \overleftarrow{\boldsymbol{h}}_j^l]$ if $l \in \{1, \cdots, L\}$. Given a test sentence, bidirectional LSTMs first obtain the L forward and backward hidden layer representations; then, ELMo embeddings are linear combinations of each layer:

$$\boldsymbol{ELMo}_j^{task} = \gamma^{task}\sum_{l=0}^{L} w_l^{task}\boldsymbol{h}_j^l \tag{4.4}$$

in which w_l^{task} determines the contribution of representations in each layer. γ^{task} specifies the importance of ELMo embeddings in the specific task.

In downstream applications, ELMo embeddings are typically employed as additional features in a supervised model for a specific text-processing task. Suppose the baseline supervised model (e.g., CNN, RNN, or feed-forward neural networks) adopted in the specific task learns the final hidden states $(\boldsymbol{h}_1^{task}, \cdots, \boldsymbol{h}_j^{task}, \cdots, \boldsymbol{h}_n^{task})$ for a test sentence $(x_1, \cdots, x_j, \cdots, x_n)$. ELMo embeddings can be leveraged in two ways to augment the baseline supervised model. On the one hand, they can be combined with the input embedding $\mathbf{x}_j$ of the baseline model, leading to $[\boldsymbol{x}_j; \boldsymbol{ELMo}_j^{task}]$ as new inputs for the supervised model. On the other hand, ELMo embeddings can be concatenated with the final representations $\boldsymbol{h}_j^{task}$ of the baseline model, resulting in $[\boldsymbol{h}_j^{task}; \boldsymbol{ELMo}_j^{task}]$, which can be directly employed to perform prediction without changing the baseline supervised architecture.

Enhanced with ELMo embeddings, remarkable performance improvements can be achieved in several text-processing tasks, such as question answering, textual entailment, semantic role labeling, coreference resolution, named entity recognition, and sentiment analysis.

4.2 GPT: Generative Pretraining

Despite the great success of ELMo, it still has some weaknesses to be addressed. First, ELMo adopted a two-layer shallow bidirectional LSTM, which makes it difficult to learn all the language regularities of the text data, and thus its potential is limited. Second, bidirectional LSTMs are not the best for capturing long-distance dependency, since they need $n - 1$ passes for the dependence modeling of the first word and the n-th word in a sequence, and their results would be further worsened by the gradient vanishing problem. Third, pretrained models are not fully exploited, since they are only being used to obtain representations that will be further employed as additional features for downstream tasks. That is, the fine-tuning model in the downstream supervised task learns from scratch and does not share the parameters of the pretraining model. Accordingly, Radford et al. (2018) propose a deep pure attention-based model GPT inspired by Transformer (Vawani et al. 2017) for both pretraining and fine-tuning. Specifically, GPT employs Transformer's decoder, which contains 12 self-attention layers, to pretrain a feed-forward language model and fine-tune the same 12-layer self-attention model for downstream tasks. This section first briefly introduces the Transformer and then gives an overview of GPT.

4.2.1 Transformer

The Transformer[3] was originally proposed to perform machine translation that automatically converts a source language sentence (token sequence $(x_0, \cdots, x_j, \cdots, x_n)$) into a target-language sentence $(y_0, \cdots, y_i, \cdots, y_m)$. It follows the encoder-decoder architecture, in which the encoder obtains the semantic representation of the source sentence and the decoder generates the target sentence token by token from left to right based on the source-side semantic representations.

[3]Model and codes can be found at https://github.com/tensorflow/tensor2tensor.

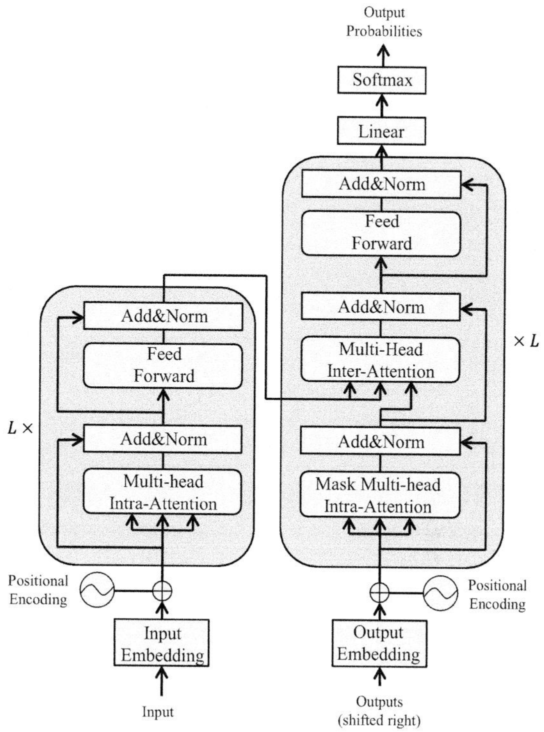

Fig. 4.2 The architecture of Transformer

The encoder includes L layers, and each layer is composed of two sublayers: the self-attention[4] sublayer followed by the feed-forward sublayer, as shown in the left part of Fig. 4.2. The decoder, as shown in the right part of Fig. 4.2, also consists of L layers. Each layer has three sublayers. The first mechanism is the masked self-attention mechanism. The second sublayer is the decoder-encoder attention sublayer, and the third sublayer is the feed-forward sublayer. Residual connection and layer normalization are performed for each sublayer in both the encoder and decoder.

Obviously, the attention mechanism is the key component. The three kinds of attention mechanisms (encoder self-attention, decoder masked self-attention, and encoder-decoder attention) can be formalized into the same formula:

$$\text{Attention}(\boldsymbol{q}, \boldsymbol{K}, \boldsymbol{V}) = \text{softmax}\left(\frac{\boldsymbol{q}\boldsymbol{K}^T}{\sqrt{d_k}}\right)\boldsymbol{V} \tag{4.5}$$

[4]The self-attention sublayer calculates the i-th representation in the upper layers by using the i-th hidden state in the current layer to attend to all the neighbors including itself, resulting in attention weights which are then employed to linearly combine all the representations in the current layer. It will be formally defined later.

where $\boldsymbol{q}$, $\boldsymbol{K}$, and $\boldsymbol{V}$ represent the query, the key list, and the value list, respectively. d_k is the dimension of the key.

For the encoder self-attention mechanism, the queries, keys, and values are from the same layer. For example, we calculate the output of the first layer in the encoder at the j-th position. Let $\boldsymbol{x}_j$ be the sum vector of the input token embedding and the positional embedding. The query is vector $\boldsymbol{x}_j$. The keys and values are the same, and both are the embedding matrix $\boldsymbol{x} = [\boldsymbol{x}_0 \cdots \boldsymbol{x}_n]$. Then, multihead attention with h heads is proposed to calculate attention in different subspaces:

$$\begin{aligned} \text{MultiHead}(q, \boldsymbol{K}, \boldsymbol{V}) &= \text{Concat}(head_1, \cdots, head_i, \cdots, head_h)\boldsymbol{W}_O \\ head_i &= \text{Attention}(\boldsymbol{q}\boldsymbol{W}_Q^i, \boldsymbol{K}\boldsymbol{W}_K^i, \boldsymbol{V}\boldsymbol{W}_V^i) \end{aligned} \tag{4.6}$$

in which Concat means that it concatenates all the head representations. $\boldsymbol{W}_Q^i$, $\boldsymbol{W}_K^i$, $\boldsymbol{W}_V^i$, and $\boldsymbol{W}_O$ denote the projection parameter matrices.

Using Eq. (4.6) followed by residential connection, layer normalization, and a feed-forward network, we can obtain the representation of the second layer. After L layers, we obtain the input contexts $\boldsymbol{C} = [\boldsymbol{h}_0, \cdots, \boldsymbol{h}_n]$.

The decoder masked self-attention mechanism is similar to that of the encoder except that the query at the i-th position can only attend to positions before i, since the predictions after the i-th position are not available in the autoregressive left-to-right unidirectional inference:

$$z_i = \text{Attention}(\boldsymbol{q}_i, \boldsymbol{K}_{\le i}, \boldsymbol{V}_{\le i}) = \text{softmax}\left(\frac{\boldsymbol{q}_i \boldsymbol{K}_{\le i}^T}{\sqrt{d_k}}\right)\boldsymbol{V}_{\le i} \tag{4.7}$$

The decoder-encoder attention mechanism calculates the source-side dynamic context that is responsible for predicting the current target-language word. The query is the output of the masked self-attention sublayer z_i, and the keys and values are the same encoder contexts $\boldsymbol{C}$. The residential connection, layer normalization, and feed-forward sublayer are then applied to yield the output of a whole layer. After L such layers, we obtain the final hidden state z_i. The softmax function is then employed to predict the output y_i, as shown in the upper right part of Fig. 4.2.

4.2.2 Pretraining the Transformer Decoder

As shown in Fig. 4.3, GPT utilizes the unidirectional Transformer decoder introduced above to pretrain the feed-forward language model on large-scale text data (e.g., the English text BookCorpus) It applies masked self-attention to attend to all

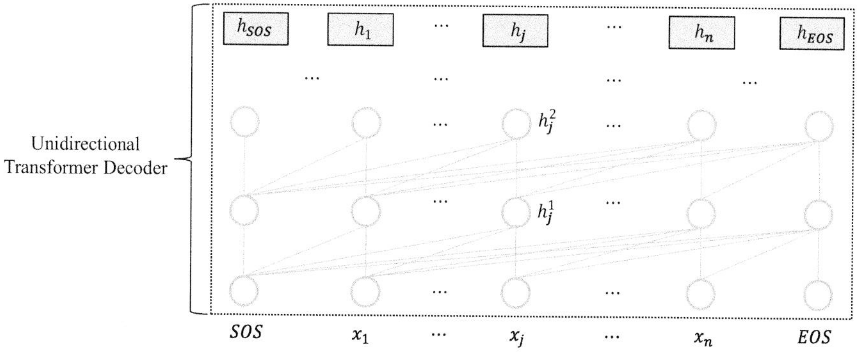

Fig. 4.3 The architecture of GPT

the preceding contexts while keeping the future contexts inaccessible. As Fig. 4.3 illustrates, when learning the representation of $\mathbf{h}_j^1$, x_j only attends to previous tokens $SOS, x_1, \cdots, x_j$. Each layer performs the same operations, leading to the hidden representation of the top layer $\mathbf{h}_j$.

GPT predicts the next token x_{j+1} with probability $p(x_{j+1}|x_0, \cdots, x_j)$ and optimizes the network parameters by maximizing the conditional log likelihood over the complete set of T training sentences:

$$\mathbb{L}_1 = \sum_{t=1}^{T} \sum_{j=0}^{n+1} \log p(x_j^{(t)}|SOS, x_1^{(t)}, \cdots, x_{j-1}^{(t)}; \theta) \tag{4.8}$$

4.2.3 *Fine-Tuning the Transformer Decoder*

When performing downstream tasks, the pretrained Transformer decoder is employed as the starting point and can be slightly adapted and further fine-tuned according to the target text-processing tasks. We know that GPT is only pretrained with the language model as the objective function and the network cannot perform specific tasks such as named entity recognition. Thus, it is necessary to fine-tune the GPT model to fit the specific tasks using task-dependent training data.

Suppose a supervised classification task contains training instances of input sequences and output labels, such as (x, y) where $x = (SOSx_1 \cdots x_j \cdots x_n EOS)$. The pretrained Transformer decoder will generate for x a sequence of final representations $(\boldsymbol{h}_{SOS}, \boldsymbol{h}_1, \cdots, \boldsymbol{h}_j, \cdots, \boldsymbol{h}_n, \boldsymbol{h}_{EOS})$ after L stacked masked self-attention layers. A linear output layer and softmax function are newly introduced to predict label probability with $\boldsymbol{h}_{EOS}$:

$$p(y|x) = p(y|SOS, x_1, \cdots, x_j, \cdots, x_n, EOS) = \text{softmax}(\boldsymbol{h}_{EOS}\boldsymbol{W}_y) \tag{4.9}$$

The network parameters of the pretrained Transformer decoder and the newly added linear projection parameter matrix $\boldsymbol{W}_y$ are then fine-tuned to maximize the following objective:

$$\mathbb{L}_2 = \sum_{(x,y)} \log p(y|x) \tag{4.10}$$

To improve the generalization ability and accelerate convergence, GPT further combines the pretrained language model objective during fine-tuning:

$$\mathbb{L} = \mathbb{L}_2 + \lambda \times \mathbb{L}_1 \tag{4.11}$$

For downstream tasks in which the input is not a single but multiple sequences, GPT simply concatenates the sequences with delimiter tokens to form a long sequence to match the pretrained Transformer decoder. For example, in the entailment task, which determines whether a premise x^1 entails hypothesis x^2, GPT uses $(x^1; \text{Delim}; x^2)$ as the final input sequence, where Delim is a delimiter token.

Radford et al. (2019) present an enhanced version of GPT-2,[5] which achieves promising performance in language generation tasks. Note that the model architecture is the same as that of GPT. The difference lies in that GPT-2 utilizes many more English texts and a much deeper Transformer decoder. The English texts contain over 8 million documents with a total of 40 GB of words. The deepest model contains 48 layers and 1542 M network parameters. Radford et al. (2019) demonstrate that only with pretraining can the model perform downstream natural language understanding and generation tasks without fine-tuning. For example, they show that the pretrained model can generate abstractive summarization quite well, achieving comparable performance with some supervised summarization models on the CNN Daily Mail dataset. Brown et al. (2020) further invent GPT-3[6], and the largest model contains up to 175 billion parameters. Surprisingly, GPT-3 shows that it can perform most of the natural language understanding and generations tasks even in few-shot or zero-shot scenarios as long as the training data is adequate and the neural network model is large enough.

4.3 BERT: Bidirectional Encoder Representations from Transformer

Although the GPT model achieves substantial progress in several natural language understanding and generation tasks, the left-to-right decoder architecture of GPT learns the semantic representation of each input x_j by relying only on the left-side

[5]The codes and models are available at https://github.com/openai/gpt-2.

[6]The models and examples are available at https://github.com/openai/gpt-3.

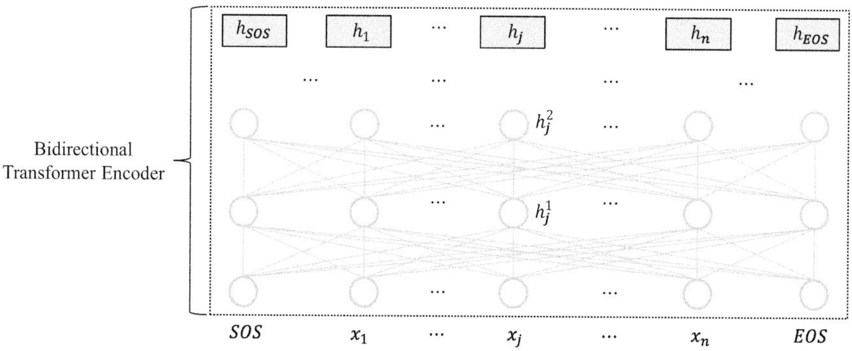

Fig. 4.4 The architecture of BERT

context $x_0, x_1, \cdots, x_{j-1}$ and cannot access the right-side context $x_{j+1}, \cdots, x_n$. It is well known that bidirectional contexts are crucial in many text-processing tasks, such as sequential labeling and question answering. Accordingly, Devlin et al. (2019) proposed a new pretraining and fine-tuning model called BERT,[7] which employs the bidirectional encoder of Transformer, as shown in Fig. 4.4, to fully exploit the contexts for semantic representation. As this figure shows, the representation of each input token $\boldsymbol{h}_j$ is learned by attending to both the left-side context $SOS, x_1, \cdots, x_{j-1}$ and the right-side context $x_{j+1}, \cdots, x_n$.

The contributions of BERT are threefold. First, BERT employs a much deeper model than GPT, and the bidirectional encoder consists of up to 24 layers with 340 million network parameters ($\text{BERT}_{\text{LARGE}}$). Second, BERT designs two novel unsupervised objective functions, including the masked language model and next sentence prediction, considering that the conventional conditional language model cannot be used for BERT. Third, BERT is pretrained on even larger text datasets (both BookCorpus with 800 million words and English Wikipedia with 2.5 billion words) than GPT. BERT is the first work to achieve breakthroughs and establish new state-of-the-art performance on 11 natural language understanding tasks, even outperforming humans on question answering tasks. Next, we briefly introduce the pretraining and fine-tuning procedure for BERT.

4.3.1 BERT: Pretraining

Both ELMo and GPT employ the conditional language model as the unsupervised pretraining objective. In contrast, the conventional language model, which conditions only in the one-sided history context, is not appropriate for BERT since BERT needs to simultaneously access the bidirectional contexts, and the

[7]Codes and pretrained models can be available at https://github.com/google-research/bert.

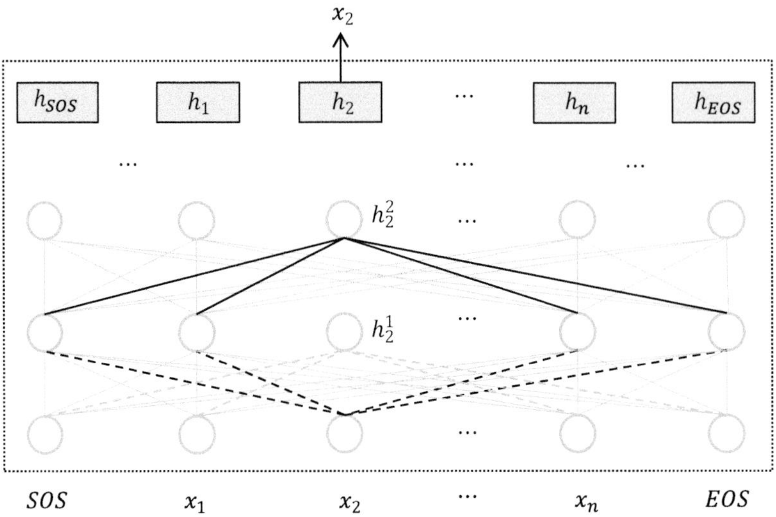

Fig. 4.5 Illustration of the problem wherein conditional language models are inappropriate for BERT

representation learning process of the multilayer encoder for a specific input token would see itself when predicting this input token. We use Fig. 4.5 to explain this problem. Suppose that we plan to use the left context (SOS, x_1) and the right context $(x_3, \cdots, x_n, EOS)$ to predict x_2, namely, calculating the probability $p(x_2|SOS, x_1, x_3, \cdots, x_n, EOS)$. At the first layer, the Transformer encoder learns the representation $\boldsymbol{h}_2^1$ by attending to all the contexts except x_2, as shown by gray dotted lines in Fig. 4.5. At the second layer, the Transformer encoder learns $\boldsymbol{h}_2^2$ in the same manner by attending to all the contexts $(\boldsymbol{h}_{SOS}^1, \boldsymbol{h}_1^1, \boldsymbol{h}_3^1, \cdots, \boldsymbol{h}_n^1, \boldsymbol{h}_{EOS}^1)$. However, since $\boldsymbol{h}_{SOS}^1, \boldsymbol{h}_1^1, \boldsymbol{h}_3^1, \cdots, \boldsymbol{h}_n^1$, and $\boldsymbol{h}_{EOS}^1$ have already considered x_2, as shown by the dotted black lines, $\boldsymbol{h}_2^2$ will contain the information of x_2 (see the word itself) through the information passing along the solid black lines in Fig. 4.5. Consequently, it is problematic if $\boldsymbol{h}_2^L$ is employed to predict x_2.

To solve this problem, two unsupervised prediction tasks are introduced to pretrain BERT. One is the masked language model, and the other is next sentence prediction.

Masked Language Model The main approach underlying the masked language model is that some percentage of the tokens in the input sequence are randomly masked, and the model is then optimized to predict only the masked tokens. For example, given an input sequence $(SOSx_1x_2 \cdots x_nEOS)$, x_2 may be randomly masked, meaning it is substituted by a special symbol MASK, as illustrated in Fig. 4.6. Then, BERT will learn semantic representations of the new sequence $(SOSx_1\text{MASK} \cdots x_nEOS)$, obtaining the final representations in the L-th layer $(\boldsymbol{h}_{SOS}, \boldsymbol{h}_1, \boldsymbol{h}_{\text{MASK}}, \cdots, \boldsymbol{h}_n, \mathbf{h}_{EOS})$. By comparing Fig. 4.6 with Fig. 4.5, it is easy to see that $\boldsymbol{h}_{\text{MASK}}$ does not contain any information of x_2 because it is absent in the

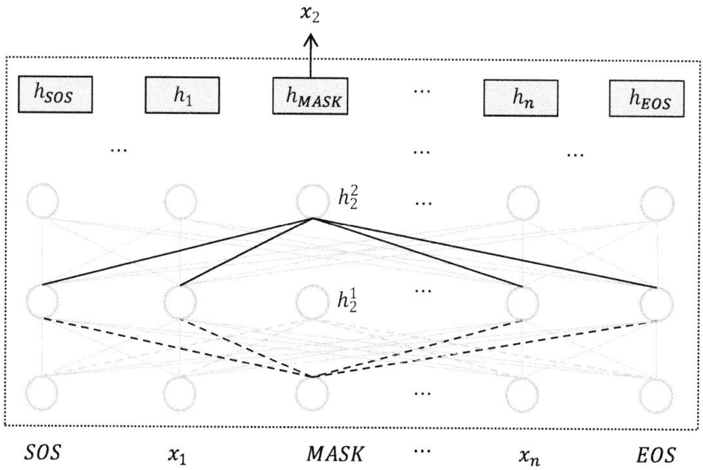

Fig. 4.6 Illustration of the masked language model for BERT

input. Finally, $\boldsymbol{h}_{\text{MASK}}$ can be utilized to predict x_2. Through the example illustration, it can be intuitively understood that the masked language model is a reasonable approach to optimizing the parameters of BERT using bidirectional contexts.

The question remains of how many and what kind of input tokens should be replaced. In practice, BERT randomly masks 15% of all the tokens in each input sequence and predicts them at the top layer. However, since the sequences are always unmasked in the test environment, a mismatch will arise between the training and testing phases. To address this issue, BERT does not always replace a token with MASK. For each of the selected 15% of tokens, 80% are replaced with MASK, 10% are replaced by another random token, and the remaining 10% remain unchanged.

Next Sentence Prediction Some text-processing tasks, such as text entailment and question answering, must deal with two sequences rather than a single sequence. For example, in the text entailment task, it must be determined whether the first sentence (premise) entails the second sentence (hypothesis). This is equivalent to predicting a label of Yes/No for the concatenation of the sentences representing a premise and a hypothesis. If BERT is only pretrained on single sentences, it will not best fit these downstream tasks. Accordingly, BERT's design includes another unsupervised training objective function that determines whether the second sequence B naturally follows the first one A. For example, A is a sentence *I am from Beijing* and B is *Beijing is the capital of China*. B is a natural sentence following A. If B is *The presidential election will be held in 2020*, B does not follow A in natural text.

The training data can be constructed easily. Each pretraining instance (A, B) is chosen according to the following strategy: in 50% of the instances, the second sequence B indeed follows A in the monolingual corpus (e.g., BookCorpus), serving as positive examples, while for the other 50%, B is a random sentence selected from the corpus and serving as negative examples.

During pretraining, A and B are concatenated into a single sequence (A[SEP]B), where [SEP] is a separator symbol between two sentences. BERT learns the L layers' semantic representations for the sequence with the bidirectional Transformer encoder. The final hidden representation of the first token (SOS is used in this manuscript, and $[CLS]$ is employed in the original paper of BERT) $\boldsymbol{h}_{SOS}^{L}$ is fed into a linear projection layer and a softmax layer to predict whether B follows A.

4.3.2 BERT: Fine-Tuning

Similar to GPT, the pretrained BERT is employed as the starting point for downstream tasks and can be slightly adapted and further fine-tuned according to the target text-processing tasks. BERT is only pretrained with a masked language model and next sentence prediction as the objective function, and the network cannot directly perform specific tasks such as text entailment and question answering. Therefore, it is necessary to fine-tune BERT to fit the specific tasks with task-dependent training data.

BERT can be applied to two kinds of downstream tasks: sequence-level classification tasks and sequence labeling tasks. For the classification tasks, the input sequence is first fed into the pretrained BERT, and the final hidden representation of the first token $\boldsymbol{h}_{SOS}^{L}$ is utilized for classification. $\boldsymbol{h}_{SOS}^{L}$ is linearly projected by a parameter matrix $\boldsymbol{W}_O$ and is then fed into the softmax layer to calculate the probability distribution of the categories. The network parameters of the pretrained BERT and the newly introduced projection matrix $\boldsymbol{W}_O$ are fine-tuned to maximize the probability $p(y|x)$ of label y on the supervised classification training set.

For the sequence labeling task, each token x_j yields a final hidden representation $\boldsymbol{h}_j^L$ through the pretrained BERT. A linear projection layer and a softmax function further operate on $\boldsymbol{h}_j^L$ to predict the label y_j. All the network parameters are fine-tuned to maximize the probability $p(y|x)$ of the label sequence y in the supervised sequential labeling training data.

4.3.3 XLNet: Generalized Autoregressive Pretraining

Although BERT has achieved great success in many text-processing tasks, there are still some shortcomings for this model. The most critical issue is that a serious discrepancy between pretraining and fine-tuning persists for BERT, since the massively used special symbol MASK during pretraining never appears in the downstream tasks during fine-tuning. Furthermore, BERT assumes that the masked tokens in the input sequence are independent of each other. According to the design of BERT, 15% the input tokens are randomly masked. For instance, after random masking, the original input

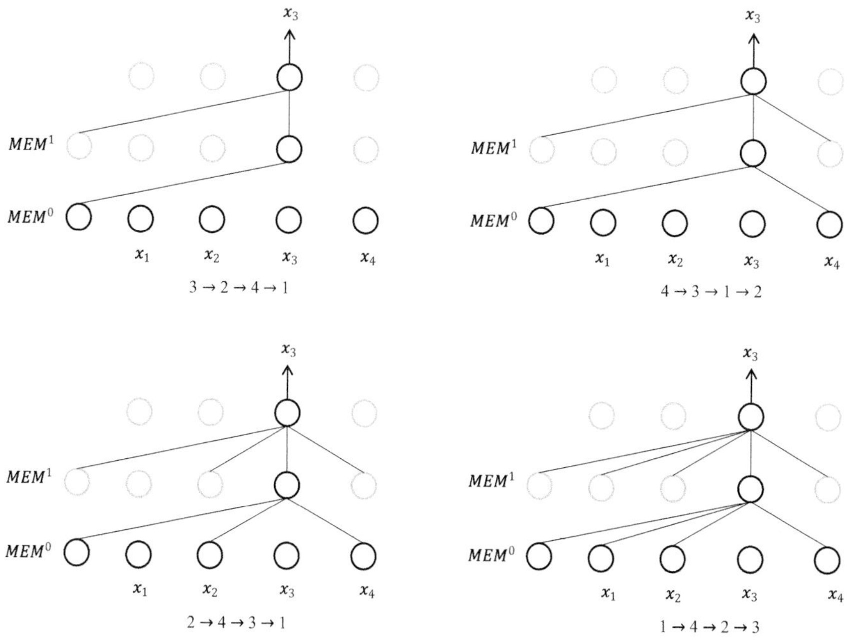

Fig. 4.7 Permutations of the sequence (x_1, x_2, x_3, x_4) and their autoregressive language model

sequence $(SOS, x_1, x_2, \cdots, x_{j-1}, x_j, x_{j+1}, \cdots, x_{n-1}, x_n, EOS)$ may become $(SOS, x_1, \text{MASK}, \cdots, x_{j-1}, \text{MASK}, x_{j+1}, \cdots, \text{MASK}, x_n, EOS)$. It is obvious that x_2 and x_{n-1} will not be used in the prediction of x_j when pretraining BERT (similarly, (x_2, x_j) not for x_{n-1} and (x_j, x_{n-1}) not for x_2). In practice, however, x_2, x_j, and x_{n-1} may depend on each other.

To overcome the above shortcomings, Yang et al. (2019) proposed a generalized autoregressive pretraining model named XLNet[8] This model aims to maintain BERT's merits in capturing the bidirectional context well without using the masking strategy. XLNet mainly includes two ideas that are novel compared to BERT: permutation language modeling and two-stream self-attention.

Permutation Language Modeling Intuitively, all the left and right contexts would have an opportunity to appear before the focal token x_j if we enumerate all the permutations of the input sequence. Take the sequence (x_1, x_2, x_3, x_4) as an example and suppose that x_3 is our focus. As shown in Fig. 4.7, different permutations will provide different contexts for x_3. The bottom right is the permutation that moves all the bidirectional contexts before x_3. Accordingly, any feed-forward (autoregressive) language model can be applied to pretrain XLNet while conditioning on bidirectional contexts.

[8]The codes and pretrained models are available at https://github.com/zihangdai/xlnet.

Let $\boldsymbol{Z}_n$ be the set of all the possible permutations of an n-token sequence, and let z_j, $z_{<j}$ be the j-th element and the preceding $j-1$ elements in a specific permutation $z \in \boldsymbol{Z}_n$. Then, XLNet is pretrained to maximize the expectation of the autoregressive language model probabilities of the permutation set:

$$\sum_{t=1}^{T}\left\{\mathbb{E}_{z\in \boldsymbol{Z}_n}\left[\sum_{j=1}^{n}\log p(x_{z_j}^{(t)}|x_{z_{<j}}^{(t)};\theta)\right]\right\} \tag{4.12}$$

Note that XLNet only permutes the factorization order (the decomposition approach to calculating the probability of $p(x)$) rather than reordering the original sequence, as Fig. 4.7 shows.

Two-Stream Self-Attention When calculating $p(x_{z_j}|x_{z_{<j}};\theta)$ in the permutation language model, the hidden representation $\boldsymbol{h}(x_{z_{<j}})$ is learned with the Transformer self-attention mechanism, and a softmax algorithm is further employed to calculate the probability distribution of the next token. It is easy to see that $\boldsymbol{h}(x_{z_{<j}})$ is not aware of the target position j when predicting x_j. Thus, $p(x_{z_j}|x_{z_{<j}};\theta)$ is computed regardless of the target position, and $p(x_{z_k}|x_{z_{<j}};\theta)$ $(k \geq j)$ shares the same probability distribution. That is, the conditional language model probability of a token will always be the same regardless of location when it is given the same history context. Obviously, this position insensitive property is undesirable because language is sensitive to word order and position. Accordingly, XLNet designs a new two-stream self-attention mechanism to address this issue.

Two types of hidden representations can be learned at time step j: content representation $\boldsymbol{h}(x_{z_{\leq j}})$ and query representation $\boldsymbol{g}(x_{z_{<j}})$:

$$\boldsymbol{h}^l(x_{z_j}) = \text{Attention}(\boldsymbol{q}_j = \boldsymbol{h}^{l-1}(x_{z_j}), \boldsymbol{K}_{\leq j}\boldsymbol{V}_{\leq j} = \boldsymbol{h}^{l-1}(x_{z_{\leq j}})) \tag{4.13}$$

$$\boldsymbol{g}^l(x_{z_j}) = \text{Attention}(\boldsymbol{q}_j = \boldsymbol{g}^{l-1}(x_{z_j}), \boldsymbol{K}_{<j}\boldsymbol{V}_{<j} = \boldsymbol{g}^{l-1}(x_{z_{<j}})) \tag{4.14}$$

Note that the content representation $\boldsymbol{h}(x_{z_{\leq j}})$ is the same as those hidden states in the conventional Transformer. The query representation $\boldsymbol{g}(x_{z_{<j}})$ is position aware, but it is learned without using the content information of the z_j-th token. The query representation at the top layer $\boldsymbol{g}^L(x_{z_{<j}})$ will be used to predict x_{z_j}. Initially, $\boldsymbol{h}^0(x_{z_j})$ is the token embedding of x_{z_j}, and $\boldsymbol{g}^0(x_{z_j})$ is a trainable vector $\boldsymbol{w}$. Figure 4.8 illustrates the main idea behind the two-stream self-attention model. Suppose the factorization order is $2 \rightarrow 4 \rightarrow 3 \rightarrow 1$ and we need to predict x_3 given (x_2, x_4). Gray solid lines denote the content representation flow (the same as in the conventional Transformer), while the black solid lines show the query representation process. The black dotted line indicates that the input serves only as the query, and its values are not used during the attention computation procedure. For example, $\boldsymbol{g}_3^1$ is a function of the weighted summation of word embeddings $\boldsymbol{x}_2$ and $\boldsymbol{x}_4$, excluding

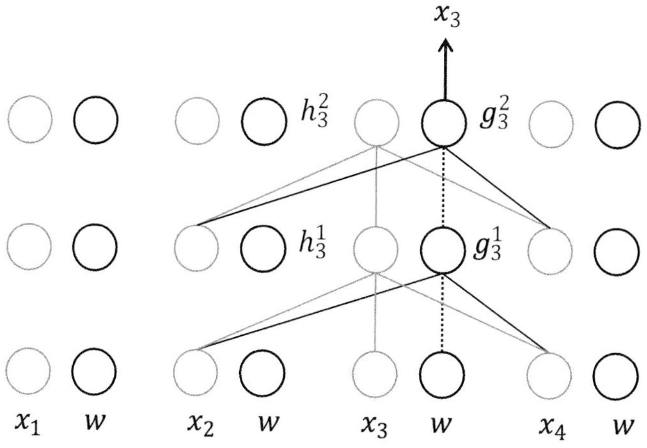

Fig. 4.8 Two-stream attention model for the perturbation $2 \rightarrow 4 \rightarrow 3 \rightarrow 1$

x_3. The weights are estimated using $\boldsymbol{g}_3^0 = \boldsymbol{w}$ as the query to attend to $\boldsymbol{x}_2$ and $\boldsymbol{x}_4$. If XLNet includes two layers, $\boldsymbol{g}_3^2$ will be adopted to predict x_3.

To speed up the convergence of the training process, XLNet only predicts the last few tokens of each sampled factorization order instead of the whole sequence. In addition, XLNet also incorporates some sophisticated techniques, such as relative position embedding and the segment recurrence mechanism from Transformer-XL (Dai et al. 2019). Finally, XLNet can outperform BERT on 20 text-processing tasks.

Interestingly, researchers at Facebook (Liu et al. 2019) find that BERT is significantly undertrained. They report that with the careful design of the key hyperparameters and training data size, BERT[9] can match or even exceed XLNet and other variants.

4.3.4 UniLM

ELMo, BERT, and XLNet aim to fully explore the bidirectional contexts of the input sequence and are mainly designed for natural language understanding tasks. GPT is appropriate for natural language generation tasks such as abstractive summarization. Nevertheless, GPT can only utilize left-side context. An interesting question is how to combine the merits of both BERT and GPT to design a pretraining model for text generation tasks.

[9]They have named the reimplementation RoBERTa; details are available at https://github.com/pytorch/fairseq.

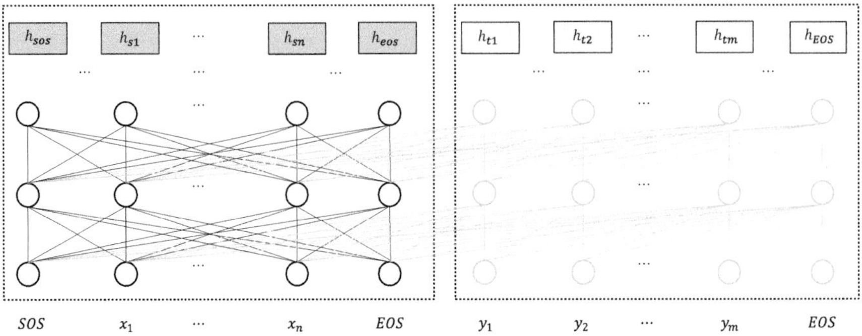

Fig. 4.9 The architecture of UniLM

Dong et al. (2019) proposed a unified pretraining language model UniLM,[10] which can adapt the Transformer model for monolingual sequence-to-sequence text generation tasks. Given a pair of two consecutive sequences (x, y) in the monolingual corpus, UniLM considers x as the input sequence and y as the output sequence. As illustrated in Fig. 4.9, UniLM applies a bidirectional Transformer encoder to the input sequence and a unidirectional Transformer decoder to the output sequence. The same masking mechanism as used for BERT is employed to pretrain the UniLM model. By pretraining the model on large-scale monolingual data, UniLM can be further fine-tuned to perform text generation tasks, such as abstractive summarization and question generation. Dong et al. (2019) report that UniLM can achieve the new state-of-the-art performance for abstractive summarization tasks on the CNN Daily Mail dataset.

To obtain a comprehensive understanding of various pretraining models, we list them in Table 4.1 and outline the key architecture and features of each model.

4.4 Further Reading

This chapter briefly introduces several popular pretraining models, including ELMo, GPT, BERT, XLNet, and UniLM. We can see that the pretraining and fine-tuning paradigms have led to major breakthroughs in many natural language understanding and generation tasks. Recently, the pretraining methodology has been developing quickly, and many improved models have been proposed, most of which focus on improving the BERT framework. These new models can be roughly divided into the following three categories. (More information about pretraining models can be found in the survey paper (Qiu et al. 2020)).

[10]Codes and models can be found at https://github.com/microsoft/unilm.

Table 4.1 Comparison of different pretraining models

Model	Architecture	Key features	Most fitted tasks
ELMo	Bidirectional LSTMs	First large-scale pretrained model that provides dynamic embedding based on test sequences for fine-tuning	Understanding
GPT	Transformer decoder	First architecture to accommodate pretraining and fine-tuning on the same model	Understanding
GPT-2	Transformer decoder	Enhanced GPT with deeper layers using more text data	Generation
GPT-3	Transformer decoder	Upgraded GPT with very deeper layers using huge scale of text data	Generation
BERT	Transformer encoder	Denoising autoencoder paradigm that employs masked language model and next sentence prediction as objectives	Understanding
XLNet	Transformer encoder	Generalized autoregressive language model using input permutations	Understanding
UniLM	Transformer	Generalized pretraining model for both natural language understanding and generation	Generation

One research direction aims at designing more sophisticated objective functions or incorporating knowledge into the BERT architecture. Sun et al. (2019) propose a model, ERNIE, that improves the masked language model by masking the entities rather than characters (subwords or words) as in the original BERT. They prove that the entity masked model works very well on many Chinese language processing tasks. They further upgrade the model to ERNIE-2.0, which incrementally learns pretraining tasks using a multitask learning framework (Sun et al. 2020). Zhang et al. (2019) present another improved model, also called ERNIE, that incorporates the representation learning of entities in knowledge graphs into the BERT pretraining process.

Another direction aims to make the pretrained model as compact as possible. Since BERT is very heavy and contains a huge number of parameters, it is computationally expensive and memory intensive, especially for the inference step. Sanh et al. (2019), Tang et al. (2019), and Jiao et al. (2019) propose using the knowledge distillation strategy to compress the big model into a small one with negligible performance drop. Lan et al. (2019) propose reducing the memory usage and speeding up the training procedure of BERT with two parameter reduction approaches, namely, factorized embedding parameterization and cross-layer sharing.

The third direction explores pretraining models for generation and cross-lingual tasks. While most of the studies tackle natural language understanding tasks by enhancing BERT, an increasing number of researchers are turning their attention to pretraining for generation tasks and cross-lingual tasks. Both UniLM and the MASS model proposed by Song et al. (2019) facilitate generation problems, but the latter uses a smart design. The MASS model masks a consecutive subsequence seq in a sentence and uses the masked sequence as input to predict the consecutive subsequence seq with a sequence-to-sequence model. Cross-lingual pretraining is also attracting increasing attention. Conneau and Lample (2019) present a model XLM for cross-lingual language model pretraining using pairs of bilingual translation sentences as input.

There is another issue, in that the inference procedure in the generation tasks always follows a left-to-right manner and cannot access future information. One promising direction is to perform synchronous bidirectional inference for generation tasks, such as the work of Zhou et al. (2019) employed in machine translation.

Exercises

4.1 Please analyze the complexity of different pretraining models including ELMo, GPT, BERT, and XLNet.

4.2 It is said that the masked language model is one type of denoising autoencoder. Please provide a detailed analysis of this claim.

4.3 Both GPT and UniLM can be employed in language generation tasks. Please comment on the difference between the two models when they are used for generation.

4.4 The task of next sentence prediction is shown to be helpful in BERT; please analyze the scenarios in which the next sentence prediction task is not necessary and give the reasons.

4.5 XLM is a cross-lingual pretraining model. Please see the details of the model and identify the kinds of downstream tasks for which it could be employed.

Chapter 5
Text Classification

5.1 The Traditional Framework of Text Classification

For simplicity of description, in the following we use "document" instead of "a piece of text" at different levels, and the text classification or text categorization problem can also be called document classification or document categorization. If not specified, the methods described below apply not only to document classification but also to text classification at other levels (e.g., sentence classification). As shown in Fig. 5.1, the goal of document classification is to divide a collection of documents into a set of predefined categories such as "technology," "sports," or "entertainment."

The traditional framework of a document classification system is represented in Fig. 5.2. The system consists of three separate components: text representation, feature selection, and classification. The literature (Sebastiani 2002) summarizes text classification techniques according to this framework. The three stages are normally separate in traditional document classification. In the following three subsections, we will introduce these three stages respectively.

In document classification, a document must be correctly and efficiently represented for subsequent classification algorithms. The representation method must truly reflect the content of the text and have sufficient ability to distinguish different types of text. We have systematically introduced text representation methods in Chap. 3. For more details on the text representation methods, particularly the traditional vector space model, readers can refer to Sect. 3.1, and we will not go into detail on these here. However, it is worth noting that the selection of a text representation method depends on the choice of classification algorithm. For example, discriminative classification models (such as ME and SVM) usually use the vector space model (VSM) for text representation. Text representation in a generative model (such as NB) is determined by the class-conditional distribution hypothesis, e.g., the multinomial distribution or the multivariable Bernoulli distribution.

There are two steps to using the vector space model for text representation: (1) generating a feature vector composed of a sequence of terms (e.g., the vocabulary)

© Tsinghua University Press 2021

C. Zong et al., *Text Data Mining*, https://doi.org/10.1007/978-981-16-0100-2_5

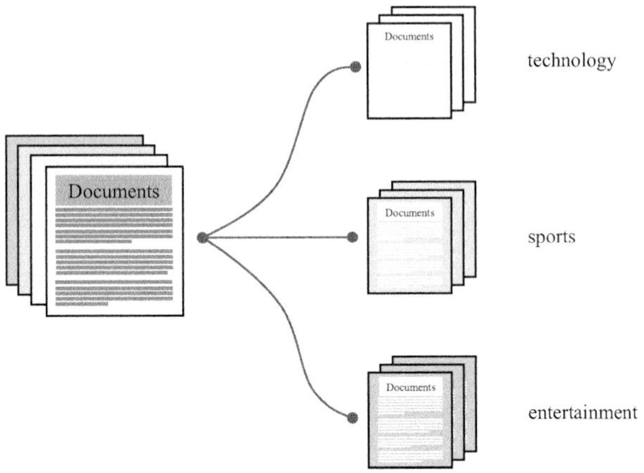

Fig. 5.1 An example of document classification

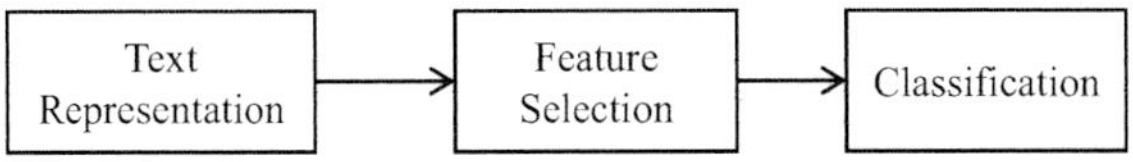

Fig. 5.2 The main components of text classification based on traditional machine learning

based on the training data and (2) assigning a weight to each term in the vector and performing some normalization for each document in the training and testing datasets. The vector space model is simple to use, but it loses too much information from the original documents.

We construct a small dataset for text classification, as shown in Table 5.1. The dataset includes two classes: "education" and "sport." Each class in the training set

Table 5.1 Text classification dataset

Id	Document	Category
$train\_d_1$	Beijing Institute of Technology was established in 1958 as one of the earliest universities that established a computer science major in China.	Education
$train\_d_2$	Students from Beijing Institute of Technology won the 4th China Computer Go Championship.	Education
$train\_d_3$	The Gymnasium of Beijing Institute of Technology is the venue for the preliminary volleyball competition of the 2008 Beijing Olympic Games in China.	Sport
$train\_d_4$	In the 5th East Asian Games, the total number of medals of China reached a new high. Both the men's and women's volleyball teams won championships.	Sport
$test\_d_1$	Beijing Institute of Technology is a national key university of China that focuses on science and engineering.	
$test\_d_2$	The Fudan University volleyball team won the volleyball championship in this year's college games.	

Table 5.2 Vocabulary for the text classification dataset given in Table 5.1

1958, 2008, 4th, 5th, Asian, Beijing, China, Olympic, champion, competition, computer, early, east, establish, game, go, gymnasium, high, institute, major, man, medal, new, number, one, preliminary, reach, science, student, team, technology, total, university, venue, volleyball, woman

includes four documents ($train\_d_1$ and $train\_d_2$ belong to the education class, and $train\_d_3$ and $train\_d_4$ belong to the sport class), and the test set consists of two documents ($test\_d_1$ and $test\_d_2$). Table 5.2 provides the vocabulary of the text classification dataset. Each document can be represented as a vector in the vector space based on this vocabulary.

5.2 Feature Selection

The traditional vector space model represents a document based on a high-dimensional vector space. To reduce the noise contained in such a high-dimensional vector and improve the computational efficiency, it is necessary to reduce its dimension before performing classification. In machine learning and pattern recognition, dimension reduction methods fall into two main categories: feature extraction and feature selection.

The purpose of feature extraction is to map the original high-dimensional sparse feature space into a low-dimensional dense feature space. The classical feature extraction methods include principal component analysis (PCA) and independent component analysis (ICA).

Feature selection is the process of selecting a subset of features for text representation and classification. In comparison with feature extraction, feature selection has been more widely discussed and used for text data. The feature selection methods for text classification normally include unsupervised and supervised methods. The former can be applied to a corpus without category annotation, but its effect is often limited. The representative approaches include term frequency (TF) and document frequency (DF), where the latter relies on category annotation, which can more effectively select a better subset of features for text classification. The commonly used supervised methods include the mutual information (MI), information gain (IG), and chi-square statistic (χ^2) methods. Yang and Pedersen (1997) and Forman (2003) systematically summarized the feature selection methods used in text classification and pointed out that a good feature selection algorithm can effectively reduce the feature space, remove redundant and noise features, and improve the efficiency of the classifier.

We introduce supervised feature selection methods in this subsection.

5.2.1 Mutual Information

In information theory, suppose that X is a discrete random variable whose probability distribution is $p(x) = P(X = x)$. The entropy of X is defined as follows:

$$H(X) = -\sum_{x} p(x) \log p(x) \tag{5.1}$$

Entropy, also known as the expectation of self-information, is used to measure the average level of "information" or "uncertainty" inherent in the variable's possible outcomes. If a random variable has greater entropy, it has greater uncertainty, and consequently, a larger amount of information is needed to represent it, while less entropy means less uncertainty and requires less information.

Suppose X and Y are a pair of discrete random variables with the joint distribution $p(x, y) = P(X = x, Y = y)$. Then, the joint entropy of X and Y is defined as:

$$H(X, Y) = -\sum_{x} \sum_{y} p(x, y) \log p(x, y) \tag{5.2}$$

The joint entropy indicates the uncertainty (i.e., the amount of information needed for representation) of a pair of random variables.

The conditional entropy describes the uncertainty of random variable Y given the value of random variable X. In other words, it indicates the amount of additional information needed to represent Y under the condition that the value of X is known. It can be defined as follows:

$$\begin{aligned} H(Y|X) &= \sum_{x} p(x) H(Y|X = x) \\ &= -\sum_{x} \sum_{y} p(x, y) \log p(y|x) \end{aligned} \tag{5.3}$$

$H(Y|X) = 0$ if and only if the value of Y is completely determined by X. Conversely, $H(Y|X) = H(Y)$ if and only if Y and X are independent of each other.

The relationship between entropy, joint entropy, and conditional entropy can be described as follows:

$$H(Y|X) = H(X, Y) - H(X) \tag{5.4}$$

Figure 5.3 displays the relationship between entropy, joint entropy, and conditional entropy. Suppose that the circle on the left represents entropy $H(X)$ and the circle on the right represents entropy $H(Y)$. The union of the two circles represents the joint entropy $H(X, Y)$, the crescent on the left represents the conditional entropy $H(X|Y)$, the crescent on the right represents the conditional entropy $H(Y|X)$, and

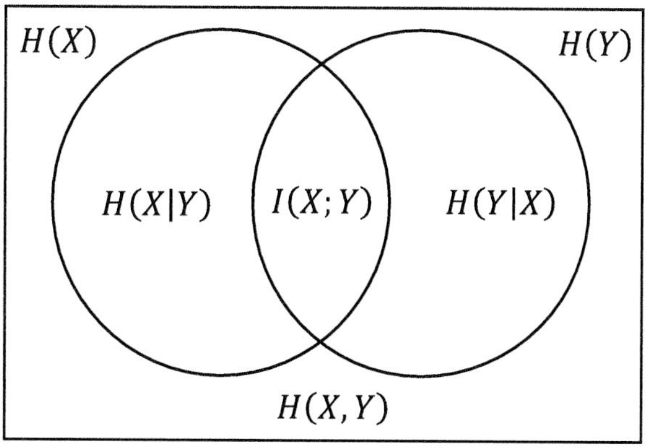

Fig. 5.3 The relationship between entropy, joint entropy, and conditional entropy

the intersection of the two circles is called the mutual information of X and Y, as defined below.

Mutual information reflects the degree to which two random variables are related to each other. For two discrete random variables X and Y, their mutual information is defined as:

$$I(X;Y) = \sum_{x,y} p(x,y) \log \frac{p(x,y)}{p(x)\,p(y)} \tag{5.5}$$

The relationship between entropy, conditional entropy, and mutual information is as follows:

$$I(X;Y) = H(Y) - H(Y|X) = H(X) - H(X|Y) \tag{5.6}$$

Mutual information is a measure of the interdependence between two random variables. It can be regarded as the amount of the reduction in uncertainty in a random variable when another random variable is known.

Let $I(x;y) = \log \frac{p(x,y)}{p(x)p(y)}$ denote the pointwise mutual information (PMI) of X and Y when they take the value (x, y). Equation (5.6) shows that MI is the expectation of PMI. In text classification, PMI measures the amount of discriminative information of class c_j provided by feature t_i.

For a given collection of documents, we first calculate the value of each term t_i and class c_j in Table 5.3. N_{t_i,c_j} indicates the document frequency of term t_i appearing in class c_j; $N_{t_i,\bar{c}_j}$ indicates the document frequency of term t_i appearing in all classes except c_j; $N_{\bar{t}_i,c_j}$ indicates the document frequency of all terms except t_i appearing in class c_j; and $N_{\bar{t}_i,\bar{c}_j}$ indicates the document frequency of all terms except t_i appearing in all classes except c_j. $N = N_{t_i,c_j} + N_{t_i,\bar{c}_j} + N_{\bar{t}_i,c_j} + N_{\bar{t}_i,\bar{c}_j}$ is the total number of documents.

Table 5.3 The document frequency statistic for each feature and class

Feature	Class	
	c_j	$\bar{c}_j$
t_i	N_{t_i,c_j}	$N_{t_i,\bar{c}_j}$
$\bar{t}_i$	$N_{\bar{t}_i,c_j}$	$N_{\bar{t}_i,\bar{c}_j}$

Based on the principle of maximum likelihood estimation, we can estimate the following probability:

$$p\left(c_j\right) = \frac{N_{t_i,c_j} + N_{\bar{t}_i,c_j}}{N} \tag{5.7}$$

$$p\left(t_i\right) = \frac{N_{t_i,c_j} + N_{t_i,\bar{c}_j}}{N} \tag{5.8}$$

$$p\left(c_j|t_i\right) = \frac{N_{t_i,c_j} + 1}{N_{t_i,c_j} + N_{t_i,\bar{c}_j} + M} \tag{5.9}$$

$$p\left(c_j|\bar{t}_i\right) = \frac{N_{\bar{t}_i,c_j} + 1}{N_{\bar{t}_i,c_j} + N_{\bar{t}_i,\bar{c}_j} + M} \tag{5.10}$$

where M denotes the number of categories and $p\left(c_j|t_i\right)$ and $p\left(c_j|\bar{t}_i\right)$ are estimated with Laplace smoothing to avoid zero probabilities.

On this basis, the mutual information $I\left(t_i; c_j\right)$ between t_i and c_j can be calculated as

$$I\left(t_i; c_j\right) = \log \frac{N_{t_i,c_j} N}{\left(N_{t_i,c_j} + N_{\bar{t}_i,c_j}\right)\left(N_{t_i,c_j} + N_{t_i,\bar{c}_j}\right)} \tag{5.11}$$

Finally, we can take either the weighted average of $I(t_i; c_j)$:

$$I_{\text{avg}}(t_i; c_j) = \sum_j p\left(c_j\right) I\left(t_i; c_j\right) \tag{5.12}$$

or the maximum value among different classes

$$I_{\max}(t_i; c_j) = \max_j \left\{I\left(t_i; c_j\right)\right\} \tag{5.13}$$

to measure the amount of discriminative information that item t_i contains for all classes.

The process of feature selection, hence, first calculates the MI score (Eq. (5.12) or Eq. (5.13)) for all terms, then ranks the terms according to their MI scores, and finally selects a subset of the top-ranked terms as the selected features.

Table 5.4 MI feature selection results for the text classification dataset

Features	MI
Computer game volleyball	0.4055
1958 2008 4th 5th Asian Olympic competition early east establish go gymnasium high major man medal new number one preliminary reach science student team total university venue woman	0.2877
China champion	0.0000
Beijing institute technology	0.1823

Table 5.4 gives the results of the feature selection for the text classification dataset (Table 5.1) based on the MI method.

5.2.2 *Information Gain*

Information gain (IG) denotes the reduction in uncertainty of the random variable Y given the condition that random variable X is observed:

$$G(Y|X) = H(Y) - H(Y|X) \tag{5.14}$$

Such a reduction in uncertainty can be represented by the difference between $H(Y)$ and $H(Y|X)$.

In the text classification task, we can regard a feature $T_i \in \{t_i, \bar{t}_i\}$ as a binary random variable that has a Bernoulli distribution (also called a 0-1 distribution) and regard the class C as a random variable that has a categorical distribution. Based on this, information gain can be defined as the difference between entropy $H(C)$ and conditional entropy $H(C|T_i)$:

$$\begin{aligned} G(T_i) &= H(C) - H(C|T_i) \\ &= -\sum_j p(c_j) \log p(c_j) - \left[\left(-\sum_j p(c_j, t_i) \log p(c_j|t_i)\right)\right. \\ &\quad \left. + \left(-\sum_j p(c_j, \bar{t}_i) \log p(c_j|\bar{t}_i)\right)\right] \end{aligned} \tag{5.15}$$

While using the same number of top-ranked features for text classification, IG performs significantly better than MI in many text classification applications because information gain takes both t_i and $\bar{t}_i$ into consideration and can be viewed as a weighted average of the pointwise mutual information $I(\bar{t}_i; c_j)$ and $I(t_i; c_j)$ (Yang and Pedersen 1997):

$$G(T_i) = \sum_j p(t_i, c_j) \cdot I(t_i, c_j) + p(\bar{t}_i, c_j) \cdot I(\bar{t}_i, c_j) \tag{5.16}$$

Table 5.5 IG feature selection results for the text classification dataset

Features	IG
Computer game volleyball	0.1308
1958 2008 4th 5th Asian Beijing Olympic competition early east establish go gymnasium high institute major man medal new number one preliminary reach science student team technology total university venue woman	0.0293
China champion	0.0000

Table 5.5 gives the feature selection results for the text classification dataset (Table 5.1) based on IG.

5.2.3 *The Chi-Squared Test Method*

The chi-square (χ^2) test is a statistical hypothesis testing method. It is widely used to test the independence of two random variables by determining whether there is a statistically significant difference between the expected frequency and the observed frequency.

As applied to text classification, suppose term $T_i \in \{t_i, \bar{t}_i\}$ and class $C_j \in \{c_j, \bar{c}_j\}$ are two binary random variables that both obey a Bernoulli distribution, where t_i and $\bar{t}_i$ represent whether t_i appears in a document or not and c_j and $\bar{c}_j$ represent whether the class of a document is c_j or not.

On this basis, we propose the null hypothesis: T_i and C_j are independent of each other. That is, $p(T_i, C_j) = p(T_i) \cdot p(C_j)$. For each term T_i and class C_j, we calculate the chi-square statistics:

$$\chi^2(T_i, C_j) = \sum_{T_i \in \{t_i, \bar{t}_i\}} \sum_{C_j \in \{c_j, \bar{c}_j\}} \frac{(N_{T_i,C_j} - E_{T_i,C_j})^2}{E_{T_i,C_j}} \tag{5.17}$$

where N_{T_i,C_j} denotes the observed document frequency defined in Table 5.3 and E_{T_i,C_j} denotes the expected document frequency based on the null hypothesis (i.e., T_i and C_j are independent of each other).

Under the null hypothesis, E_{t_i,c_j} can be estimated based on Eqs. (5.17) and (5.18) as follows:

$$\begin{aligned} E_{t_i,c_j} &= N \cdot p(t_i, c_j) = N \cdot p(t_i) \cdot p(c_j) \\ &= N \cdot \frac{N_{t_i,c_j} + N_{t_i,\bar{c}_j}}{N} \cdot \frac{N_{t_i,c_j} + N_{\bar{t}_i,c_j}}{N} \end{aligned} \tag{5.18}$$

Table 5.6 Chi-square feature selection results for the text classification dataset

Features	χ^2
Computer game volleyball	4.0000
1958 2008 4th 5th Asian Beijing Olympic competition early east establish go gymnasium high institute major man medal new number one preliminary reach science student team technology total university venue woman	1.3333
China champion	0.0000

Similarly, we can estimate $E_{\bar{t}_i,c_j}$, $E_{t_i,\bar{c}_j}$, and $E_{\bar{t}_i,\bar{c}_j}$. Finally, by bringing the above results into Eq. (5.17), the chi-square statistic can be written as:

$$\chi^2(T_i, C_j) = \frac{N \cdot \left(N_{t_i,c_j} N_{\bar{t}_i,\bar{c}_j} - N_{\bar{t}_i,c_j} N_{t_i,\bar{c}_j}\right)^2}{\left(N_{t_i,c_j} + N_{\bar{t}_i,c_j}\right) \cdot \left(N_{t_i,c_j} + N_{t_i,\bar{c}_j}\right) \cdot \left(N_{t_i,\bar{c}_j} + N_{\bar{t}_i,\bar{c}_j}\right) \cdot \left(N_{\bar{t}_i,c_j} + N_{\bar{t}_i,\bar{c}_j}\right)} \tag{5.19}$$

The higher the $\chi^2(T_i, C_j)$ value, the less valid the null hypothesis is, and the higher the correlation between T_i and C_j.

Similar to MI and CHI, the weighted average or maximum $\chi^2(T_i, C_j)$ across all classes can measure the amount of discriminative information contained by term T_i, and the top-ranked terms can be used as the selected features:

$$\chi^2_{\max}(T_i) = \max_{j=1,\ldots,M} \left\{\chi^2(T_i, C_j)\right\} \tag{5.20}$$

$$\chi^2_{\text{avg}}(T_i) = \sum_{j=1}^{M} p(c_j)\chi^2(T_i, C_j) \tag{5.21}$$

Table 5.6 shows the results from the feature selection for the text classification dataset (Table 5.1) using the χ^2 method.

5.2.4 *Other Methods*

Nigam et al. (2000) proposed a weighted log-likelihood ratio (WLLR) method to measure the correlation between term t_i and class c_j for feature selection for text classification:

$$\begin{aligned} \text{WLLR}(t_i, c_j) &= p(t_i|c_j) \cdot \log \frac{p(t_i|c_j)}{p(t_i|\bar{c}_j)} \\ &= \frac{N_{t_i,c_j}}{N_{t_i,c_j} + N_{\bar{t}_i,c_j}} \cdot \log \frac{N_{t_i,c_j}(N_{t_i,\bar{c}_j} + N_{\bar{t}_i,\bar{c}_j})}{N_{t_i,\bar{c}_j}(N_{t_i,c_j} + N_{\bar{t}_i,c_j})} \end{aligned} \tag{5.22}$$

Table 5.7 The text classification dataset (Table 5.1) after feature selection

Id	Document after dimension reduction	Category
$train\_d_1$	computer university	Education
$train\_d_2$	computer	Education
$train\_d_3$	game volleyball	Sport
$train\_d_4$	game medal volleyball	Sport
$test\_d_1$	university	
$test\_d_2$	game university volleyball volleyball	

Li et al. (2009a) further analyzed six kinds of feature selection methods (MI, IG, χ^2, WLLR, and so on). They found that frequency $p\left(t_i|c_j\right)$ and odd ratio $\frac{p(t_i|c_j)}{p(t_i|\bar{c}_j)}$ are two basic components in feature selection, and the above feature selection methods can be formulated as the combination of the two basic components. Based on this, they proposed a general feature selection method called general weighted frequency and odd (WFO) for text classification:

$$\mathrm{WFO}\left(t_i, c_j\right) = p\left(t_i|c_j\right)^{\lambda}\left(\log\frac{p\left(t_i|c_j\right)}{p\left(t_i|\bar{c}_j\right)}\right)^{1-\lambda}$$

$$= \left(\frac{N_{t_i,c_j}}{N_{t_i,c_j}+N_{\bar{t}_i,c_j}}\right)^{\lambda}\left(\log\frac{N_{t_i,c_j}(N_{t_i,\bar{c}_j}+N_{\bar{t}_i,\bar{c}_j})}{N_{t_i,\bar{c}_j}(N_{t_i,c_j}+N_{\bar{t}_i,c_j})}\right)^{1-\lambda} \tag{5.23}$$

We assume that the feature space obtained after feature selection is {computer, volleyball, game, medal, university}. Based on the reduced feature set, the text representations of the documents in Table 5.1 are shown in Table 5.7.

5.3 Traditional Machine Learning Algorithms for Text Classification

After text representation and feature selection, the next step is to employ a classification algorithm to predict the class label of the documents. Early text classification algorithms included the Rocchio approach, the K-nearest neighbor classifier, and decision trees. The most widely used text classification algorithms in traditional machine learning are naïve Bayes, maximum entropy (ME), and support vector machines (SVM).

5.3.1 *Naïve Bayes*

The Bayesian model is a kind of generative algorithm that models the joint distribution $p(\boldsymbol{x}, y)$ of the observation $\boldsymbol{x}$ and its class y. In practice, the joint distribution is transformed into the product of the class-prior distribution $p(y)$ and the class-conditional distribution $p(\boldsymbol{x}|y)$:

$$p\left(\boldsymbol{x}, y\right) = p(y) \times p(\boldsymbol{x}|y) \tag{5.24}$$

The Bernoulli distribution or the categorical distribution can be used to model the former for binary and multiclass classifications, respectively. The remaining problem is how to estimate the class-conditional distribution $p(\boldsymbol{x}|y)$ for different applications.

In text classification, to solve the above problem, it is necessary to further simplify the class-conditional distribution of documents. A simple way is to ignore the word order relationships in the document and assume that a document is a bag of words where the individual words are interchangeable. Mathematically, such simplification can be described as an assumption that the class-conditional distributions of words are independent of each other. Based on this assumption, the class-conditional distribution of a document can be written as the product of multiple class-conditional distributions of words. Such a bag-of-words assumption is consistent with the discriminant model where a document can be represented based on a vector space model. The Bayesian model under this assumption is called the naïve Bayes model.

There are two main hypotheses for the class-conditional distribution, known as the multinomial distribution and the multivariate Bernoulli distribution (McCallum et al. 1998). The multivariate Bernoulli distribution only captures the presence of words in a document and ignores their frequency. In comparison, the multinomial distribution is used more often and has generally better classification performance. In this section, we will introduce the naïve Bayes model based on the multinomial distribution.

First, we represent a document $\boldsymbol{x}$ as a sequence of words:

$$\boldsymbol{x} = \left[w_1, w_2, \cdots, w_{|\boldsymbol{x}|}\right] \tag{5.25}$$

Under the bag-of-words assumption, $p(\boldsymbol{x}|y)$ has the form of a multinomial distribution:

$$\begin{aligned} p(\boldsymbol{x}|c_j) &= p([w_1, w_2, \cdots, w_{|\boldsymbol{x}|}]|c_j) \\ &= \prod_{i=1}^{V} p(t_i|c_j)^{N(t_i, \boldsymbol{x})} \end{aligned} \tag{5.26}$$

where V is the dimension of the vocabulary, t_i is the i-th term in the vocabulary, $\theta_{i|j} = p(t_i|c_j)$ is the probability of occurrence of t_i in class c_j, and $N(t_i, \boldsymbol{x})$ is the term frequency of t_i in document $\boldsymbol{x}$.

We take the multiclass classification problem as an example for description. We assume that the class y obeys the categorical distribution:

$$p\left(y = c_j\right) = \pi_j \tag{5.27}$$

According to the assumption of a multinomial distribution, the joint distribution of $p\left(\boldsymbol{x}, y\right)$ can be written as

$$p(\boldsymbol{x}, y = c_j) = p(c_j) \cdot p(\boldsymbol{x}|c_j) = \pi_j \prod_{i=1}^{V} \theta_{i|j}^{N(t_i, \boldsymbol{x})} \tag{5.28}$$

Naïve Bayes learns the parameters $(\boldsymbol{\pi}, \boldsymbol{\theta})$ based on the principle of maximum likelihood estimation (MLE). Given the training set $\{\boldsymbol{x}_k, y_k\}_{k=1}^{N}$, the optimization objective is to maximize the log-likelihood function $L\left(\boldsymbol{\pi}, \boldsymbol{\theta}\right) = \log \prod_{k=1}^{N} p\left(\boldsymbol{x}_k, y_k\right)$. By solving the MLE problem, we obtain the estimated value of the parameters:

$$\pi_j = \frac{\sum_{k=1}^{N} I(y_k = c_j)}{\sum_{k=1}^{N} \sum_{j'=1}^{C} I(y_k = c_{j'})} = \frac{N_j}{N} \tag{5.29}$$

$$\theta_{i|j} = \frac{\sum_{k=1}^{N} I(y_k = c_j) N(t_i, \boldsymbol{x}_k)}{\sum_{k=1}^{N} I(y_k = c_j) \sum_{i'=1}^{V} N(t_{i'}, \boldsymbol{x}_k)} \tag{5.30}$$

It can be seen that the estimated value of the class-prior probability π_j is the document frequency of the j-th class in the training set, and the estimated value of the class-conditional probability of term t_i in class c_j is also the frequency of t_i in the documents of class c_j over all the terms in the vocabulary.

To prevent the occurrence of zero probabilities, a Laplace smoothing technique is often applied to Eq. (5.30):

$$\theta_{i|j} = \frac{\sum_{k=1}^{N} I(y_k = c_j) N(t_i, \boldsymbol{x}_k) + 1}{\sum_{i'=1}^{V} \sum_{k=1}^{N} I(y_k = c_j) N(t_{i'}, \boldsymbol{x}_k) + V} \tag{5.31}$$

We train a multinomial naïve Bayes model on the dimension-reduced training set (Table 5.7). Let t_1 = computer, t_2 = volleyball, t_3 = game, t_4 = medal, t_5 = university, and $y = 1$ for the class "education" and $y = 0$ for the class "sport." The parameter estimation results are shown in Table 5.8.

Table 5.8 Naïve Bayes parameter estimation on the dimension-reduced text classification dataset (Table 5.7)

$p(y)$	$p(y = 1) = 0.5$	$p(y = 0) = 0.5$
$p(t_i\|y)$	$p(t_1\|y = 1) = 0.375$	$p(t_1\|y = 0) = 0.1$
	$p(t_2\|y = 1) = 0.125$	$p(t_2\|y = 0) = 0.3$
	$p(t_3\|y = 1) = 0.125$	$p(t_3\|y = 0) = 0.3$
	$p(t_4\|y = 1) = 0.125$	$p(t_4\|y = 0) = 0.2$
	$p(t_5\|y = 1) = 0.25$	$p(t_5\|y = 0) = 0.1$

We classify the test documents in Table 5.7 based on the above model. Suppose the representation of test document $test\_d_1$ is $\boldsymbol{x}_1$. The joint probabilities of $\boldsymbol{x}_1$ and each class are

$$p(\boldsymbol{x}_1, y = 1) = p(y = 1) \cdot p(t_5|y = 1) = 0.125$$
$$p(\boldsymbol{x}_1, y = 0) = p(y = 0) \cdot p(t_5|y = 0) = 0.05$$

According to Bayes' theorem, the posterior probabilities of $\boldsymbol{x}_1$ belonging to each class are

$$p(y = 1|\boldsymbol{x}_1) = \frac{0.125}{0.125 + 0.05} = 0.714$$
$$p(y = 0|\boldsymbol{x}_1) = 0.286$$

Thus, $test\_d_1$ belongs to the "education" class.

Similarly, the joint probabilities of the test document $test\_d_2$ belonging to each class are

$$p(\boldsymbol{x}_2, y = 1) = p(y = 1) \cdot p(t_2|y = 1)^2 \cdot p(t_3|y = 1) \cdot p(t_5|y = 1) = 0.00024$$
$$p(\boldsymbol{x}_2, y = 0) = p(y = 0) \cdot p(t_2|y = 0)^2 \cdot p(t_3|y = 0) \cdot p(t_5|y = 0) = 0.00135$$

The posterior probabilities are

$$p(y = 1|\boldsymbol{x}_2) = 0.153$$
$$p(y = 0|\boldsymbol{x}_2) = 0.847$$

Thus, $test\_d_2$ belongs to the "sport" class.

5.3.2 *Logistic/Softmax and Maximum Entropy*

Logistic regression is a classification algorithm, although its name contains the term "regression." It is a linear classification model that is widely used for binary classification. Softmax regression is its extension from binary classification to multiclass classification. In natural language processing, there is also a commonly used model called maximum entropy (ME). Although softmax regression and maximum entropy were proposed in different ways, their essence is the same.

We first introduce the three models with an emphasis on logistic regression.

We begin by introducing the sigmoid function $\sigma(z) = \frac{1}{1+e^{-z}}$, which can convert the range of real numbers (from negative infinity to positive infinity) to the range of probability (from 0 to 1). Thus, it is often used to approximate a distribution. Its derivative is

$$\frac{\mathrm{d}\sigma(z)}{\mathrm{d}z} = \sigma(z)(1 - \sigma(z)) \tag{5.32}$$

For a binary classification problem, let $y \in \{0, 1\}$ denote its class, $\boldsymbol{x}$ denote the feature vector, and $\boldsymbol{\theta}$ denote the weight vector. Logistic regression defines the posterior probability of $y \in \{0, 1\}$ given $\boldsymbol{x}$ as follows:

$$\begin{cases} p(y = 1|\boldsymbol{x}; \boldsymbol{\theta}) = h_{\boldsymbol{\theta}}(\boldsymbol{x}) = \sigma(\boldsymbol{\theta}^{\mathrm{T}}\boldsymbol{x}) \\ p(y = 0|\boldsymbol{x}; \boldsymbol{\theta}) = 1 - h_{\boldsymbol{\theta}}(\boldsymbol{x}) \end{cases} \tag{5.33}$$

where the probability $p(y = 1|x)$ is defined by a logistic function.

The two equations above can be written in a unified form:

$$\begin{aligned} p(y|\boldsymbol{x}; \boldsymbol{\theta}) &= (h_{\boldsymbol{\theta}}(\boldsymbol{x}))^{y}(1 - h_{\boldsymbol{\theta}}(\boldsymbol{x}))^{(1-y)} \\ &= \left(\frac{1}{1 + \mathrm{e}^{-\boldsymbol{\theta}^{\mathrm{T}}\boldsymbol{x}}}\right)^{y}\left(1 - \frac{1}{1 + \mathrm{e}^{-\boldsymbol{\theta}^{\mathrm{T}}\boldsymbol{x}}}\right)^{(1-y)} \end{aligned} \tag{5.34}$$

For the hypothesis given by Eq. (5.34), logistic regression estimates the parameters based on the principle of maximum likelihood estimation. Given a training set $\{(\boldsymbol{x}_i, y_i)\}, i = 1, \cdots, N$, the log-likelihood of the model is

$$l(\boldsymbol{\theta}) = \sum_{i=1}^{n} y_i \log h_{\boldsymbol{\theta}}(\boldsymbol{x}_i) + (1 - y_i)\log(1 - h_{\boldsymbol{\theta}}(\boldsymbol{x}_i)) \tag{5.35}$$

First-order optimization methods such as gradient ascent and stochastic gradient ascent are usually used to solve this optimization problem. In addition, quasi-Newton methods such as BFGS (Broyden–Fletcher–Goldfarb–Shanno) and L-BFGS (limited-memory BFGS) are also used to increase learning efficiency for large-scale training data.

Softmax regression is the extension of logistic regression from binary classification to multiclass classification, and it is also called multiclass logistic regression. Logistic regression can also be viewed as a special case of softmax regression where the number of classes is two. Softmax regression is the most widely used classification algorithm in traditional machine learning and is often used as the last layer of deep neural networks to perform classification.

Given the feature vector $\boldsymbol{x}$, the posterior probability of class $y = c_j$ is defined in terms of a softmax function as follows:

$$\begin{aligned} p(y = c_j|\boldsymbol{x}; \boldsymbol{\Theta}) &= h_j(\boldsymbol{x}) \\ &= \frac{\exp(\boldsymbol{\theta}_j^{\mathrm{T}}\boldsymbol{x})}{\sum_{l=1}^{C} \exp(\boldsymbol{\theta}_l^{\mathrm{T}}\boldsymbol{x})}, \quad j = 1, 2, \cdots, C \end{aligned} \tag{5.36}$$

where the parameters of the model are $\boldsymbol{\Theta} = \{\boldsymbol{\theta}_j\}, j = 1, \cdots, C$.

Given the training set $\{(\boldsymbol{x}_1, y_1), \ldots, (\boldsymbol{x}_N, y_N)\}$, the log-likelihood of softmax regression is

$$L(\boldsymbol{\Theta}) = \sum_{i=1}^{N}\sum_{j=1}^{C} I(y_i = c_j)\log h_j(\boldsymbol{x}_i) \tag{5.37}$$

Note that the negative log-likelihood of softmax regression is also called the cross-entropy loss function and is widely used in classification. It is worth noting that softmax regression and naïve Bayes can be seen as a pair of discriminant generative models (Ng and Jordan 2002).

In comparison with softmax regression, maximum entropy is a widely used model in NLP classification tasks. Maximum entropy assigns the joint probability to observation and label pairs $(\boldsymbol{x}, y)$ based on a log-linear model that is quite similar to softmax regression:

$$\underset{\boldsymbol{\theta}}{p}(\boldsymbol{x}, y) = \frac{\exp(\boldsymbol{\theta} \cdot f(\boldsymbol{x}, y))}{\sum_{\boldsymbol{x}', y'} \exp(\boldsymbol{\theta} \cdot f(\boldsymbol{x}', y'))} \tag{5.38}$$

where $\boldsymbol{\theta}$ is a vector of weights, as mentioned earlier, and f is a function that maps pairs $(\boldsymbol{x}, y)$ to a binary-value feature vector.

The feature vector in softmax regression is defined based on the vector space model of the observation $\boldsymbol{x}$. In maximum entropy, it is defined by the following feature function, which describes the known relationships between the observation $\boldsymbol{x}$ and the class label y:

$$f_i(\boldsymbol{x}, y) = \begin{cases} 1, \boldsymbol{x} \text{ satisfies a certain fact, and } y \text{ belongs to a category} \\ 0, \text{ others} \end{cases} \tag{5.39}$$

By using the text classification dataset in Table 5.7, for example, we can construct one feature function of $(\boldsymbol{x}, y)$ as follows: whether the category is "education" when the document contains the word "university." When the feature template is consistent with the definition of the vector space model of softmax regression, the two models are equivalent. It has also been proven that the parameter estimation principles of the two methods (i.e., maximum entropy and maximum likelihood) are also identical.

5.3.3 *Support Vector Machine*

Support vector machine (SVM) is a kind of supervised discriminative learning algorithm for binary classification. It is one of the most popular and widely discussed algorithms in traditional machine learning. There are two core ideas in SVM: first, if the data points are linearly separable, a good separation in terms of a hyperplane is the one that has the largest distance to the nearest training data points on both sides; second, if the data points are linearly nonseparable, SVM can transform the data to high-dimension space, where the data points may become linearly separable through nonlinear transformation based on kernel functions. The linear SVM is widely used for text classification.

The logistic regression model mentioned above is also a kind of linear binary classification model that uses maximum likelihood as its learning criterion. The learning criterion used in linear SVM is called the maximum margin.

For a linear classification hypothesis

$$f(\boldsymbol{x}) = \boldsymbol{w}^{\mathrm{T}}\boldsymbol{x} + b \tag{5.40}$$

Its classification hyperplane is $\boldsymbol{w}^{\mathrm{T}}\boldsymbol{x} + b = 0$. The maximum margin criterion can be expressed as

$$\begin{aligned} &\max_{\boldsymbol{w},b} \frac{1}{2}\|\boldsymbol{w}\|^2 \\ &\text{s.t.} \quad y_i\left(\boldsymbol{w}^{\mathrm{T}}\boldsymbol{x}_i + b\right) \geq 1, \quad i = 1, \cdots, N \end{aligned} \tag{5.41}$$

As a quadratic optimization problem, it can be solved with any off-the-shelf quadratic programming optimization package. However, instead of directly solving the primal optimization problem, SVM tends to solve the following dual problem based on the Lagrange multiplier method:

$$\begin{aligned} &\max_{\boldsymbol{\alpha}} \sum_{i=1}^{N} \alpha_i - \frac{1}{2}\sum_{i,j=1}^{m} y_i y_j \alpha_i \alpha_j \langle \boldsymbol{x}_i, \boldsymbol{x}_j \rangle \\ &\text{s.t.} \quad \alpha_i \geq 0, i = 1, \cdots, N \\ &\sum_{i=1}^{N} \alpha_i y_i = 0 \end{aligned} \tag{5.42}$$

where $\alpha_i \geq 0$ is the Lagrange multiplier. The solutions of the dual problem satisfy the Karush–Kuhn–Tucker (KKT) conditions. According to the KKT conditions, only the weights of data points on the boundaries of the margin are positive ($\alpha_i > 0$), and the weights of the remaining data points are all zero ($\alpha_i = 0$). It can be further inferred that the classification hyperplane is only supported by the data points on the boundaries. This is the main reason why the model is called a "support vector" machine.

The hard-margin SVM can work only when the data are completely linearly separable. In the case of noise points or outliers, the margin of SVM will become smaller or even negative. To solve this problem, the soft-margin SVM was proposed, which introduces slack variables into the primary problem:

$$\begin{aligned}
&\max_{w,b} \frac{1}{2}\|w\|^2 + C\sum_{i=1}^{N}\xi_i \\
&\text{s.t.} \quad y_i\left(\boldsymbol{w}^{\mathrm{T}}\boldsymbol{x}_i + b\right) \geq 1 - \xi_i \\
&\qquad \xi_i \geq 0, i = 1, \ldots, N
\end{aligned} \tag{5.43}$$

where ξ_i is the slack variable and C is the parameter that determines the tradeoff between increasing the margin size and ensuring that the points lie on the correct side of the margin boundary. The corresponding dual problem of soft-margin SVM is

$$\begin{aligned}
&\max_{\boldsymbol{\alpha}} \sum_{i=1}^{N}\alpha_i - \frac{1}{2}\sum_{i,j=1}^{m} y_i y_j \alpha_i \alpha_j \langle \boldsymbol{x}_i, \boldsymbol{x}_j \rangle \\
&\text{s.t.} \quad 0 \leq \alpha_i \leq C, i = 1, \cdots, N \\
&\qquad \sum_{i=1}^{N}\alpha_i y_i = 0
\end{aligned} \tag{5.44}$$

Meanwhile, to address the linearly nonseparable classification problem in low-dimensional space, SVM introduces the kernel function, which allows the algorithm to fit the maximum-margin hyperplane in a transformed high-dimensional feature space. Although the problem is linearly nonseparable in the original input space, it might be linearly separable in the transformed feature space.

The kernel function is defined as the inner product of kernel data in the transformed space:

$$K(\boldsymbol{x}, z) = \varphi(\boldsymbol{x})^{\mathrm{T}}\varphi(z) \tag{5.45}$$

According to Eq. (5.44), the operations involved in SVM are all inner product operations toward $\boldsymbol{x}$. We can therefore use a nonlinear kernel function to replace the dot product in the transformed feature space without needing to know the exact mapping function. The resulting dual problem is formally similar to the previous problem:

$$\begin{aligned}
&\max_{\boldsymbol{\alpha}} W(\boldsymbol{\alpha}) = \sum_{i=1}^{N}\alpha_i - \frac{1}{2}\sum_{i,j=1}^{m} y_i y_j \alpha_i \alpha_j K(\boldsymbol{x}_i, \boldsymbol{x}_j) \\
&\text{s.t.} \quad 0 \leq \alpha_i \leq C, i = 1, \cdots, N \\
&\qquad \sum_{i=1}^{N}\alpha_i y_i = 0
\end{aligned} \tag{5.46}$$

and the decision function is accordingly

$$
\begin{aligned}
f(\boldsymbol{x}) &= \sum_{i=1}^{N} \alpha_i^* y_i \langle \varphi(\boldsymbol{x}_i), \varphi(\boldsymbol{x}) \rangle + b^* \\
&= \sum_{i=1}^{N} \alpha_i^* y_i K(\boldsymbol{x}_i, \boldsymbol{x}) + b^*
\end{aligned} \tag{5.47}
$$

The commonly used kernel functions include:

- Linear kernel: $K(\boldsymbol{x}, z) = \boldsymbol{x}^{\mathrm{T}} z$,
- Polynomial kernel: $K(\boldsymbol{x}, z) = \left(\boldsymbol{x}^{\mathrm{T}} z + c\right)^d$,
- Radial basis function: $K(\boldsymbol{x}, z) = \exp\left(-\frac{|x-z|^2}{2\delta^2}\right)$

as well as some other kernels, such as the sigmoid kernel, pyramid kernel, string kernel, and tree kernel functions. The linear kernel function is mostly used in text classification because the feature space representing a document is usually high-dimensional and linearly separable.

Thus far, we have introduced how to convert the primal problem of SVM into the dual problem shown in Eq. (5.46), but we still need to solve the dual problem and obtain the optimal parameters $\boldsymbol{\alpha}^*$ and b^*. A representative method for this task is the sequential minimal optimization (SMO) algorithm. Interested readers can refer to Platt (1998) for more details.

As a representative classification algorithm in traditional machine learning, SVM has been widely used in text classification since the 1990s. According to the comparative study of Yang and Liu (1999), SVM's performance on topic-based text classifications is significantly better than those of NB, linear least squares fit, and a three-layer feed-forward neural network and is equivalent to or slightly better than the k-nearest neighbor classifier. For the sentiment classification task (Pang et al. 2002), it was also reported that SVM performed better than NB and ME on the movie review corpus.

5.3.4 Ensemble Methods

The pursuit of ensemble methods has been motivated by the intuition that the appropriate integration of different participants might leverage distinct strengths. In traditional machine learning, ensemble methods mostly combine multiple learning algorithms to obtain better predictive performance than any of the base learning algorithms alone. There are three main methods for generating multiple base learning algorithms: (1) training on different data subsets; (2) training on different feature sets; and (3) adopting different classification algorithms.

Bagging (bootstrap aggregating) and boosting algorithms belong to the first category. Bagging, proposed by Breiman (1996), involves training each base classifier based on a randomly extracted subset of the training set and obtaining the ensemble predication by voting on multiple base classifiers. Boosting involves incrementally building an ensemble model by training each base classifier iteratively to emphasize the training instances that previous base classifiers misclassified. AdaBoost is one of the representative (Freund et al. 1996) variants of the boosting algorithm.

Ensemble learning has been successfully applied to text classification. An early study (Larkey and Croft 1996) combined different types of machine learning algorithms to obtain an ensemble classifier with better performance for text classification. Schapire and Singer (2000) proposed BoosTexter, a text classification system based on boosting, which performed better than traditional algorithms. Xia et al. (2011) performed a comparative study of the effectiveness of the ensemble technique for sentiment classification by integrating different feature sets and classification algorithms to synthesize a more accurate sentiment classification procedure.

5.4 Deep Learning Methods

Traditional text representation and classification algorithms rely on manually designed features, which have many shortcomings, such as the high-dimensional problem, data-sparsity problem, and poor representation learning ability. In recent years, deep learning techniques, represented by deep neural networks, have made great breakthroughs in speech recognition, image processing, and text mining. Because of its powerful representation learning ability and the end-to-end learning framework, deep learning has been widely applied to and made great progress in many text mining tasks, including text classification.

In the following, we will introduce several representative deep learning methods for text classification.

5.4.1 *Multilayer Feed-Forward Neural Network*

A multilayer feed-forward neural network is a forward-structured artificial neural network that maps a set of input vectors to a set of output vectors in a multilayer fully connected manner. Compared to the linear classification algorithm, a multilayer forward neural network adds a hidden layer and an activation function for nonlinear transformation. Ideally, a multilayer feed-forward network can approximate any nonlinear functions.

Figure 5.4 shows the structure of a three-layer feed-forward network. Suppose $\boldsymbol{x} \in \mathbb{R}^M, \boldsymbol{h} \in \mathbb{R}^S, \boldsymbol{y} \in \mathbb{R}^C$ are the input layer, hidden layer, and output layer, respectively. The nodes between two adjacent layers are fully connected. For example, the hidden node b_h is connected with all input nodes, x_1, &, x_i, &, x_M, and

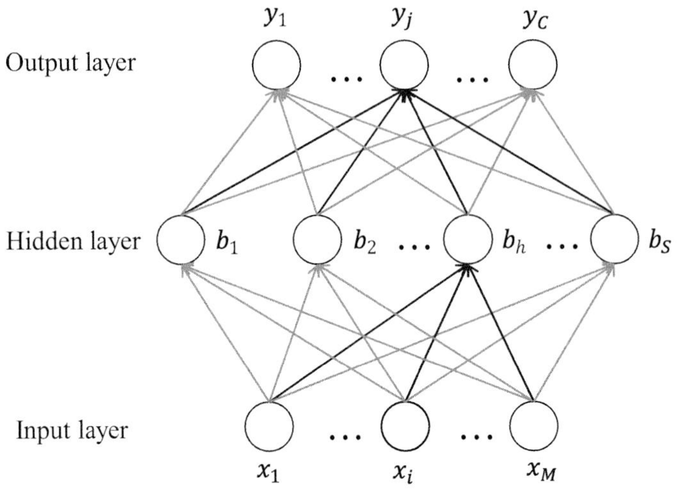

Fig. 5.4 The structure of a three-layer feed-forward neural network

the output node y_j is connected with all hidden nodes, b_1, &, b_h, &, b_S. $\boldsymbol{W} \in \mathbb{R}^{M\times S}$ represents the weight matrix between the hidden layer and the output layer, where w_{hj} is the weight of the connection between b_h and y_j. $\boldsymbol{V} \in \mathbb{R}^{S\times C}$ represents the weight matrix between the input layer and the hidden layer, where v_{ih} is the weight of the connection between x_i and b_h. The network structure can be formulated as follows:

$$b_h = \sigma\left(\alpha_h\right) = \sigma\left(\sum_{i=1}^{M} v_{ih}x_i + \gamma_h\right) \tag{5.48}$$

$$\hat{y}_j = \sigma\left(\beta_j\right) = \sigma\left(\sum_{h=1}^{S} w_{hj}b_h + \theta_j\right) \tag{5.49}$$

where $\sigma(\cdot)$ is a nonlinear activation function such as sigmoid.

Given a training set $D = \left\{\left(\boldsymbol{x}_1, \boldsymbol{y}_1\right), \left(\boldsymbol{x}_2, \boldsymbol{y}_2\right), \ldots, \left(\boldsymbol{x}_N, \boldsymbol{y}_N\right)\right\}$, define the following least mean squares loss function:

$$E = \frac{1}{2}\sum_{k=1}^{N}\sum_{j=1}^{C}\left(\hat{\boldsymbol{y}}_{k_j} - \boldsymbol{y}_{k_j}\right)^2 \tag{5.50}$$

Learning or training can be viewed as the process of optimizing the loss function, i.e., determining the optimal parameters of the model that best fits the training data according to the loss function. For training, a feed-forward neural network uses the back-propagation (BP) algorithm, which is essentially a stochastic gradient descent optimization.

Artificial neural networks were investigated in the early research on text classification (Yang and Liu 1999) but have not been widely used because of the computational inefficiency at that time. Moreover, neural networks were only used as a classifier module rather than an end-to-end joint framework of feature representation learning and classification widely used now. A document was first represented as a sparse feature vector $\boldsymbol{x} = [x_1, x_2, x_3, \cdots]^{\mathrm{T}}$ based on the manually designed vector space and then sent to the feed-forward neural network for classification only, similar to the process for traditional classification algorithms such as naïve Bayes and SVM.

Recently, with the development of representation learning ability and the application of the end-to-end learning method, artificial neural network models, renamed deep learning, have achieved great success in many text data mining fields, including text classification. The representative deep learning algorithms include convolutional neural networks and recurrent neural networks.

5.4.2 *Convolutional Neural Network*

Convolutional neural network (CNN) is a special kind of feed-forward neural network in which the hidden layers consist of a series of convolutional and pooling layers. In comparison with multilayer feed-forward neural networks, a CNN has the characteristics of local connection, shared weights, and translation invariance.

Figure 5.5 shows the basic structure of a convolutional neural network for text classification, which consists of an input layer, a convolutional layer, a pooling layer, a fully connected layer, and an output layer.

A text classification model based on CNN usually has the following steps:

(1) The input text is normally subjected to morphological processing (e.g., tokenization for English or word segmentation for Chinese) and converted to a word sequence, and then the word embedding is used for the initialization of the network.
(2) Feature extraction is then performed through the convolutional layer. Taking Fig. 5.5 as an example, there are three different sizes of convolution kernels, 2×5, 3×5, and 4×5, and two convolution kernels for each size. It should be noted that when computing the convolutional representation matrix of the input text, the two-dimensional convolution is usually performed only in one direction (i.e., the width of the convolution kernel and the dimension of the word vector are maintained), and the step of the convolution operation is set to 1. Each convolution kernel is used to operate on the representation matrix of the input text, and each convolution kernel will obtain a vector representation of the input text.

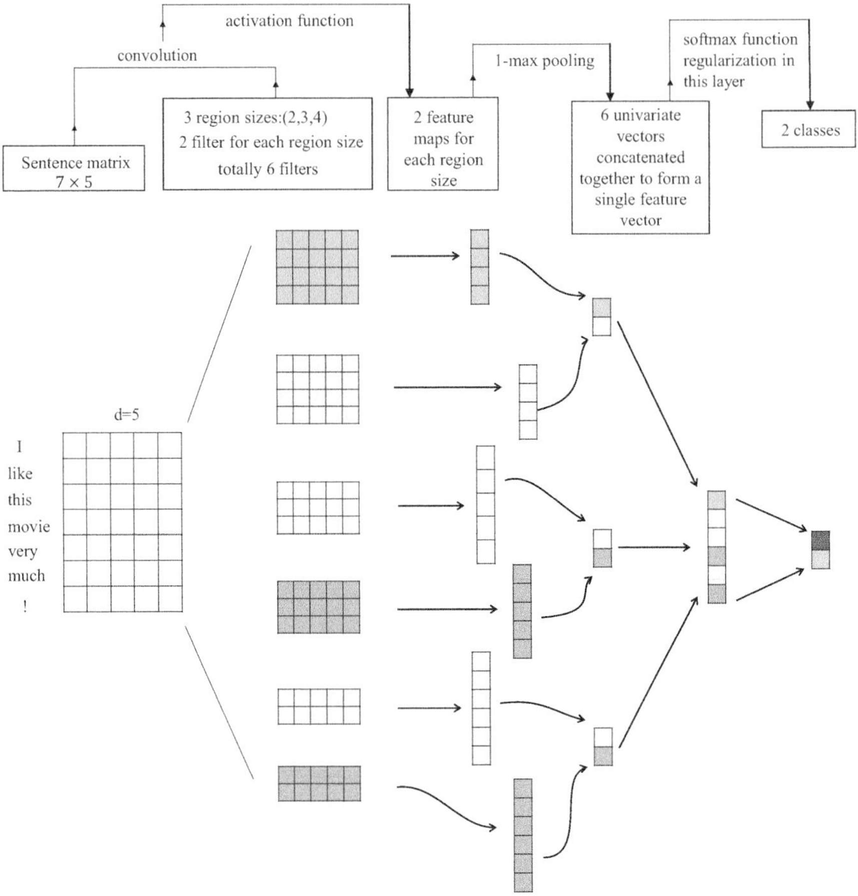

Fig. 5.5 The basic structure of CNN for text classification

(3) The pooling layer downsamples the feature vectors outputted by the convolutional layer and obtains an abstract text representation whose dimension is equal to the number of convolution kernels. The vector representation outputted by the pooling layer is then fed into the softmax layer (i.e., a fully connected layer plus a normalization layer) for classification.

Kim (2014) first proposed convolutional neural networks for text classification and found that they achieved significantly better performance than classical machine learning methods in both topic and sentiment classification tasks. Kalchbrenner et al. (2014) proposed a dynamic convolutional neural network that used the dynamic k-max pooling operation to downsample and several of the most important features to represent local features after two-dimensional convolutions. Zhang et al. (2015) proposed a character-level CNN that represents the text and performs convolution operations at a finer granularity (e.g., characters) and achieved better or competitive results in comparison with word-level CNN and RNN.

5.4.3 *Recurrent Neural Network*

(1) RNN, LSTM, Bi-LSTM, and GRU

Recursive neural network is a kind of deep neural network created by applying the same set of weights recursively over a structured input. It has been widely used in learning sequences and tree structures in natural language processing. Usually, the recursive neural network over time (i.e., a modeling sequence) is called a recurrent neural network. In the following, RNN refers to recurrent neural network if not stated specifically otherwise.

The structure of a recurrent neural network is shown in Fig. 5.6. The left side is the structure that runs recurrently over time, and the right side is the structure that is expanded to a sequence. Suppose $\boldsymbol{x}_t$ is the input at time step t and $\boldsymbol{o}_t$ is the output of the model. It can be seen that $\boldsymbol{o}_t$ is related to not only $\boldsymbol{x}_t$ but also the hidden layer state of the previous time step $\boldsymbol{s}_{t-1}$. $\boldsymbol{o}_t$ can be described as follows:

$$\boldsymbol{s}_t = f\left(\boldsymbol{U}\boldsymbol{x}_t + \boldsymbol{W}\boldsymbol{s}_{t-1}\right) \tag{5.51}$$

$$\boldsymbol{o}_t = \boldsymbol{V}\boldsymbol{s}_t \tag{5.52}$$

where $\boldsymbol{U} \in \mathbb{R}^{h \times d}$, $\boldsymbol{W} \in \mathbb{R}^{h \times h}$, and $\boldsymbol{V} \in \mathbb{R}^{c \times h}$ are the weight matrices of the input node to the hidden node, the current hidden node to the next hidden node, and the hidden node to the output node, respectively. d, h, and c are the dimensions of the input layer, the hidden layer, and the output, respectively. f is a nonlinear activation function (e.g., tanh). By feeding $\boldsymbol{o}_t$ to a softmax layer, we can perform classification for each node or the entire sequence:

$$\boldsymbol{p}_t = \text{softmax}\left(\boldsymbol{o}_t\right) \tag{5.53}$$

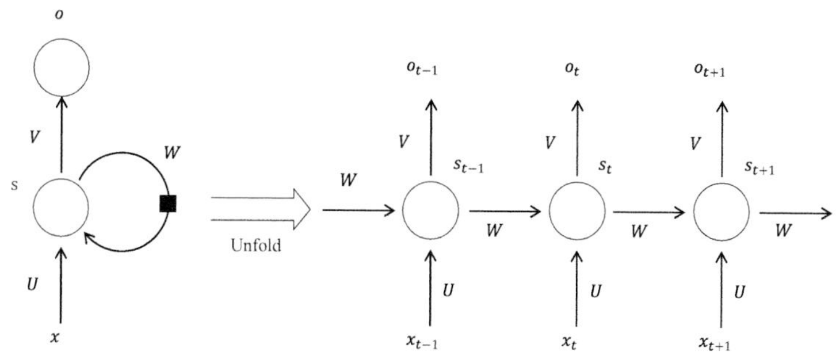

Fig. 5.6 The structure of a recurrent neural network

RNN learns the model parameters by the back-propagation through time (BPTT) algorithm, which is a generalization of the back-propagation algorithm of feed-forward neural networks.

To address the problems of vanishing gradients and exploding gradients when processing long sequence data, Hochreiter and Schmidhuber (1997) proposed the long short-term memory (LSTM) model, which was further improved and promoted by Gers et al. (2002). Schuster and Paliwal (1997) proposed the bidirectional RNN to make better use of the forward and backward context information. Graves et al. (2013) employed bidirectional LSTM (Bi-LSTM) in speech recognition to encode the sequence from front to back and back to front. To address the complexity and redundancy of LSTM, Cho et al. (2014) proposed a gated recurrent unit (GRU) based on LSTM. GRU simplifies the structure of LSTM by combining the forget gate and the input gate into an update gate while merging the cell state and the hidden layer.

When using an RNN to model sequence data, one can learn from the attention mechanism of the human brain, which adaptively selects some key information from a large number of input signals. This approach can improve the performance and efficiency of the model. Inspired by this, the attention mechanism was proposed to differentiate the importance of component units in sequence in semantic composition. For example, the representation of a sentence will be the weighted sum of the representations of the words it contains, and furthermore, the representation of a document will be the weighted sum of the representations of the sentences it contains.

More details of LSTM, GRU, and the attention mechanism can be found in Chap. 3 of this book.

(2) Sentence-Level Classification Model Based on RNN

In this section, we take sentence-level sentiment classification as an example to introduce how to apply RNN to text classification. Let us assume that the input sentence is "I like this movie" and the class label is "positive."

As shown in Fig. 5.7, we first obtain the initial representation of the sentence with the pretrained word vector $[x_1, x_2, \ldots, x_T]$. Each word embedding $\boldsymbol{x}_t$ is sent to Bi-LSTM according to the word order:

$$\overrightarrow{\boldsymbol{c}}_t, \overrightarrow{\boldsymbol{h}}_t = \text{LSTM}\left(\overrightarrow{\boldsymbol{c}}_{t-1}, \overrightarrow{\boldsymbol{h}}_{t-1}, \boldsymbol{w}_t\right) \tag{5.54}$$

$$\overleftarrow{\boldsymbol{c}}_t, \overleftarrow{\boldsymbol{h}}_t = \text{LSTM}\left(\overleftarrow{\boldsymbol{c}}_{t+1}, \overleftarrow{\boldsymbol{h}}_{t+1}, \boldsymbol{w}_t\right) \tag{5.55}$$

The hidden vector is

$$\boldsymbol{h}_t = \left[\overrightarrow{\boldsymbol{h}}_t, \overleftarrow{\boldsymbol{h}}_t\right] \tag{5.56}$$

After preprocessing all words, the hidden states are $[\boldsymbol{h}_1, \boldsymbol{h}_2, \ldots, \boldsymbol{h}_T]$.

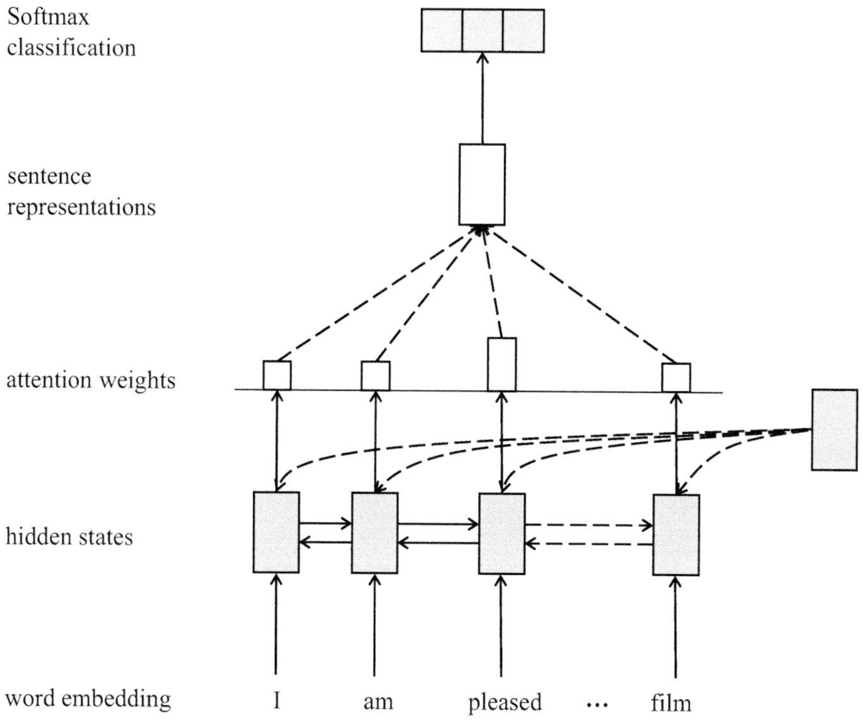

Fig. 5.7 The basic structure of RNN for sentence-level text classification

Then, we calculate the attention weight α_t according to the attention mechanism:

$$\alpha_t = \text{softmax}\left(\boldsymbol{u}_t^{\mathrm{T}}\boldsymbol{q}\right) \tag{5.57}$$

where $\boldsymbol{u}_t = \tanh(\boldsymbol{W}\boldsymbol{h}_t + \boldsymbol{b})$ and $\boldsymbol{q}$ is the query vector. The final sentence representation vector is obtained in the form of the weighted sum of the hidden state of each word in the sentence:

$$\boldsymbol{r} = \sum_t \alpha_t \boldsymbol{h}_t \tag{5.58}$$

The prediction is finally obtained by feeding r to a softmax layer:

$$\boldsymbol{p} = \text{softmax}(\boldsymbol{W}_c\boldsymbol{r} + \boldsymbol{b}_c) \tag{5.59}$$

where $\boldsymbol{W}_c$ and $\boldsymbol{b}_c$ are the weight matrix and the bias term, respectively.

The cross-entropy E between the ground truth $\boldsymbol{y}$ and the prediction distribution $\boldsymbol{p}$ is used as the loss function:

$$E = -\sum_{j=1}^{C} y_j \log p_j \tag{5.60}$$

The model parameters are learned through the BPTT algorithm.

(3) Hierarchical Document-Level Text Classification Model

Document-level text classification refers to text classification for the entire document, where each document is assigned a class label. A simple method for document-level text classification is to treat the document as a long sentence and employ an RNN to encode and classify it. However, this approach does not consider the hierarchical structure of the document.

A document usually contains multiple sentences, and each sentence contains multiple words. Therefore, a document can be modeled according to such a "word-sentence-document" hierarchy. Tang et al. (2015a) first employed CNN (or LSTM) to encode word sequences in a sentence and then used a gated RNN to encode the sentence sequences in the document. Yang et al. (2016) furthermore proposed a hierarchical attention GRU model that consists of five parts: the word-level encoding layer, the word-level attention layer, the sentence-level encoding layer, the sentence-level attention layer, and the softmax layer, as shown in Fig. 5.8.

- Word-level encoding layer: For each sentence, we send the initialized word embedding to Bi-GRU and obtain the forward hidden state $\overrightarrow{\boldsymbol{h}}_{it}$ and the backward hidden state $\overleftarrow{\boldsymbol{h}}_{it}$ of each word. Their concatenation is used as the representation of each word $\boldsymbol{h}_{it} = \left[\overrightarrow{\boldsymbol{h}}_{it}, \overleftarrow{\boldsymbol{h}}_{it}\right]$.
- Word-level attention layer: We first calculate the weight according to $\alpha_{it} = \frac{\exp(\boldsymbol{u}_{it}^{\mathrm{T}}\boldsymbol{u}_w)}{\sum_t \exp(\boldsymbol{u}_{it}^{\mathrm{T}}\boldsymbol{u}_w)}$, where $\boldsymbol{u}_{it} = \tanh(\boldsymbol{W}_w\boldsymbol{h}_{it} + \boldsymbol{b}_w)$ and $\boldsymbol{u}_w$ is a query vector that measures the importance of each word in the sentence. It can be seen as a high-level representation of the query statement "Which word is more important?" It is randomly initialized in the model and trained with the other parameters of the model jointly. Finally, the weighted sum of the hidden representation of each word is used as the representation of the sentence $\boldsymbol{s}_i = \sum_t \alpha_{it}\boldsymbol{h}_{it}$.
- Sentence-level encoding layer: After word-level encoding and attention, each sentence obtains its representation. A document consists of multiple sentences. Similar to the word-level encoding layer, the representation of each sentence is sent to the Bi-GRU to obtain the forward embedding vector $\overrightarrow{\boldsymbol{h}}_i$ and the backward embedding vector $\overleftarrow{\boldsymbol{h}}_i$. The concatenation is used as the hidden representation of each sentence $\boldsymbol{h}_i = \left[\overrightarrow{\boldsymbol{h}}_i, \overleftarrow{\boldsymbol{h}}_i\right]$.

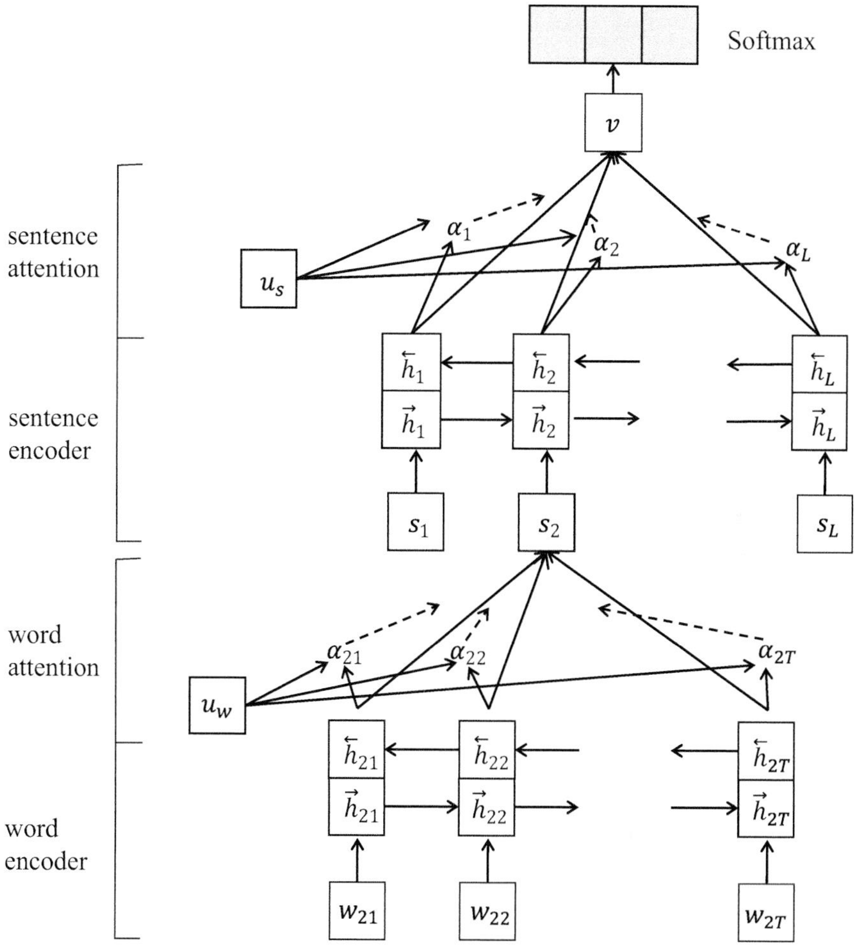

Fig. 5.8 The hierarchical structure of RNN for document-level text classification

- Sentence-level attention layer: We introduce the attention mechanism again to distinguish the importance of different sentences for document representation. The weight for each sentence is $\alpha_i = \frac{\exp(\boldsymbol{u}_i^{\mathrm{T}}\boldsymbol{u}_s)}{\sum_i \exp(\boldsymbol{u}_i^{\mathrm{T}}\boldsymbol{u}_s)}$, where $\boldsymbol{u}_i = \tanh(\boldsymbol{W}_s\boldsymbol{h}_i + \boldsymbol{b}_s)$. The final document representation is also a weighted sum of the representations of all sentences $\boldsymbol{v} = \sum_i \alpha_i \boldsymbol{h}_i$.
- Softmax layer: The document representation $\boldsymbol{v}$ is sent to a softmax layer for document classification: $\boldsymbol{p} = \mathrm{softmax}(\boldsymbol{W}_c\boldsymbol{v} + \boldsymbol{b}_c)$, where $\boldsymbol{W}_c$ and $\boldsymbol{b}_c$ are the weight matrix and bias, respectively.

5.5 Evaluation of Text Classification

Assume that there are M categories in a text classification task, represented by $C_1, \ldots, C_M$. For each of the classes, we calculate statistics on the number of documents with respect to the following four cases:

(1) True positive (TP): where the system correctly predicts it as a positive example (i.e., both the prediction and the ground truth belong to this class).
(2) True negative (TN): where the system correctly predicts it as a negative example (i.e., both the prediction and the ground truth do not belong to this class).
(3) False positive (FP): where the system incorrectly predicts it as a positive example (i.e., the prediction belongs to this class, but the ground truth does not belong to this class).
(4) False negative (FN): where the system incorrectly predicts it as a negative example (i.e., the ground truth belongs to this class, but the prediction does not belong to this class).

After obtaining TP, TN, FP, and FN for each class, we can obtain the microlevel statistics as shown in Table 5.9.

I. Recall, Precision, and F_1 Score

By using $j \in \{1, 2, \ldots, M\}$ to denote the class index, we define the following metrics for each class:

(1) Recall is defined as the proportion of examples being correctly predicted as this class among all examples whose ground truth is this class:

$$R_j = \frac{\mathrm{TP}_j}{\mathrm{TP}_j + \mathrm{FN}_j} \times 100\% \tag{5.61}$$

(2) Precision is defined as the proportion of examples being correctly predicted as the current class among all examples being predicted as the current class:

$$P_j = \frac{\mathrm{TP}_j}{\mathrm{TP}_j + \mathrm{FP}_j} \times 100\% \tag{5.62}$$

Table 5.9 The microlevel statistics of text classification

Category	TP	FP	FN	TN
C_1	TP_1	FP_1	FN_1	TN_1
C_2	TP_2	FP_2	FN_2	TN_2
$\vdots$	$\vdots$	$\vdots$	$\vdots$	$\vdots$
C_M	TP_M	FP_M	FN_M	TN_M

(3) An ideal system with high precision and high recall will return many results, with all results labeled correctly. A system with high recall but low precision returns many positive predictions, but most of them are incorrect in comparison with the ground truth labels. A system with high precision but low recall returns very few positive predictions, but most of its predicted labels are correct when compared with the ground truth labels. We ideally hope that a classification system has both high recall and high precision, but the two are often contradictory. We therefore define the harmonic average of precision and recall as the F_1 score to comprehensively evaluate the effects of these two aspects.

To distinguish the importance of recall and precision, we can also define a more general F_β score:

$$F_\beta = \frac{(\beta^2 + 1)PR}{\beta^2 P + R} \times 100\% \tag{5.63}$$

When $\beta = 1$, F_β becomes the standard F_1 score.

II. Accuracy, Macroaverage, and Microaverage

Recall, precision, and F score can only evaluate the classification performance for a certain class. To measure the performance on the entire classification task, we define the classification accuracy as follows:

$$\text{Acc} = \frac{\#\text{Correct}}{N} \times 100\% \tag{5.64}$$

where N is the number of all examples and #Correct is the number of examples that are correctly predicted.

In addition to classification accuracy, we can also use the macroaverage and microaverage of previous class-oriented measures across all classes to evaluate the performance of the entire classification task.

The recall, precision, and F_1 score based on the macroaverage are defined as follows:

$$\text{Macro\_P} = \frac{1}{C}\sum_{j=1}^{C} \frac{\text{TP}_i}{\text{TP}_i + \text{FP}_i} \tag{5.65}$$

$$\text{Macro\_R} = \frac{1}{C}\sum_{j=1}^{C} \frac{\text{TP}_i}{\text{TP}_i + \text{FN}_i} \tag{5.66}$$

$$\text{Macro\_F}_1 = \frac{2 \times \text{Macro\_P} \times \text{Macro\_R}}{\text{Macro\_P} + \text{Macro\_R}} \tag{5.67}$$

Table 5.10 An example of binary classification results

Prediction/ground truth	Positive(+)	Negative(−)	Total
Positive(+)	250	20	270
Negative(−)	50	180	230
Total	300	200	500

Table 5.11 The evaluation of the classification results in Table 5.10

	TP	FP	FN	TN	Recall	Precision	F_1	Acc
Positive(+)	250	20	50	180	0.8333	0.9259	0.8772	0.8600
Negative(−)	180	50	20	250	0.9000	0.7826	0.8372	
Macroaverage					0.8667	0.8543	0.8605	
Microaverage					0.8600	0.8600	0.8600	

The recall, precision, and F_1 score based on the microaverage are defined as follows:

$$\text{Micro\_P} = \frac{\sum_{j=1}^{C} \text{TP}_i}{\sum_{j=1}^{C} (\text{TP}_i + \text{FP}_i)} \tag{5.68}$$

$$\text{Micro\_R} = \frac{\sum_{j=1}^{C} \text{TP}_i}{\sum_{j=1}^{C} (\text{TP}_i + \text{FN}_i)} \tag{5.69}$$

$$\text{Micro\_F}_1 = \frac{2 \times \text{Micro\_P} \times \text{Micro\_R}}{\text{Micro\_P} + \text{Micro\_R}} \tag{5.70}$$

According to the classification results of a binary classification problem as shown in Table 5.10, we calculate all the aforementioned measures in Table 5.11.

III. P-R Curve and ROC Curve

In a classification problem, predictions are made based on the comparison of the prediction score and a predefined prediction threshold. For example, the threshold value of logistic regression is normally set to be 0.5. When the positive probability is greater than 0.5, we predict it as positive; when the positive probability is less than 0.5, we predict it as negative.

To evaluate the performance of classification models more comprehensively under different recall scores, we can adjust the prediction threshold of the classifier and observe the corresponding precision-recall (P-R) curve by using recall as the x-axis and precision as the y-axis. The P-R curve shows the tradeoff between precision and recall for different thresholds, and the area under the P-R curve can be used to measure the general performance of a classification system. A high area under the curve represents both high recall and high precision, where high precision relates to a low false-positive rate, and high recall relates to a low false-negative rate. High scores for both show that the classifier returns accurate results (high precision) and returns a majority of all positive results (high recall). The mean average precision (mAP) is a metric that can be viewed as a simplification of the area under the P-R curve. It computes the average precision for recall over 0 to 1. For example, we can set the recall to be 0, 0.1, 0.2, 0.3, 0.4, 0.5, 0.6, 0.7, 0.8, 0.9, and 1.0. Then, we use the 11-point average precision for evaluation.

Similar to the P-R curve, we can also plot the ROC (receiver operating characteristic) curve by using the false-positive rate as the x-axis and the true positive rate (i.e., recall) as the y-axis. The area under the ROC curve is called AUC (area under the ROC curve). The higher the AUC value is, the better the general classification performance of the classifier.

The ROC curve summarizes the tradeoff between recall and the false-positive rate. The P-R curve summarizes the tradeoff between precision and recall. The ROC curve is appropriate when the observations are balanced between each class, whereas the P-R curve is more suitable for imbalanced datasets.

5.6 Further Reading

Classification algorithms based on statistical machine learning can be roughly divided into two categories: the discriminative model and the generative model. In general, a discriminative model models the decision boundary between the classes (i.e., learns the decision function $y = f(\boldsymbol{x})$ or the posterior probability $p(y|\boldsymbol{x})$ directly); a generative model explicitly models the distribution of each class $p(\boldsymbol{x}|y)$ as well as the joint distribution of observation and class label (i.e., $p(\boldsymbol{x}, y) = p(y)p(\boldsymbol{x}|y)$). With application to text classification, the typical generative model is the naïve Bayes model, and the typical discriminative models include logistic/softmax regression, maximum entropy model, support vector machine, and artificial neural networks.

The classification models introduced in this chapter are all designed for classification on the entire document, and the discussion has not involved structure predication in the document. Given a piece of text $\boldsymbol{x}$ that consists of multiple nodes $\boldsymbol{x}_t$, it is a classification task to predict the label of $\boldsymbol{x}$ and a sequence labeling task to predict the labels of all nodes $\boldsymbol{x}_t$ in $\boldsymbol{x}$. In a sequence labeling task, each node $\boldsymbol{x}_t$ has a label y_t. Typical sequence labeling models include hidden Markov models (HMMs) and conditional random fields (CRFs). HMM can be viewed as an extension of the naïve Bayes model from classification to sequence labeling. In addition to modeling the relationship between $\boldsymbol{x}_t$ and y_t, HMMs also use the state transition probability to model the relationship of y_{t-1} and y_t. Similarly, the CRF model is the extension of maximum entropy from classification to sequence labeling. The CRF model adopts the log-linear model hypothesis of the maximum entropy model and defines similar feature functions. In addition, the CRF model also defines a state transition feature function to learn the structural relationships in a sequence. Interested readers can refer to Zong (2013); Li (2019) for more details about the sequence labeling models.

Recurrent neural networks naturally have the ability to handle both classification and sequence labeling problems. In the RNN structure shown in Fig. 5.6, if we perform prediction on each node of the sequence, it is a sequence labeling problem; if we obtain the representation of the entire document via semantic composition (e.g., attention) and only perform classification for the document, it is a classification problem. Such a high degree of flexibility is also a major advantage of deep neural networks for text modeling in comparison with traditional machine learning models.

Exercises

5.1 Please derive the naïve Bayes model under the assumption of multivariable Bernoulli distribution according to (McCallum et al. 1998).

5.2 What are the main differences between the softmax regression model and the maximum entropy model?

5.3 Why is the linear kernel more widely used than the other nonlinear kernels when using SVM for text classification?

5.4 What are the main differences between the multilayer feed-forward neural network and the convolutional neural network?

5.5 Can a convolutional neural network capture n-gram grammatical features in text? Why?

5.6 What are the main differences between recurrent neural networks and convolutional neural networks? Which do you think is more suitable for document classification? Why?

Chapter 6
Text Clustering

6.1 Text Similarity Measures

Different clustering algorithms can produce different results by adopting different perspectives, but almost all of them are performed based on similarity measures. Therefore, the key problem of text clustering is how to effectively measure the similarity of texts.

In text clustering, a cluster is represented by a collection of similar documents, and there are three main types of text similarities:

- Similarity between two documents;[1]
- Similarity between two document collections;
- Similarity between a document and a document collection.

We will introduce the three kinds of similarity measures below.

6.1.1 *The Similarity Between Documents*

(1) Distance-Based Similarity

In a vector space model, a document is represented as a vector in the vector space. The simplest way to measure document similarity is to use the distance between two vectors in vector space. The smaller the distance between two vectors, the higher the similarity of the two documents. The commonly used distance metrics include Euclidean distance, Manhattan distance, Chebyshev distance, Minkowski distance, Mahalanobis distance, and Jaccard distance.

[1] For the simplicity of description, we use "document" to refer to a piece of text at different levels (e.g., sentence, document, etc.).

© Tsinghua University Press 2021

C. Zong et al., *Text Data Mining*, https://doi.org/10.1007/978-981-16-0100-2_6

Let $\boldsymbol{a}$ and $\boldsymbol{b}$ be the vector representations of two documents, and the following distances are defined as follows.

a. Euclidean distance

$$d(\boldsymbol{a}, \boldsymbol{b}) = \left(\sum_{k=1}^{M}(a_k - b_k)^2\right)^{1/2} \tag{6.1}$$

b. Manhattan distance

$$d(\boldsymbol{a}, \boldsymbol{b}) = \sum_{k=1}^{M} |a_k - b_k| \tag{6.2}$$

c. Chebyshev distance

$$d(\boldsymbol{a}, \boldsymbol{b}) = \max_{k} |a_k - b_k| \tag{6.3}$$

d. Minkowski distance

$$d(\boldsymbol{a}, \boldsymbol{b}) = \left(\sum_{k=1}^{M}(a_k - b_k)^p\right)^{1/p} \tag{6.4}$$

(2) Cosine Similarity

Cosine similarity computes the similarity between two vectors by calculating the cosine of the angle between the two vectors:

$$\cos(\boldsymbol{a}, \boldsymbol{b}) = \frac{\boldsymbol{a}^{\mathrm{T}}\boldsymbol{b}}{\|\boldsymbol{a}\|\|\boldsymbol{b}\|} \tag{6.5}$$

The range of cosine similarity is $[-1, 1]$. The smaller the angle between the two vectors is, the higher the cosine similarity. When the angle between two vectors is 0° (i.e., the same direction), the cosine similarity is 1; when the angle between two vectors is 90° (i.e., orthogonal direction), the cosine similarity is 0; when the angle between two vectors is 180° (i.e., opposite direction), the cosine similarity is -1.

The inner product of two vectors is proportional to the cosine similarity. The inner product of two vectors after L-2 normalization (see Chap. 3) is equivalent to the cosine similarity: $\boldsymbol{a} \cdot \boldsymbol{b} = \boldsymbol{a}^{\mathrm{T}}\boldsymbol{b}$ (Fig. 6.1).

Distance-based similarity measures the absolute distance between two vectors in the vector space. Cosine similarity measures the angle of vectors in the vector space and is the most widely used method for measuring the similarity of texts.

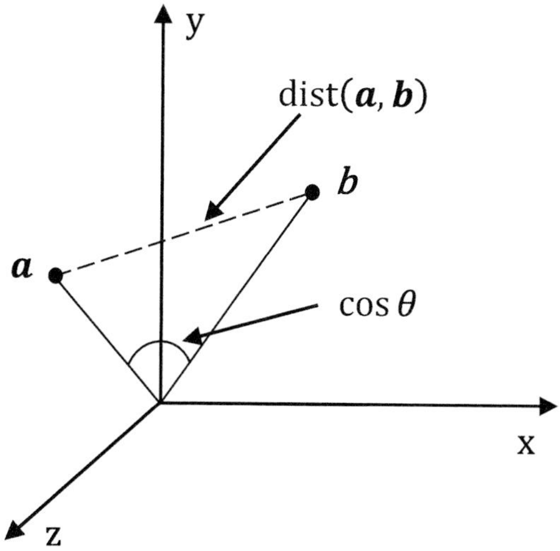

Fig. 6.1 Distance measurement samples in the vector space model

(3) Distribution-Based Similarity

The previous two kinds of similarity measures are performed based on the vector space. However, a document is sometimes represented by a distribution rather than a vector space model, especially in generative models. In this case, the statistical distance can be used to measure the similarity between two documents.

Statistical distance measures the difference between two distributions. A commonly used metric is the Kullback–Leibler (K-L) distance (also called K-L divergence). Based on the BOW assumption, a document can be represented by a categorical distribution over terms. Suppose P and Q are two categorical distributions, and the K-L distance of P and Q is defined as

$$D_{\mathrm{KL}}(P \| Q) = \sum_i P(i) \log \frac{P(i)}{Q(i)} \tag{6.6}$$

The K-L distance is not symmetrical, that is, $D_{\mathrm{KL}}(P \| Q) \neq D_{\mathrm{KL}}(Q \| P)$. A symmetrical K-L distance can therefore be used instead:

$$D_{\mathrm{SKL}}(P, Q) = D_{\mathrm{KL}}(P \| Q) + D_{\mathrm{KL}}(Q \| P) \tag{6.7}$$

It is worth noting that when a document is of short length, it is meaningless to use a categorical distribution to represent it and use the K-L distance to measure the similarity of two documents. In fact, such distribution-based metrics are more suitable for measuring the similarity between two collections of texts than that between two short pieces of texts.

(4) Other Measures

There are other methods for similarity measures. For example, the Jaccard similarity coefficient is another widely used metric that measures the similarity between two sets; it is defined as the size of the intersection divided by the size of the union of the two sets:

$$J\left(\boldsymbol{x}_i, \boldsymbol{x}_j\right) = \frac{|\boldsymbol{x}_i \cap \boldsymbol{x}_j|}{|\boldsymbol{x}_i \cup \boldsymbol{x}_j|} \tag{6.8}$$

where a document is represented by a set of words.

Note that the aforementioned similarity measures can be used not only in text clustering but also in other text data mining tasks.

6.1.2 The Similarity Between Clusters

A cluster is a collection of similar documents. The similarity between two clusters can be computed based on the similarities of the documents contained in them. Suppose $d(C_m, C_n)$ denotes the distance between clusters C_m and C_n, $d(\boldsymbol{x}_i and \boldsymbol{x}_j)$ denotes the distance between documents $\boldsymbol{x}_i$ and $\boldsymbol{x}_j$. There are several ways to measure the similarity between the two clusters as follows.

(1) A single linkage denotes the shortest distance between two documents extracted from two clusters respectively:

$$d(C_m, C_n) = \min_{\boldsymbol{x}_i \in C_m, \boldsymbol{x}_j \in C_n} d(\boldsymbol{x}_i, \boldsymbol{x}_j) \tag{6.9}$$

(2) A complete linkage denotes the longest distance between two documents extracted from two clusters respectively:

$$d(C_m, C_n) = \max_{\boldsymbol{x}_i \in C_m, \boldsymbol{x}_j \in C_n} d(\boldsymbol{x}_i, \boldsymbol{x}_j) \tag{6.10}$$

(3) The average linkage denotes the average distance between two documents extracted from two clusters respectively:

$$d(C_m, C_n) = \frac{1}{|C_m| \cdot |C_n|} \sum_{\boldsymbol{x}_i \in C_m} \sum_{\boldsymbol{x}_j \in C_n} d(\boldsymbol{x}_i, \boldsymbol{x}_j) \tag{6.11}$$

(4) The centroid method is the distance between the centroid of two clusters:

$$d(C_m, C_n) = d(\bar{x}(C_m), \bar{x}(C_n)) \tag{6.12}$$

where $\bar{x}(C_m)$ and $\bar{x}(C_n)$ denote the centroids of the clusters C_m and C_n, respectively.

(5) Ward's method. For each cluster, we first define the within-cluster variance as the sum of squares of the distance between each document and the cluster centroid. The increase in total within-cluster variance after merging the two clusters can therefore be used as a cluster distance metric:

$$\begin{aligned} d(C_m, C_n) = & \sum_{\boldsymbol{x}_k \in C_m \cup C_n} d(\boldsymbol{x}_k, \bar{\boldsymbol{x}}(C_m \cup C_n)) \\ & - \sum_{\boldsymbol{x}_i \in C_m} d(\boldsymbol{x}_i, \bar{\boldsymbol{x}}(C_m)) - \sum_{\boldsymbol{x}_j \in C_n} d(\boldsymbol{x}_j, \bar{\boldsymbol{x}}(C_n)) \end{aligned} \tag{6.13}$$

where $d(\boldsymbol{a}, \boldsymbol{b}) = \|\boldsymbol{a} - \boldsymbol{b}\|^2$.

Ward's method is a criterion applied in hierarchical clustering. It minimizes the total within-cluster variance by finding the pair of clusters at each step that leads to a minimum increase in total within-cluster variance after they are merged.

In addition to the five abovementioned methods, the K-L divergence can also be used for calculating the distance between two clusters. The equation for the K-L divergence is shown as Eq. (6.6). The difference is that the categorical distributions P and Q are estimated by a cluster rather than a document.

6.2 Text Clustering Algorithms

There are extensive types of text clustering methods, including partition-based methods, hierarchy-based methods, density-based methods, grid-based methods, and graph-based methods, each of which contains some typical algorithms. In the following, we introduce several representative text clustering algorithms.

6.2.1 *K-Means Clustering*

The K-means algorithm, proposed by MacQueen in 1967, is a widely used partition-based clustering algorithm.

For a given dataset $\{\boldsymbol{x}_1, \boldsymbol{x}_2, \ldots, \boldsymbol{x}_N\}$, the goal of K-means clustering is to divide the N samples into K $(K \leq N)$ clusters to minimize the sum of the squared distances within each cluster, which is called the within-cluster sum of squares (WCSS):

$$\arg\min_{C} \sum_{k=1}^{K} \sum_{\boldsymbol{x} \in C_k} \|\boldsymbol{x} - \boldsymbol{m}_k\|^2 \tag{6.14}$$

To achieve this objective, the standard K-means clustering algorithm (also called the Lloyd–Forgy method) uses the iterative optimization method. In each iteration step, the distances between each sample and the K centroids (i.e., the means) of the cluster are first calculated. The samples are then assigned to the clusters with the nearest centroid, and the centroids of existing clusters are updated. This process is repeated until the minimum WCSS is reached.

Formally, given the initial centroids of the K clusters $\boldsymbol{m}_1^{(0)}, \boldsymbol{m}_2^{(0)}, \ldots, \boldsymbol{m}_K^{(0)}$, the algorithm iterates in the following two steps:

(1) Assignment: Assign each sample into the cluster that minimizes the sum of squares within clusters:

$$C^{(t)}(\boldsymbol{x}_i) = \arg\min_{k=1,\ldots,K} ||\boldsymbol{x}_i - \boldsymbol{m}_k^{(t-1)}||^2 \tag{6.15}$$

where t denotes the steps of the iterations and $C(\boldsymbol{x})$ denotes the index of the cluster to which $\boldsymbol{x}$ is assigned.

(2) Updating: Update the centroids for each of the K clusters:

$$\boldsymbol{m}_k^{(t+1)} = \frac{1}{|C_k^{(t)}|} \sum_{\boldsymbol{x}_i \in C_k^{(t)}} \boldsymbol{x}_i \tag{6.16}$$

The two steps are iteratively performed until the algorithm converges to a local minimum. But such an alternated iterative optimization cannot guarantee the global minimum of the WCSS.

In practice, we can also choose different distance metrics. For example, in text clustering, the cosine similarity is more often used:

$$d\left(\boldsymbol{x}, \boldsymbol{m}_k^{(t)}\right) = \frac{\boldsymbol{x} \cdot \boldsymbol{m}_k^{(t)}}{\|\boldsymbol{x}\| \left\|\boldsymbol{m}_k^{(t)}\right\|} \tag{6.17}$$

However, it should be noted that the above iterative optimization can ensure the decrease in WCSS only under the Euclidean distance metric. If different distance metrics are used, there is a risk that the algorithm may not converge.

In summary, the K-means clustering algorithm is described as follows.

Table 6.1 displays a small text clustering dataset that contains ten short documents extracted from the domains of education, sports, technology, and literature.

Algorithm 1: K-means clustering algorithm

Input : dataset $\mathscr{D} = \{\boldsymbol{x}_1, \boldsymbol{x}_2, \ldots, \boldsymbol{x}_N\}$, number of clusters K;
Output: clusters $\{C_1, C_2, \ldots, C_K\}$.

Randomly select K samples in $\mathscr{D}$ as the initial mean vectors $\{\boldsymbol{m}_1, \boldsymbol{m}_2, \ldots, \boldsymbol{m}_K\}$;
while *not converged* **do**
  for $i = 1, \ldots, N$ **do**
    for $k = 1, \ldots, K$ **do**
      calculate the distance $d(\boldsymbol{x}_i, \boldsymbol{m}_k) = \|\boldsymbol{x}_i - \boldsymbol{m}_k\|^2$ between $\boldsymbol{x}_i$ and $\boldsymbol{m}_k$;
    end
    divide sample $\boldsymbol{x}_i$ into the cluster of nearest mean vector $\arg\min_k\{d(\boldsymbol{x}_i, \boldsymbol{m}_k)\}$
  end
  for $i = 1, \ldots, K$ **do**
    update the mean vector of each cluster: $\boldsymbol{m}_k^{\text{new}} = \frac{1}{|C_k|} \sum_{\boldsymbol{x}_i \in C_k} \boldsymbol{x}_i$.
  end
end

Table 6.1 Text clustering dataset

ID	Sentence
$\boldsymbol{x}_1$	Beijing Institute of Technology was established in 1958 as one of the earliest universities that established a computer science major in China.
$\boldsymbol{x}_2$	Students from Beijing Institute of Technology won the 4th China Computer Go Championship.
$\boldsymbol{x}_3$	The Gymnasium of Beijing Institute of Technology is the venue for the preliminary volleyball competition of the 2008 Beijing Olympic Games in China.
$\boldsymbol{x}_4$	In the 5th East Asian Games, the total number of medals of China reached a new high. Both the men's and women's volleyball teams won championships.
$\boldsymbol{x}_5$	Artificial intelligence, also known as machine intelligence, refers to the intelligence represented by an artificially produced system.
$\boldsymbol{x}_6$	Artificial intelligence is a branch of computer science that attempts to produce an intelligent machine that can react in a manner similar to human intelligence.
$\boldsymbol{x}_7$	The three Go competitions between artificial intelligence AlphaGo and human champion Jie Ke end with the human's thorough defeat.
$\boldsymbol{x}_8$	The first sparrow of spring! The year beginning with youngest hope than ever!
$\boldsymbol{x}_9$	The brooks sing carols and glees to the spring. The symbol of youth, the grass blade, like a long green ribbon, streams from the sod into the summer.
$\boldsymbol{x}_{10}$	The grass flames up on the hillsides like a spring fire, not yellow but green is the color of its flame.

Let $D = \{\boldsymbol{x}_1, \boldsymbol{x}_2, \ldots, \boldsymbol{x}_{10}\}$ denote this clustering dataset, in which $\boldsymbol{x}_i$ corresponds to the i-th document. Before text clustering, we first perform feature selection. The dataset includes 118 words. Due to the small scale of the corpus, we have not chosen supervised feature selection methods (such as MI and IG) for feature selection. Instead, we use an unsupervised feature selection method, term frequency, to select those features with a frequency of no less than two in this corpus. This method results in a simplified vocabulary that contains 22 words:

Table 6.2 Dimension-reduced text clustering dataset

ID	Sentence
$\boldsymbol{x}_1$	Beijing institute technology university computer science China
$\boldsymbol{x}_2$	Beijing institute technology China computer champion
$\boldsymbol{x}_3$	Beijing institute technology volleyball competition game China
$\boldsymbol{x}_4$	game China volleyball win champion
$\boldsymbol{x}_5$	artificial intelligence machine intelligence intelligence
$\boldsymbol{x}_6$	artificial intelligence computer science intelligent machine human intelligence
$\boldsymbol{x}_7$	artificial intelligence go competition human champion
$\boldsymbol{x}_8$	spring young
$\boldsymbol{x}_9$	spring young grass green
$\boldsymbol{x}_{10}$	grass spring green

"volleyball," "Beijing," "China," "institute," "win," "go," "champion," "computer," "science," "technology," "human," "race," "university," "artificial," "intelligence," "machine," "game," "competition," "spring," "young," "green," "grass."

The dimension-reduced dataset is shown in Table 6.2.

We perform K-means clustering on the corpus dimension-reduced dataset by setting $K = 3$ and use the Euclidean distance as the similarity measure. We use principal component analysis (PCA) to reduce the dimension of the feature space and take the top two components as the x-axis and y-axis to visualize the clustering process:

(i) Initialization: The initial clusters are $\{C_1 : \{\boldsymbol{x}_4\}, C_2 : \{\boldsymbol{x}_5\}, C_3 : \{\boldsymbol{x}_8\}\}$;

(ii) The first iteration: Calculate the distance of each document to the centroid of each cluster. Taking $\boldsymbol{x}_3$ as an example, its distances to the three centroids $\boldsymbol{x}_4, \boldsymbol{x}_5, and \boldsymbol{x}_8$ are 2.45, 3.16, and 3, respectively. Thus, $\boldsymbol{x}_3$ is assigned to its nearest cluster C_1. After assignment for each document, the updated clusters become $\{C_1 : \{\boldsymbol{x}_2, \boldsymbol{x}_3, \boldsymbol{x}_4\}, C_2 : \{\boldsymbol{x}_5, \boldsymbol{x}_6, \boldsymbol{x}_7\}, C_3 : \{\boldsymbol{x}_1, \boldsymbol{x}_8, \boldsymbol{x}_9, \boldsymbol{x}_{10}\}\}$, as shown in Fig. 6.2a.

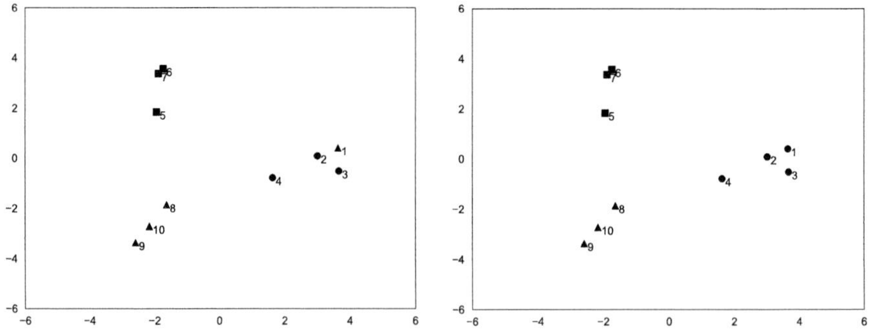

Fig. 6.2 Clustering text with K-means algorithm ($K = 3$)

(iii) The second iteration: Calculate the distance of each document to the centroid of each cluster after the first iteration. Taking $\boldsymbol{x}_1$ as an example, its distances to the three centroids are 2.08, 3.02, and 2.29. Thus, $\boldsymbol{x}_6$ is assigned to its nearest cluster C_1. After assignment for each document, the updated clusters become $\{C_1 : \{\boldsymbol{x}_1, \boldsymbol{x}_2, \boldsymbol{x}_3, \boldsymbol{x}_4\}, C_2 : \{\boldsymbol{x}_5, \boldsymbol{x}_6, \boldsymbol{x}_7\}, C_3 : \{\boldsymbol{x}_8, \boldsymbol{x}_9, \boldsymbol{x}_{10}\}\}$, as shown in Fig. 6.2b.

(iv) The third iteration: According to the distance between each document and the centroid of each cluster after the third iteration, the cluster assignments no longer need to be changed, and the algorithm converges. The final clusters are $\{C_1 : \{\boldsymbol{x}_1, \boldsymbol{x}_2, \boldsymbol{x}_3, \boldsymbol{x}_4\}, C_2 : \{\boldsymbol{x}_5, \boldsymbol{x}_6, \boldsymbol{x}_7\}, C_3 : \{\boldsymbol{x}_8, \boldsymbol{x}_9, \boldsymbol{x}_{10}\}\}$, as shown in Fig. 6.2b.

Although the K-means algorithm is widely used because of its simplicity and efficiency, it still has several shortcomings: ① it remains difficult to determine the value of clustering number K, and ② the result depends on the selected initial centroids or metric selection. For example, if documents $\boldsymbol{x}_2$, $\boldsymbol{x}_5$, and $\boldsymbol{x}_8$ are selected as initial centroids of three clusters, the algorithm will terminate within one iteration, and the final clustering results will be $\{C_1 : \{\boldsymbol{x}_1, \boldsymbol{x}_2, \boldsymbol{x}_3, \boldsymbol{x}_4\}, C_2 : \{\boldsymbol{x}_5, \boldsymbol{x}_6, \boldsymbol{x}_7\}, C_3 : \{\boldsymbol{x}_8, \boldsymbol{x}_9, \boldsymbol{x}_{10}\}\}$.

6.2.2 *Single-Pass Clustering*

In comparison with K-means, single-pass clustering is an even simpler and more efficient clustering algorithm, as it only needs to traverse a collection of documents once to perform the clustering. In the initial stage, the algorithm takes a document from the corpus and constructs a cluster with this document. It then iteratively processes a new document and computes the similarity between this document and each existing cluster. If the similarity is lower than a predefined threshold, a new cluster will be generated; otherwise, it will be assigned to the cluster with the highest similarity. This process repeats until all the documents in the dataset have been processed.

Single-pass clustering involves a similarity computation between a document and a cluster, the methods for which are summarized in Sect. 6.2. In standard single-pass clustering, the similarity between the document and the mean vector of the cluster is employed.

The detailed algorithm is described as follows.

We perform single clustering on the dimension-reduced dataset shown in Table 6.2. The opposite value of the Euclidean distance is used as the similarity metric, and the threshold T is set to be -2.3. All documents are processed in sequence. The clustering process is as follows:

(i) Read the first document $\boldsymbol{x}_1$, establish an initial cluster C_1, and assign $\boldsymbol{x}_1$ to C_1. The initial cluster is $\{C_1 : \{\boldsymbol{x}_1\}\}$;

Algorithm 2: Single-pass clustering algorithm

Input : dataset $\mathscr{D} = \{\boldsymbol{x}_1, \boldsymbol{x}_2, \ldots, \boldsymbol{x}_N\}$, similarity threshold T;
Output: clusters $\{C_1, C_2, \ldots, C_M\}$.

$M = 1$; $C_1 = \{x_1\}$; $\boldsymbol{m}_1 = \boldsymbol{x}_1$
for $i = 2, \ldots, N$ **do**
  for $k = 1, \ldots, M$ **do**
    calculate the similarity $d(\boldsymbol{x}_i, \boldsymbol{m}_k)$ between $\boldsymbol{x}_i$ and $\boldsymbol{m}_k$
  end
  select the cluster of highest similarity $k^* = \arg\max_k \{d(\boldsymbol{x}_i, \boldsymbol{m}_k)\}$
end
if $d(\boldsymbol{x}_i, \boldsymbol{m}_{k^*}) > T$ **then**
  add x_i into cluster C_{k^*}: $C_{k^*} \leftarrow (C_{k^*} \cup \boldsymbol{x}_i)$
  update the mean vector of C_{k^*}: $\boldsymbol{m}_{k^*} = \frac{1}{|C_{k^*}|} \sum_{\boldsymbol{x}_j \in C_{k^*}} \boldsymbol{x}_j$
end
else
  $M += 1$; $C_M = \{\boldsymbol{x}_i\}$
end

(ii) Process document $\boldsymbol{x}_2$. Because the similarity between $\boldsymbol{x}_2$ and the centroid of C_1 is -2.18, which is higher than T, we assign $\boldsymbol{x}_1$ to C_1. The updated clusters are $\{C_1 : \{\boldsymbol{x}_1, \boldsymbol{x}_2\}\}$;

(iii) Process document $\boldsymbol{x}_3$. The similarity between $\boldsymbol{x}_3$ and the centroid of the existing clusters C_1 is -2.18, which is higher than T; therefore, we assign $\boldsymbol{x}_3$ to C_1. The updated clustering result is $\{C_1 : \{\boldsymbol{x}_1, \boldsymbol{x}_2, \boldsymbol{x}_3\}\}$;

(iv) Process document $\boldsymbol{x}_4$. The similarity between $\boldsymbol{x}_4$ and the centroid of the existing cluster C_1 is -2.47. The highest similarity is lower than T; therefore, we assign $\boldsymbol{x}_4$ to C_2. The updated clustering result is $\{C_1 : \{\boldsymbol{x}_1, \boldsymbol{x}_2, \boldsymbol{x}_3\}, C_2 : \{\boldsymbol{x}_4\}\}$;

(v) Process document $\boldsymbol{x}_5$. The similarities between $\boldsymbol{x}_5$ and the centroids of existing clusters C_1 and C_2 are -2.85 and -2.83, respectively. The highest similarity is lower than T; therefore, we establish a new cluster C_3 and assign $\boldsymbol{x}_5$ to it. The updated clustering result is $\{C_1 : \{\boldsymbol{x}_1, \boldsymbol{x}_2, \boldsymbol{x}_3\}, C_2 : \{\boldsymbol{x}_4\}, C_3 : \{\boldsymbol{x}_5\}\}$;

(vi) Process document $\boldsymbol{x}_6$. The similarities between $\boldsymbol{x}_6$ and the centroids of existing clusters C_1, C_2 and C_3 are -3.02, -3.32, and -1.73 respectively. The highest similarity is higher than T (with C_3); therefore, we assign $\boldsymbol{x}_6$ to C_3. The updated clustering result is $\{C_1 : \{\boldsymbol{x}_1, \boldsymbol{x}_2, \boldsymbol{x}_3\}, C_2 : \{\boldsymbol{x}_4\}, C_3 : \{\boldsymbol{x}_5, \boldsymbol{x}_6\}\}$;

(vii) Process document $\boldsymbol{x}_7$. The similarities between $\boldsymbol{x}_7$ and the centroids of existing clusters C_1, C_2 and C_3 are -3.13, -3.0, -2.18 respectively. The highest similarity is higher than T (with C_3); therefore, we assign $\boldsymbol{x}_7$ to C_3. The updated clustering result is $\{C_1 : \{\boldsymbol{x}_1, \boldsymbol{x}_2, \boldsymbol{x}_3\}, C_2 : \{\boldsymbol{x}_4\}, C_3 : \{\boldsymbol{x}_5, \boldsymbol{x}_6, \boldsymbol{x}_7\}\}$;

(viii) Process document $\boldsymbol{x}_8$. The similarities between $\boldsymbol{x}_8$ and the centroids of existing clusters C_1, C_2 and C_3 are -2.67, -2.65, -2.33 respectively. The highest similarity is lower than T; therefore, we establish a new cluster C_4 and assign $\boldsymbol{x}_8$ to it. The updated clustering result is $\{C_1 : \{\boldsymbol{x}_1, \boldsymbol{x}_2, \boldsymbol{x}_3\}, C_2 : \{\boldsymbol{x}_4\}, C_3 : \{\boldsymbol{x}_5, \boldsymbol{x}_6, \boldsymbol{x}_7\}, , C_4 : \{\boldsymbol{x}_8\}\}$;

(ix) Process document $\boldsymbol{x}_9$. The similarities between $\boldsymbol{x}_9$ and the centroids of existing clusters C_1, C_2, C_3, C_4, and C_5 are -3.02, -3.0, -2.73 and -1.41 respectively. The highest similarity is higher than T (with C_4); therefore, we assign $\boldsymbol{x}_9$ to C_4. The updated clustering result is $\{C_1 : \{\boldsymbol{x}_1, \boldsymbol{x}_2, \boldsymbol{x}_3\}, C_2 : \{\boldsymbol{x}_4\}, C_3 : \{\boldsymbol{x}_5, \boldsymbol{x}_6, \boldsymbol{x}_7\}, , C_4 : \{\boldsymbol{x}_8, \boldsymbol{x}_9\}\}$;

(x) Process document $\boldsymbol{x}_{10}$. The similarities between $\boldsymbol{x}_{10}$ and the centroids of existing clusters C_1, C_2, C_3 and C_4 are -2.85, -2.83, -2.53 and -1.22 respectively. The highest similarity is higher than T (with C_4); therefore, we assign $\boldsymbol{x}_{10}$ to C_4. The updated clustering result is $\{C_1 : \{\boldsymbol{x}_1, \boldsymbol{x}_2, \boldsymbol{x}_3\}, C_2 : \{\boldsymbol{x}_4\}, C_3 : \{\boldsymbol{x}_5, \boldsymbol{x}_6, \boldsymbol{x}_7\}, C_4 : \{\boldsymbol{x}_8, \boldsymbol{x}_9, \boldsymbol{x}_{10}\}\}$;

Thus, all documents in the corpus are processed. The final clustering result is shown in Fig. 6.3.

Because of its simplicity and efficiency, the single-pass clustering algorithm is suitable for scenarios including large-scale and real-time streaming data, such as topic detection and tracking, which we will introduce in Chap. 9. However, it also contains some inherent flaws. For example, its performance greatly depends on the order of processed documents, and the threshold is sometimes hard to determine in advance.

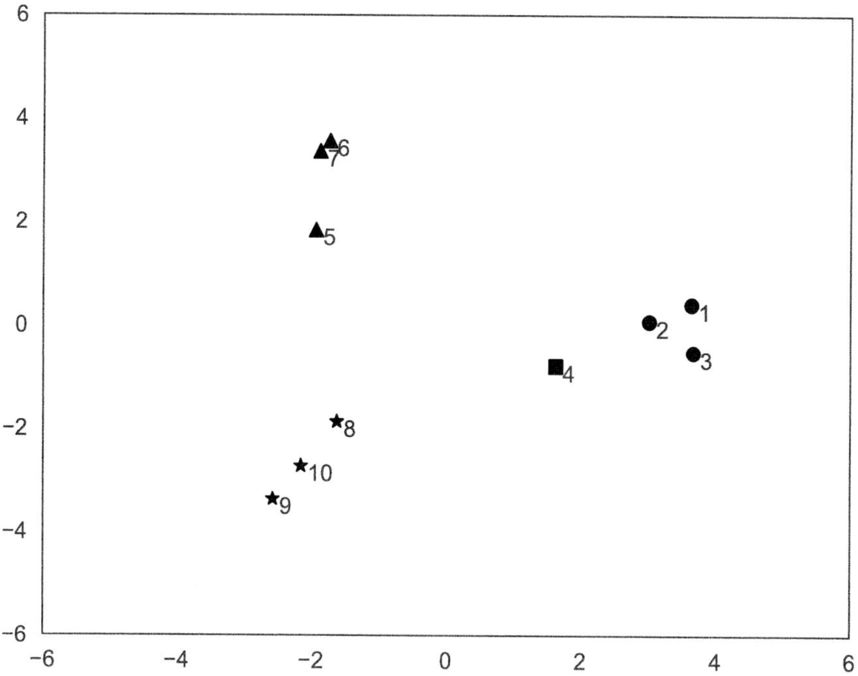

Fig. 6.3 Clustering result with single-pass clustering algorithm

6.2.3 Hierarchical Clustering

Hierarchical clustering is a class of cluster analysis methods that seek to build a hierarchy of clusters. It can be divided into two main types:

(1) Agglomerative hierarchical clustering: This is a bottom-up approach where each element starts in its own cluster and similar pairs of clusters are merged as we move up the hierarchy.
(2) Divisive hierarchical clustering: This is a top-down approach where all elements start in one cluster and splits are performed recursively as we move down the hierarchy.

In agglomerative hierarchical clustering, each document is initially considered as an individual cluster, and the most similar two clusters are merged together in each iteration until one cluster or K clusters are formed.

In the clustering process, the similarity between two clusters needs to be calculated. The commonly used measures, including single linkage, complete linkage, average linkage, and Ward's method, are described in detail in Sect. 6.2.

Algorithm 3: Agglomerative hierarchical clustering algorithm

Input : dataset $\mathscr{D} = \{\boldsymbol{x}_1, \boldsymbol{x}_2, \ldots, \boldsymbol{x}_N\}$, number of clusters K;
Output: clusters $\{C_1, C_2, \ldots, C_K\}$.

for $i = 1, \ldots, N$ **do**
    $C_i = \{\boldsymbol{x}_i\}$
end
for $i = 1, \ldots, N$ **do**
    for $j = 1, \ldots, N$ **do**
        calculate the similarity between two clusters $d(C_i, C_j)$
    end
end
while $size(\mathscr{C}) > K$ **do**
    find the nearest two clusters C_{i^*} and C_{j^*}.
    for $h = 1, \ldots, size(\{C_k\})$ **do**
        if $h \neq i^*$ *and* $h \neq j^*$ **then**
            update the similarity $d(C_h, C_{i^*} \cup C_{j^*})$
        end
        delete C_{i^*} and C_{j^*} from $\mathscr{C}$
        add $C_{i^*} \cup C_{j^*}$ to $\mathscr{C}$
        update the index of each cluster and record samples in each cluster.
    end
end

The results of hierarchical clustering can be represented by a dendrogram, which is a tree-like diagram that records the sequences of merges or splits, as shown in Fig. 6.4. Each leaf node represents a document, and each intermediate node has two subnodes, indicating that the two component clusters merged into one cluster. The

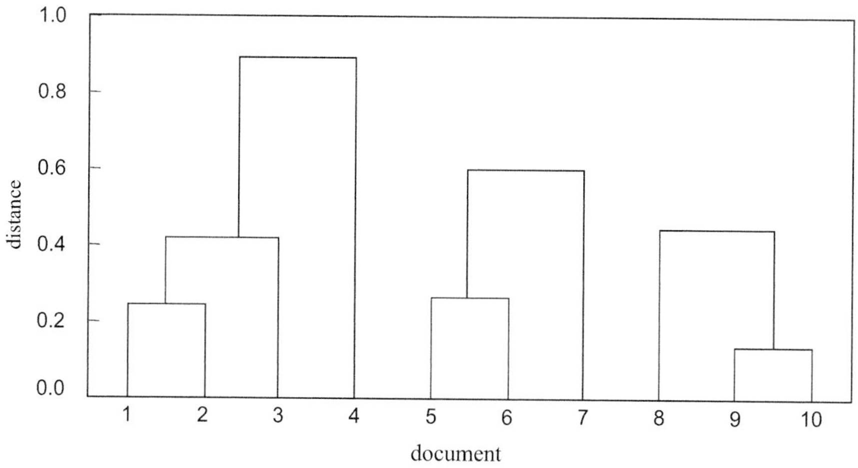

Fig. 6.4 The agglomerative hierarchical clustering results (dendrogram)

height of the leaf nodes is 0, and the height of each intermediate node represents the distance of its two subnodes and is inversely proportional to their similarity. Cutting the tree horizontally at a given height yields partitioning clustering results at a selected level.

We perform agglomerative hierarchical clustering on the dimension-reduced clustering dataset shown in Table 6.2 by using cosine to measure the similarity between documents and average linkage to measure the similarity between clusters and setting the expected number of clusters K as 3. The clustering process is as follows:

(i) Initialize a cluster for each document. This results in ten clusters in our task. The initial clusters are $\{C_1 : \{\boldsymbol{x}_1\}, C_2 : \{\boldsymbol{x}_2\}, C_3 : \{\boldsymbol{x}_3\}, C_4 : \{\boldsymbol{x}_4\}, C_5 : \{\boldsymbol{x}_5\}, C_6 : \{\boldsymbol{x}_6\}, C_7 : \{\boldsymbol{x}_7\}, C_8 : \{\boldsymbol{x}_8\}, C_9 : \{\boldsymbol{x}_9\}, C_{10} : \{\boldsymbol{x}_{10}\}\}$.

(ii) Compute the similarities between each cluster pair. Because the similarity between clusters C_9 and C_{10} is the highest (0.87), the two clusters are merged. The updated clusters are $\{C_1 : \{\boldsymbol{x}_1\}, C_2 : \{\boldsymbol{x}_2\}, C_3 : \{\boldsymbol{x}_3\}, C_4 : \{\boldsymbol{x}_4\}, C_5 : \{\boldsymbol{x}_5\}, C_6 : \{\boldsymbol{x}_6\}, C_7 : \{\boldsymbol{x}_7\}, C_8 : \{\boldsymbol{x}_8\}, C_9 : \{\boldsymbol{x}_9, \boldsymbol{x}_{10}\}\}$.

(iii) Compute the similarities between each cluster pair and merge the two clusters C_1 and C_2, which have the highest similarity. The updated clustering result is $\{C_1 : \{\boldsymbol{x}_1, \boldsymbol{x}_2\}, C_3 : \{\boldsymbol{x}_3\}, C_4 : \{\boldsymbol{x}_4\}, C_5 : \{\boldsymbol{x}_5\}, C_6 : \{\boldsymbol{x}_6\}, C_7 : \{\boldsymbol{x}_7\}, C_8 : \{\boldsymbol{x}_8\}, C_9 : \{\boldsymbol{x}_9, \boldsymbol{x}_{10}\}\}$.

(iv) Compute the similarities between each cluster pair and merge the two clusters C_5 and C_6, which have the highest similarity. The updated clustering result is $\{C_1 : \{\boldsymbol{x}_1, \boldsymbol{x}_2\}, C_3 : \{\boldsymbol{x}_3\}, C_4 : \{\boldsymbol{x}_4\}, C_5 : \{\boldsymbol{x}_5, \boldsymbol{x}_6\}, C_7 : \{\boldsymbol{x}_7\}, C_8 : \{\boldsymbol{x}_8\}, C_9 : \{\boldsymbol{x}_9, \boldsymbol{x}_{10}\}\}$.

(v) Compute the similarities between each cluster pair and merge the two clusters C_1 and C_3, which have the highest similarity. The updated clustering result is $\{C_1 : \{\boldsymbol{x}_1, \boldsymbol{x}_2, \boldsymbol{x}_3\}, C_4 : \{\boldsymbol{x}_4\}, C_5 : \{\boldsymbol{x}_5, \boldsymbol{x}_6\}, C_7 : \{\boldsymbol{x}_7\}, C_8 : \{\boldsymbol{x}_8\}, C_9 : \{\boldsymbol{x}_9, \boldsymbol{x}_{10}\}\}$.

(vi) Compute the similarities between each cluster pair and merge the two clusters C_8 and C_9, which have the highest similarity. The updated clustering result is $\{C_1 : \{\boldsymbol{x}_1, \boldsymbol{x}_2, \boldsymbol{x}_3\}, C_4 : \{\boldsymbol{x}_4\}, C_5 : \{\boldsymbol{x}_5, \boldsymbol{x}_6\}, C_7 : \{\boldsymbol{x}_7\}, C_8 : \{\boldsymbol{x}_8, \boldsymbol{x}_9, \boldsymbol{x}_{10}\}\}$.

(vii) Compute the similarities between each cluster pair and merge the two clusters C_5 and C_7, which have the highest similarity. The updated clustering result is $\{C_1 : \{\boldsymbol{x}_1, \boldsymbol{x}_2, \boldsymbol{x}_3\}, C_4 : \{\boldsymbol{x}_4\}, C_5 : \{\boldsymbol{x}_5, \boldsymbol{x}_6, \boldsymbol{x}_7\}, C_8 : \{\boldsymbol{x}_8, \boldsymbol{x}_9, \boldsymbol{x}_{10}\}\}$.

(viii) Compute the similarities between each cluster pair and merge the two clusters C_1 and C_4, which have the highest similarity. The updated clustering result is $\{C_1 : \{\boldsymbol{x}_1, \boldsymbol{x}_2, \boldsymbol{x}_3, \boldsymbol{x}_4\}, C_5 : \{\boldsymbol{x}_5, \boldsymbol{x}_6, \boldsymbol{x}_7\}, C_8 : \{\boldsymbol{x}_8, \boldsymbol{x}_9, \boldsymbol{x}_{10}\}\}$.

At this point, the number of clusters reaches the preset value ($K = 3$), and the hierarchical clustering ends. The clustering results in terms of the dendrogram are shown in Fig. 6.4.

The top-down divisive hierarchical clustering process follows the opposite process as the bottom-up clustering process. Initially, all the documents are contained in one cluster, and the documents that are not similar are separated iteratively from the cluster until all documents are divided into different clusters.

6.2.4 *Density-Based Clustering*

In density-based clustering, clusters are defined as areas of higher density than the remainder of the data. The basic concept is that the densely distributed data points in the data space are separated by the sparsely distributed data points; the connected high-density regions are the target clusters we are looking for.

Density-based spatial clustering of applications with noise (DBSCAN) is a representative algorithm of density-based clustering. Given a set of data points in the data space, the points that are closely connected (points with many nearby neighbors) will be grouped together and marked as high-density regions, and the points that lie alone in low-density regions (whose nearest neighbors are too far away) will be marked as outliers.

Let r denote the radius of the neighborhood and n denote the minimum number of data points required to construct a high-density region. On this basis, the following basic concepts are defined.

- r-neighborhood: The r-neighborhood of a sample P refers to the circular domain with P at the center and r as the radius.
- Core point: Point P is a core point if P's r-neighborhood contains at least n points.
- Directly reachable: Point Q is directly reachable from P if Q is in the r-neighborhood of P.
- Reachable: If there exists a sequence of data points $P_1, P_2, \ldots, P_T$ and P_{t+1} is directly reachable from P_t for any $t = 1, \ldots, T - 1$, we say that point

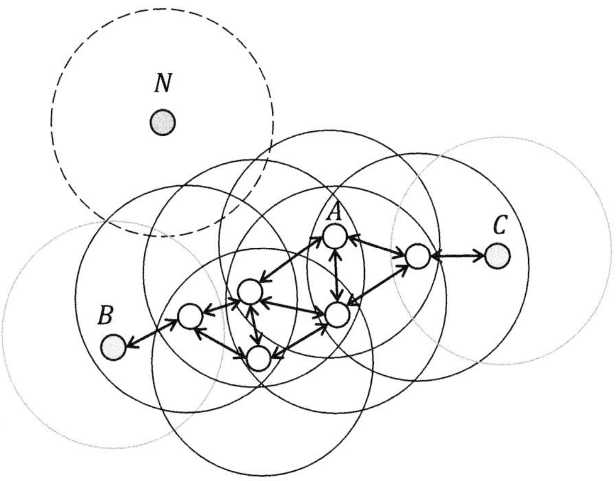

Fig. 6.5 An illustration of the DBSCAN algorithm

P_T is reachable from P_1. According to the definition of direct reachability, $P_1, P2, \ldots, P_{T-1}$ in the sequence are all core points.

- Density-connected: Two points Q_1 and Q_2 are density-connected if both Q_1 and Q_2 can be reachable from a core point P.

The DBSCAN algorithm supposes that for any core point P, the points in the dataset that are reachable from P belong to the same cluster. Figure 6.5 gives an example of the DBSCAN algorithm where $n = 4$. Point A and other hollow points are core samples, and boundary points B and C are non-core points. Points B and C are reachable from point A, that is, B and C are density-connected; therefore, together with the core points, they construct a cluster. Point N is a noise point that is not density-connected to A, B, or C.

Starting from a core point, the DBSCAN algorithm expands continuously to reachable regions to obtain a maximum region containing core points and boundary points. In this region, any two points are connected with each other and aggregated into a cluster. The process is repeated for each unlabeled core point until all core points in the dataset are processed. The points that are not included in any clusters are called noise points and grouped in a noise cluster.

We perform DBSCAN clustering on the dimension-reduced clustering dataset shown in Table 6.2, using cosine distance with $r = 0.6$ and $n = 3$. The clustering process is as follows.

(i) Initially, mark all data points as unvisited. Select $\boldsymbol{x}_1$ first and mark it as visited. The r-neighborhood of $\boldsymbol{x}_1$ includes points $\boldsymbol{x}_1$, $\boldsymbol{x}_2$ and $\boldsymbol{x}_3$. Because its size is not smaller than n, make the connected high-density region $\{\boldsymbol{x}_1, \boldsymbol{x}_2, \boldsymbol{x}_3\}$. The clustering result is $\{C_1 : \{\boldsymbol{x}_1, \boldsymbol{x}_2, \boldsymbol{x}_3\}\}$;

(ii) Select an unvisited point $\boldsymbol{x}_4$ and mark it as visited. The r-neighborhood of $\boldsymbol{x}_4$ includes $\boldsymbol{x}_1, \boldsymbol{x}_2$, and $\boldsymbol{x}_3$. The updated clustering result is $\{C_1 : \{\boldsymbol{x}_1, \boldsymbol{x}_2, \boldsymbol{x}_3, \boldsymbol{x}_4\}\}$;

(iii) Select an unvisited point $\boldsymbol{x}_5$ and mark it as visited. The r-neighborhood of $\boldsymbol{x}_5$ includes $\boldsymbol{x}_5$, $\boldsymbol{x}_6$, and $\boldsymbol{x}_7$, the size of which is not smaller than n.

Algorithm 4: DBSCAN algorithm

```
Input  : dataset 𝒟, radius r, the number of samples n required to construct a high-density
         region;
Output: set of clusters 𝒞.
𝒞 = ∅
for P in 𝒟 do
    if P has been visited then
        continue
    end
    find a set R_P of all samples in the r-neighborhood of P
    if |R_P| < n then
        mark P as a noise sample
    end
    else
        add sample P to a new cluster C
        find a set S_P of directly reachable samples from P
        for Q in S_P do
            if Q is a noise sample then
                add Q to cluster C
            end
            if Q has not been visited then
                add Q to cluster C
            end
            find a set R_Q of samples within the r-neighborhood of Q
            if |R_P| ≥ n then
                S_P = S_P ∪ R_Q
            end
            add C to 𝒞
        end
    end
end
```

Therefore, make the connected high-density region $\{\boldsymbol{x}_5, \boldsymbol{x}_6, \boldsymbol{x}_7\}$ a new cluster. The updated clustering result is $\{C_1 : \{\boldsymbol{x}_1, \boldsymbol{x}_2, \boldsymbol{x}_3, \boldsymbol{x}_4\}, C_2 : \{\boldsymbol{x}_5, \boldsymbol{x}_6, \boldsymbol{x}_7\}\}$;

(iv) Select an unvisited point $\boldsymbol{x}_8$ and mark it as visited. The r-neighborhood of $\boldsymbol{x}_8$ includes $\boldsymbol{x}_8$, $\boldsymbol{x}_9$, and $\boldsymbol{x}_{10}$, the size of which is smaller than n. Therefore, make the connected high-density region $\{\boldsymbol{x}_8, \boldsymbol{x}_9, \boldsymbol{x}_{10}\}$ a new cluster. The updated clustering result is $\{C_1 : \{\boldsymbol{x}_1, \boldsymbol{x}_2, \boldsymbol{x}_3, \boldsymbol{x}_4\}, C_2 : \{\boldsymbol{x}_5, \boldsymbol{x}_6, \boldsymbol{x}_7\}, C_3 : \{\boldsymbol{x}_8, \boldsymbol{x}_9, \boldsymbol{x}_{10}\}\}$;

(v) At this point, all points in the dataset are marked as visited, and clustering is finished. The final clustering result is $\{C_1 : \{\boldsymbol{x}_1, \boldsymbol{x}_2, \boldsymbol{x}_3, \boldsymbol{x}_4\}, C_2 : \{\boldsymbol{x}_5, \boldsymbol{x}_6, \boldsymbol{x}_7\}, C_3 : \{\boldsymbol{x}_8, \boldsymbol{x}_9, \boldsymbol{x}_{10}\}\}$, as shown in Fig. 6.6.

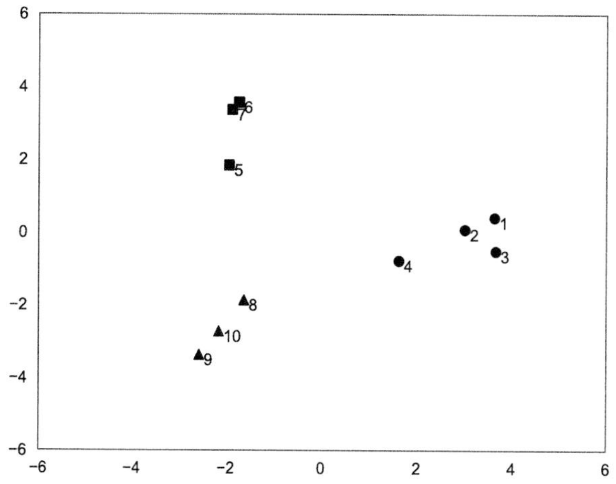

Fig. 6.6 The clustering result with the DBSCAN clustering algorithm

6.3 Evaluation of Clustering

The evaluation of clustering is also called cluster validity analysis. There are two main categories of methods for evaluating clustering: external criteria and internal criteria. The main difference between them is whether external information is used for clustering validation.

6.3.1 External Criteria

In external criteria, the quality of clustering is measured by the consistency between the clustering result and a clustering reference, which is considered the ground truth. The clustering reference is usually manually labeled.

For a dataset $\mathscr{D} = \{\boldsymbol{d}_1, \boldsymbol{d}_2, \ldots, \boldsymbol{d}_n\}$, assume that the clustering reference is denoted by $\mathscr{P} = \{P_1, P_2, \ldots, P_m\}$, where P_i represents the i-th cluster in the clustering reference, and the clustering result is $\mathscr{C} = \{C_1, C_2, \ldots, C_k\}$, where C_i is a model-obtained cluster. For any two different samples $\boldsymbol{d}_i$ and $\boldsymbol{d}_j$ in $\mathscr{D}$, define the following four relationships based on their co-occurrences in $\mathscr{C}$ and $\mathscr{P}$, respectively:

(1) SS: $\boldsymbol{d}_i$ and $\boldsymbol{d}_j$ belong to the same cluster in $\mathscr{C}$ and the same cluster in $\mathscr{P}$;
(2) SD: $\boldsymbol{d}_i$ and $\boldsymbol{d}_j$ belong to the same cluster in $\mathscr{C}$ but different clusters in $\mathscr{P}$;
(3) DS: $\boldsymbol{d}_i$ and $\boldsymbol{d}_j$ belong to different clusters in $\mathscr{C}$ but the same cluster in $\mathscr{P}$;
(4) DD: $\boldsymbol{d}_i$ and $\boldsymbol{d}_j$ belong to different clusters in $\mathscr{C}$ and different clusters in $\mathscr{P}$;

Let a, b, c, d denote the number of SS, SD, DS, and DD, respectively. The following evaluation measures can be defined:

(a) Rand index

$$\mathrm{RS} = \frac{a+d}{a+b+c+d} \tag{6.18}$$

(b) Jaccard index

$$\mathrm{JC} = \frac{a}{a+b+c} \tag{6.19}$$

(c) Fowlkes and Mallows index

$$\mathrm{FMI} = \sqrt{\frac{a}{a+b} \cdot \frac{a}{a+c}} \tag{6.20}$$

The range of the above three indices is [0, 1]. The larger the value of the index is, the higher the similarity of $\mathscr{C}$ and $\mathscr{P}$ and the better the performance of the clustering result $\mathscr{C}$.

6.3.2 Internal Criteria

The internal criteria are based on internal information (such as distribution and structure) and evaluate a cluster without reference to external information. Cohesion and separation are two key factors for evaluating the clustering performance in internal criteria. Generally, internal criteria prefer clusters with high similarity within a cluster (high cohesion) and low similarity between clusters (high separation).

The typical internal criteria include the silhouette coefficient, I index, Davies–Bouldin index, Dunn index, Calinski–Harabasz index, Hubert's Γ statistic, and the cophenetic correlation coefficient. Most of these metrics include factors of both cohesion and separation. In the following, we will introduce the representative measure: the silhouette coefficient. Readers can refer to (Liu et al. 2010) for the details of other methods.

The silhouette coefficient was first proposed by Peter J. Rousseeuw in 1986 and has become a commonly used internal criterion for clustering evaluation. Assuming $\boldsymbol{d}$ is a sample belonging to cluster C_m, we first calculate the average distance between $\boldsymbol{d}$ and the other samples in C_m as:

$$a(\boldsymbol{d}) = \frac{\sum\limits_{\boldsymbol{d}' \in C_m, \boldsymbol{d} \neq \boldsymbol{d}'} \mathrm{dist}(\boldsymbol{d}, \boldsymbol{d}')}{|C_m| - 1} \tag{6.21}$$

We then calculate the minimum average distance between $\boldsymbol{d}$ and the samples in the other clusters:

$$b(\boldsymbol{d}) = \min_{C_j:1\leq j\leq k, j\neq m} \left\{ \frac{\sum_{\boldsymbol{d}'\in C_j} \mathrm{dist}(\boldsymbol{d}, \boldsymbol{d}')}{|C_j|} \right\} \tag{6.22}$$

Among them, $a(\boldsymbol{d})$ reflects the degree of cohesion in the cluster to which $\boldsymbol{d}$ belongs; $b(\boldsymbol{d})$ reflects the degree of separation between $\boldsymbol{d}$ and the other clusters.

On this basis, the silhouette coefficient with respect to $\boldsymbol{d}$ is defined as follows:

$$\mathrm{SC}(\boldsymbol{d}) = \frac{b(\boldsymbol{d}) - a(\boldsymbol{d})}{\max\{a(\boldsymbol{d}), b(\boldsymbol{d})\}} \tag{6.23}$$

The overall silhouette coefficient is then defined as the average silhouette coefficient across all samples in the dataset:

$$\mathrm{SC} = \frac{1}{N}\sum_{i=1}^{N} \mathrm{SC}(\boldsymbol{d}_i) \tag{6.24}$$

The range of the silhouette coefficient is $[-1, 1]$. The higher the silhouette coefficient is, the better the clustering performance.

6.4 Further Reading

The performance of text clustering depends on the quality of the text representation. Traditional text clustering methods mainly use the vector space model for text representation. This type of representation has some inherent shortcomings, including high-dimensional and sparsity problems, which are inefficient for similarity calculation and text clustering.

In text classification, supervised feature selection methods (e.g., MI and IG) are widely used to improve the quality of text representation. However, because the labels of documents are unknown in text clustering, we can only use unsupervised feature selection methods (e.g., document frequency and term frequency). The unsupervised feature extraction algorithms (e.g., PCA, ICA) are also options for dimension reduction in text clustering. In addition, topic models such as latent semantic analysis (LSA), probabilistic latent semantic analysis (PLSA), and latent Dirichlet distribution (LDA) also provide a way to represent a document by transforming the high-dimensional sparse vectors of words into low-dimensional dense vectors of topics. In addition, some studies also attempt to use the concepts in a knowledge base (such as WordNet, HowNet, Wikipedia, etc.) to guide text representation, similarity calculation, and clustering.

In recent years, with the rise of deep learning, distributed representations such as word embedding have been widely used in text data mining. For example, as introduced in Chap. 3, a piece of text at different levels (e.g., word, phrase, sentence, and document) can be represented by a densely distributed low-dimensional vector. Another advantage of representation learning is that it can learn a task-related representation. Both advantages bring new perspectives to text clustering.

In addition to the clustering methods we described above, there are some special clustering algorithms, such as suffix tree clustering (STC), that are specific to text processing. As a type of data structure, a suffix tree was first proposed to support effective matches and queries for strings. By using the suffix tree structure to represent and process text, the suffix tree clustering algorithm regards text as a sequence of words rather than a set of words and captures more word order information.

Clustering text streams is a special problem of text clustering, which has been widely used in the fields of topic detection and tracking and social media mining. Unlike traditional text clustering, the text data in these fields often appear in the form of online text streams, which creates challenges for text clustering. The single-pass clustering algorithm is a widely used method for real-time large-scale text stream clustering. We will also see in Chap. 9 that some online variants of the traditional clustering algorithms, such as group-average agglomerative clustering (Allan et al. 1998a; Yang et al. 1998), have also been proposed to address these challenges.

Exercises

6.1 Please point out the similarities and differences between the classification and clustering problems.

6.2 What is the relationship between Euclidean distance and cosine similarity when measuring the similarity of two documents?

6.3 Is KL divergence suitable for the similarity calculation of short documents? In addition to KL divergence, can you think of other distribution-based similarity calculation methods?

6.4 Please give the detailed K-means clustering process for the clustering dataset in Table 6.2 when document $\boldsymbol{x}_1$, $\boldsymbol{x}_5$, and $\boldsymbol{x}_8$ are selected as the initial centroids.

6.5 What is the single-pass clustering results if the document order for processing is reversed in Table 6.2?

6.6 Please try to perform divisive hierarchical clustering on the clustering dataset in Table 6.2.

Chapter 7
Topic Model

7.1 The History of Topic Modeling

The following text is extracted from *Walden*, a book by Henry David Thoreau, where the words with wavy lines are related to animals, the words with dots are related to locations, the words with double underscores are related to plants, and the words with underscores are related to colors.

> The first sparrow of spring! The year beginning with younger hope than ever! The faint silvery warblings heard over the partially bare and moist fields from the bluebird, the song sparrow, and the red-wing, as if the last flakes of winter tinkled as they fell! What at such a time are histories, chronologies, traditions, and all written revelations? The brooks sing carols and glees to the spring. The marsh hawk, sailing low over the meadow, is already seeking the first slimy life that awakes. The sinking sound of melting snow is heard in all dells, and the ice dissolves apace in the ponds. The grass flames up on the hillsides like a spring fire—"etprimitus oritur herba imbribus primoribus evocata"—as if the earth sent forth an inward heat to greet the returning sun; not yellow but green is the color of its flame;—the symbol of perpetual youth, the grass-blade, like a long green ribbon, streams from the sod into the summer, checked indeed by the frost, but anon pushing on again, lifting its spear of last year's hay with the fresh life below. It grows as steadily as the rill oozes out of the ground. It is almost identical with that, for in the growing days of June, when the rills are dry, the grass-blades are their channels, and from year to year the herds drink at this perennial green stream, and the mower draws from it betimes their winter supply. So our human life but dies down to its root, and still puts forth its green blade to eternity.

© Tsinghua University Press 2021

C. Zong et al., *Text Data Mining*, https://doi.org/10.1007/978-981-16-0100-2_7

Table 7.1 The topic in the above paragraph

Animal	Plant	Location	Color
Sparrow	Grass	Meadow	Silvery
Bluebird	Grass-Blade	Dells	Green
Marsh hawk	Hay	Ponds	Yellow

By extracting these words, we can obtain a set of topics, each of which is denoted by a set of representative words, as shown in Table 7.1.

The idea of topic modeling is derived from the field of information retrieval. Susan Dumais et al. proposed latent semantic indexing (LSI), using singular value decomposition (SVD) technology to map document vectors from high-dimensional word space to a low-dimensional semantic space (i.e., topic space). This method can discover the implicit topic information in texts, such as the linguistic phenomena of polysemy and synonymy, without relying on any prior knowledge, and ultimately provide search results that match the user's query not only at the lexical level but also at the semantic level.

The LSI model is based on the framework of algebra, while the probabilistic latent semantic indexing (PLSI) model proposed by Thomas Hofmann simulates the process of generating words in documents through the probabilistic generative model and extends the LSI model to the framework of probability. LSI and PLSI are also called latent semantic analysis (LSA) and probabilistic latent semantic analysis (PLSA), respectively, and have been widely used in information retrieval, natural language processing, and text data mining.

The PLSA model can only fit a limited collection of documents in terms of the training dataset. The parameter space of PLSA increases linearly with the number of documents contained in the training set, which makes it either prone to overfitting or challenges its ability to infer meaning from unseen documents. To solve these problems, David Blei et al. proposed the latent Dirichlet allocation (LDA) model, which introduced the prior distribution of parameters on the basis of PLSA and replaced the MLE used in PLSA with Bayesian estimation. LDA can be used not only as a text representation method but also for dimension reduction and has been successfully applied in many downstream text data mining tasks.

7.2 Latent Semantic Analysis

In 1988, Susan Dumais et al. proposed using LSA for distributional semantic representation (Dumais et al. 1988; Deerwester et al. 1990). The goal is to represent a piece of text by a set of implicit semantic concepts rather than the explicit terms in the vector space model.

LSA assumes that words with similar semantics are more likely to appear in similar pieces of text. Unlike the high-dimensional and sparse text representation method VSM, LSA uses SVD technology to map high-dimensional representations of texts into a low-dimensional latent semantic space. This low-dimensional representation reveals the semantic relationship among words (documents). Such latent semantic concepts are called topics.

7.2.1 Singular Value Decomposition of the Term-by-Document Matrix

Given a set of documents, the following term-by-document matrix can be constructed based on the vector space model:

$$\boldsymbol{X} = \begin{bmatrix} x_{1,1} & \dots & x_{1,n} \\ \vdots & \ddots & \vdots \\ x_{m,1} & \dots & x_{m,n} \end{bmatrix}$$

where $x_{i,j}$ denotes the weight of the i-th term in the j-th document, m is the number of terms, n is the number of documents. Thus, $\left[x_{i,1}, \cdots, x_{i,n}\right]$ denotes the representation of the i-th term across all documents, and each column $\left[x_{1,j}, \cdots, x_{m,j}\right]^{\mathrm{T}}$ denotes the representation of the j-th document in the vector space model.

Perform SVD on $\boldsymbol{X}$:

$$\boldsymbol{X} = \boldsymbol{T}\boldsymbol{\Sigma}\boldsymbol{D}^{\mathrm{T}} \tag{7.1}$$

where $\boldsymbol{\Sigma} = \operatorname{diag}(\sigma_1, \ldots, \sigma_r)$ is an r-order rectangular diagonal matrix with nonnegative real numbers $\sigma_1, \cdots, \sigma_r (\sigma_1 \geq \sigma_2 \geq \cdots \geq \sigma_r > 0)$ on the diagonal. The column vectors of $\boldsymbol{T}$ $(\boldsymbol{t}_1, \boldsymbol{t}_2, \cdots, \boldsymbol{t}_r)$ and the column vectors of $\boldsymbol{D}$ $(\boldsymbol{d}_1, \boldsymbol{d}_2, \cdots, \boldsymbol{d}_r)$ construct a set of unit orthogonal vectors that satisfies $\boldsymbol{T}^{\mathrm{T}}\boldsymbol{T} = \boldsymbol{I}_r$ and $\boldsymbol{D}^{\mathrm{T}}\boldsymbol{D} = \boldsymbol{I}_r$.

The above formula can also be written in the form of the sum of r-rank matrices:

$$\boldsymbol{X} = \sigma_1 \boldsymbol{t}_1 \boldsymbol{d}_1^{\mathrm{T}} + \cdots + \sigma_r \boldsymbol{t}_r \boldsymbol{d}_r^{\mathrm{T}} \tag{7.2}$$

where the singular value $\sigma_1, \cdots, \sigma_r$ reflects the strength of r independent concepts implied in $\boldsymbol{X}$.

In text representation, the traditional term-by-document matrix is usually very sparse because of the high dimension of the feature space and the short length of documents. Meanwhile, there is a high linear correlation among the high-dimensional features. LSA decomposes the term-by-document matrix $\boldsymbol{X}$ by truncated SVD, as shown in Formula (7.2), keeping the k $(k < r)$ largest singular values. The orthogonal space composed of the k singular values and the corresponding singular vectors is regarded as the latent semantic space of the text. This means that the dimension of text representation can be reduced from m to k by selecting k latent topics instead of m explicit terms as the basis for text representation. The

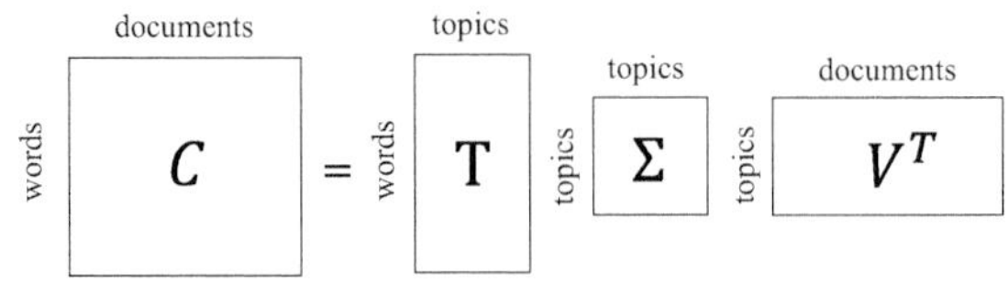

Fig. 7.1 Matrix decomposition of LSA model

k-rank approximation of the original matrix $\boldsymbol{X}$ is then obtained:

$$\hat{\boldsymbol{X}} = \sigma_1 \boldsymbol{t}_1 \boldsymbol{d}_1^{\mathrm{T}} + \cdots + \sigma_k \boldsymbol{t}_k \boldsymbol{d}_k^{\mathrm{T}} \tag{7.3}$$

written in the form of matrices:

$$\hat{\boldsymbol{X}} = \boldsymbol{T}_k \boldsymbol{\Sigma}_k \boldsymbol{D}_k^{\mathrm{T}} \tag{7.4}$$

where $\boldsymbol{T}_k = [\boldsymbol{t}_1 \cdots \boldsymbol{t}_k]$ is called the term-by-topic matrix and $\boldsymbol{D}_k = [\boldsymbol{d}_1 \cdots \boldsymbol{d}_k]$ is called the document-by-topic matrix. The above matrix decomposition process is illustrated in Fig. 7.1.

Such a low-rank approximation is expected to merge the dimensions associated with terms that have similar meanings, for example, {car, truck, flower} → {1.38 × car + 0.52 × truck, flower}.

7.2.2 *Conceptual Representation and Similarity Computation*

After obtaining the low-rank approximation of the term-by-document matrix, we are concerned with the following five issues.

(1) Conceptual representation of terms

Each row in $\hat{\boldsymbol{X}}$ corresponds to one term represented by a vector of its term weights in different documents.

The conceptual representation of terms t_j can be obtained by extracting the j-th row of $\hat{\boldsymbol{X}}$ in Eq. (7.8):

$$\begin{aligned} \left[x_{j,1} \cdots x_{j,n}\right] &= [t_{j,1} \cdots t_{j,k}] \boldsymbol{\Sigma}_k \boldsymbol{D}_k^{\mathrm{T}} \\ &= [\sigma_1 t_{j,1} \quad \cdots \quad \sigma_k t_{j,k}] \boldsymbol{D}_k^{\mathrm{T}} \end{aligned} \tag{7.5}$$

If we treat $\boldsymbol{\Sigma}_k \boldsymbol{D}_k^{\mathrm{T}}$ as the coordinate system of the latent semantic space, the conceptual representation of the j-th term is $[t_{j,1} \ldots t_{j,k}]$.

The similarity between different terms can therefore be measured by the inner product of two row vectors of $\hat{\boldsymbol{X}}$. For this purpose, a quadratic symmetric matrix

$\hat{X}\hat{X}^{\mathrm{T}}$ is constructed to contain the inner product of all terms:

$$\begin{aligned}\hat{X}\hat{X}^{\mathrm{T}} &= T_k \Sigma_k D_k^{\mathrm{T}} D_k \Sigma_k T_k^{\mathrm{T}} \\ &= T_k \Sigma_k \left(T_k \Sigma_k\right)^{\mathrm{T}}\end{aligned} \tag{7.6}$$

The similarity between the i-th and j-th terms, i.e., the elements of the i-th row and the j-th column in $\hat{X}\hat{X}^{\mathrm{T}}$, is equal to the inner product of the corresponding row vectors in the matrix $T_k \Sigma_k$.

(2) Conceptual representation of documents

Extracting the i-th column of the matrix in Eq. (7.8), we can obtain the conceptual representation of the i-th document:

$$\begin{aligned}x_i &= \left[x_{1,i} \cdots x_{m,i}\right]^{\mathrm{T}} \\ &= T_k \Sigma_k \left[d_{1,i} \cdots d_{k,i}\right]^{\mathrm{T}} \\ &= T_k \left[\sigma_1 d_{1,i} \cdots \sigma_k d_{k,i}\right]^{\mathrm{T}}\end{aligned} \tag{7.7}$$

Similarly, the inner product of two column vectors is used to measure the similarity between two documents. Constructing another quadratic symmetric matrix:

$$\begin{aligned}\hat{X}^{\mathrm{T}}\hat{X} &= D_k \Sigma_k T_k^{\mathrm{T}} T_k \Sigma_k D_k^{\mathrm{T}} \\ &= \left(D_k \Sigma_k\right) \left(D_k \Sigma_k\right)^{\mathrm{T}}\end{aligned} \tag{7.8}$$

The similarity between the i-th and the j-th document, i.e., the element of the i-th row and the j-th column in $\hat{X}^{\mathrm{T}}\hat{X}$, is equal to the inner product of the corresponding row vectors of the $D_k \Sigma_k$ matrix.

(3) The correlation between terms and documents

The term-by-document approximation matrix $\hat{X}$ reflects the relevance between terms and documents. Formula (7.4) can be rewritten as:

$$\begin{aligned}\hat{X} &= T_k \Sigma_k D_k^{\mathrm{T}} \\ &= T_k \Sigma_k^{1/2} \Sigma_k^{1/2} D_k^{\mathrm{T}}\end{aligned} \tag{7.9}$$

By taking $T_k \Sigma_k^{1/2}$ as the coordinate system, the conceptual representation of the j-th term can be written as $[\sqrt{\sigma_1} t_{j,1} \cdots \sqrt{\sigma_k} t_{j,k}]^{\mathrm{T}}$. By taking $D_k \Sigma_k^{1/2}$ as the coordinate system, the conceptual representation of the i-th document is expressed as $[\sqrt{\sigma_1} d_{1,i} \ldots \sqrt{\sigma_k} d_{k,i}]^{\mathrm{T}}$. Thus, the correlation between the j-th term and the i-th document is derived as follows:

$$
\begin{aligned}
\left[\sqrt{\sigma_1}t_{j,1}\cdots\sqrt{\sigma_k}t_{j,k}\right]\left[\sqrt{\sigma_1}d_{1,i}\cdots\sqrt{\sigma_k}d_{k,i}\right]^{\mathrm{T}} &= \sum_{h=1}^{k}\sigma_h t_{j,h}d_{h,i} \\
&= [\hat{X}]_{j,i}
\end{aligned} \tag{7.10}
$$

(4) Conceptual representation of new documents

We have described the conceptual representations of the documents contained in the training dataset, and now, we explain how to obtain their conceptual representation. Suppose $\boldsymbol{x}'$ is a new document. By taking the column vector $\delta_1\boldsymbol{t}_1, \delta_2\boldsymbol{t}_2, \cdots, \delta_k\boldsymbol{t}_k$ of $T_k\Sigma_k$ as the coordinate system, we can describe the relationship between $\boldsymbol{x}'$ and its representation in the new coordinate system $\boldsymbol{d}'$ as follows:

$$
\boldsymbol{x}' = \boldsymbol{T}_k\boldsymbol{\Sigma}_k\boldsymbol{d}' \tag{7.11}
$$

After applying a left multiplication to both sides of the equation by $\boldsymbol{\Sigma}_k^{-1}\boldsymbol{T}_k^{\mathrm{T}}$, we obtain $\boldsymbol{d}'$ as follows:

$$
\begin{aligned}
\boldsymbol{d}' &= \boldsymbol{\Sigma}_k^{-1}\boldsymbol{T}_k^{\mathrm{T}}\boldsymbol{x}' \\
&= \boldsymbol{F}\boldsymbol{x}'
\end{aligned} \tag{7.12}
$$

where $\boldsymbol{F} = \boldsymbol{\Sigma}_k^{-1}\boldsymbol{T}_k^{\mathrm{T}}$ is called the folding-in matrix, which represents a linear transformation from term space to concept space.

7.3 Probabilistic Latent Semantic Analysis

LSA is based on the framework of algebra, but algebra lacks insight from the perspective of probability. In addition, the difficulty of using SVD on a large scale also restricts the application of LSA. In 1999, Thomas Hoffmann proposed the probabilistic latent semantic analysis (PLSA) model (Hofmann 1999), which extended latent semantic analysis's algebra framework to include probability.

7.3.1 Model Hypothesis

PLSA is a probabilistic graphical model that models the process of text generation based on probabilistic graphs. Figure 7.2 shows the plate diagram of PLSA, where random variables $\boldsymbol{d}$, $\boldsymbol{w}$, and z represent documents, terms, and topics, respectively. $\boldsymbol{d}$ and $\boldsymbol{w}$ are observable variables, and z are hidden variables that cannot be directly

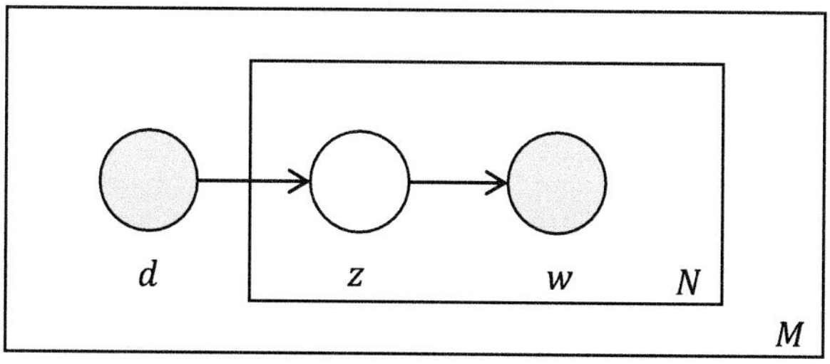

Fig. 7.2 The plate diagram of PLSA

observed. M, N and K denote the number of terms, documents, and topics. In PLSA, the document-by-topic matrix $\boldsymbol{D}$ and term-by-topic matrix $\boldsymbol{T}$ in LSA, as shown in Fig. 7.1, are modeled by the document-conditional topic distributions $p(z|\boldsymbol{d})$ and topic-conditional term distributions $p(\boldsymbol{w}|z)$, where $p(z_k|d_i)$ denotes the probability of topic z_k conditioned on document d_i, and $p\left(w_j|z_k\right)$ denotes the probability of term w_j conditioned on topic z_k.

The PLSA model assumes that a document is generated by the following process:

For each document:
  Choose a document d_i according to the probability $p(d_i)$;
  For each word position in each document:
    Choose a topic z_k, according to the probability $p\left(z_k|d_i\right)$;
    Choose a term w_j according to the topic z_k and the probability $p\left(w_j|z_k\right)$.

Accordingly, the joint distribution of observed variables $\left(d_i, w_j\right)$ can be written as:

$$
\begin{aligned}
p\left(d_i, w_j\right) &= p\left(d_i\right) p\left(w_j|d_i\right) \\
&= p\left(d_i\right) \sum_{k=1}^{K} p(w_j|z_k) p(z_k|d_i)
\end{aligned}
\tag{7.13}
$$

where $p(w_j|z_k)$ and $p\left(z_k|d_i\right)$ are parameters to be estimated.

7.3.2 *Parameter Learning*

Given a training corpus containing multiple documents, PLSA estimates the parameters $p(w_j|z_k)$ and $p\left(z_k|d_i\right)$ based on the principle of maximum likelihood estimation (MLE). If the training corpus is regarded as a sequence of documents, each document consists of a sequence of terms, and the log-likelihood of the joint distribution of observed variables can be written as:

$$\begin{aligned}\mathscr{L} &= \log\prod_{i=1}^{N}\prod_{j=1}^{M} p\left(d_i, w_j\right)^{n(d_i,w_j)} \\ &= \sum_{i=1}^{N}\sum_{j=1}^{M} n\left(d_i, w_j\right)\log p\left(d_i\right)\sum_{k=1}^{K} p\left(w_j|z_k\right) p\left(z_k|d_i\right)\end{aligned} \tag{7.14}$$

where $n\left(d_i, w_j\right)$ is the frequency of term w_j in document d_i.

Due to the existence of hidden variables z_k, the log-likelihood contains the log of sum, and it is difficult to perform MLE directly. Instead, the expectation maximization (EM) algorithm can be used to solve the MLE problem. This book omits the detailed derivations of the EM algorithm for PLSA and only displays its main steps. Interested readers can refer to (Mei and Zhai 2001) for more details.

The main steps of the EM algorithm for PLSA are as follows:

(1) Parameter Initialization: $\boldsymbol{\Theta}^{(0)} = \{p\left(w_j|z_k\right)^{(0)}, p\left(z_k|d_i\right)^{(0)}\}$.
(2) E-step: Calculate the posterior probability of the latent variables given the observed variables based on current parameter $\boldsymbol{\Theta}^{(t)} = \{p\left(w_j|z_k\right)^{(t)}, p\left(z_k|d_i\right)^{(t)}\}$:

$$p\left(z_k|d_i, w_j\right) = \frac{p\left(w_j|z_k\right) p\left(z_k|d_i\right)}{\sum_{h=1}^{K} p\left(w_j|z_h\right) p\left(z_h|d_i\right)} \tag{7.15}$$

(3) M-step: Perform the traditional MLE to maximize the lower bound of $\mathscr{L}$ under parameter $\boldsymbol{\Theta}^{(t)}$ and obtain the updated parameters $\boldsymbol{\Theta}^{(t+1)}$.

$$\begin{cases} p\left(w_j|z_k\right)^{(t+1)} = \dfrac{\sum_{i=1}^{N} n\left(d_i, w_j\right) p\left(z_k|d_i, w_j\right)}{\sum_{j=1}^{M}\sum_{i=1}^{N} n\left(d_i, w_j\right) p\left(z_k|d_i, w_j\right)} \\ p\left(z_k|d_i\right)^{(t+1)} = \dfrac{\sum_{j=1}^{M} n\left(d_i, w_j\right) p\left(z_k|d_i, w_j\right)}{n\left(d_i\right)} \end{cases} \tag{7.16}$$

where $n\left(d_i\right)$ denotes the total number of terms contained in document d_i.

(4) Repeat E-step and M-step until the algorithm converges.

Based on the above equations, we can easily obtain the conditional topic distribution given a document in the training corpus. For an unseen document d', we can keep parameter $p(\boldsymbol{w}|z)$ learned from the training corpus fixed, continuously run the EM algorithm based on d', and update $p(z|d')$ iteratively until the algorithm converges.

7.4 Latent Dirichlet Allocation

Let $\boldsymbol{\theta}_i = [p(z_1|d_i), p(z_2|d_i), \ldots, p(z_K|d_i)]$ denote the conditional topic distribution given the i-th document and $\boldsymbol{\varphi}_k = [p(w_1|z_k), p(w_2|z_k), \ldots, p(w_V|z_k)]$ denote the conditional term distribution given the k-th topic. In PLSA, $\boldsymbol{\theta}_i$ and $\boldsymbol{\varphi}_k$ are viewed as deterministic but unknown variables in advance and can be estimated by performing MLE on the training corpus. The size of parameters $\boldsymbol{\theta}_i$ increases linearly with the number of documents, and such linear growth in parameters suggests that the model is prone to overfitting.

To address these problems, David Blei, Andrew Ng, and Michael Jordan proposed a more generalized topic model based on PLSA, the LDA (Blei et al. 2003). LDA introduces a Dirichlet distribution as the priors to the document-conditional topic distribution and the topic-conditional term distribution. The parameters $\boldsymbol{\theta}_i$ and $\boldsymbol{\varphi}_k$ are considered random variables and are drawn from two Dirichlet distributions rather than, as in PLSA, being treated as deterministic variables.

The Dirichlet distribution and the categorical distribution are a pair of conjugate distributions, and the learning algorithm of the model is also changed from maximum likelihood estimation to Bayesian estimation, accordingly. In comparison, LDA is a more well-defined generative model whose parameters do not grow with the size of the training corpus and are able to generalize easily to new documents. Girolami and Kabán (2003) proved that PLSA is essentially a special type of LDA based on MAP estimation by taking a uniform Dirichlet distribution $\text{Dir}(\mathbf{1})$ as the prior distribution.

7.4.1 Model Hypothesis

The plate diagram of LDA is shown in Fig. 7.3, where double circles denote observable variables, single circles denote latent variables, arrows denote conditional dependency between two variables, and boxes denote repeated operations.

The main notations in LDA are shown in Table 7.2. In LDA, the generation process of a document is assumed as follows.

For each topic:

  Choose $\boldsymbol{\varphi}_k \sim \text{Dir}(\boldsymbol{\beta})$, where $\boldsymbol{\varphi}_k$ is the parameter of the term distribution given the k-th topic.

For each document:

  Choose $\boldsymbol{\theta}_m \sim \text{Dir}(\boldsymbol{\alpha})$, where $\boldsymbol{\theta}_m$ is the parameter of the topic distribution given the m-th document.

  For each word position in each document:

    Choose a topic for the current position: $z_{m,n} \sim \text{Cat}(\boldsymbol{\theta}_m)$.

    Based on the topic $z_{m,n}$, choose a term for the current position $w_{m,n} \sim \text{Cat}(\boldsymbol{\varphi}_{z_{m,n}})$.

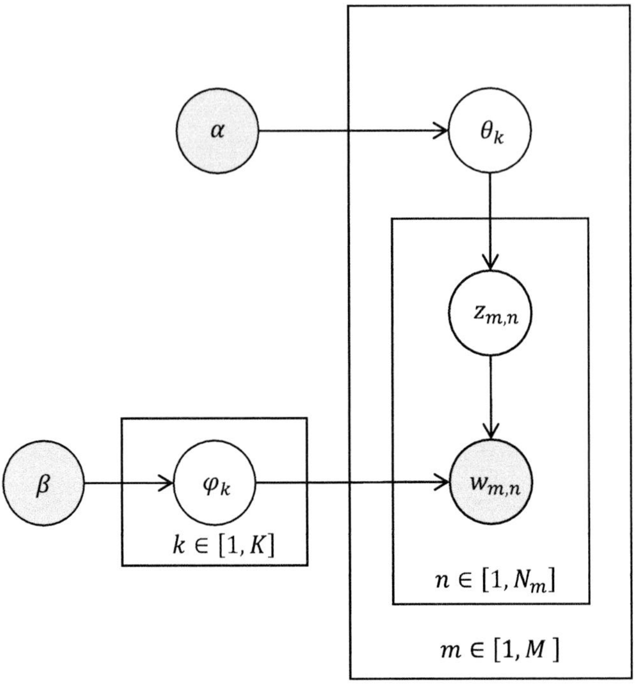

Fig. 7.3 The plate diagram of LDA

Table 7.2 The main parameters of the LDA model

Parameters	Meaning
M	The number of documents
K	The number of topics
V	The number of terms(i.e., The dimensions of the vocabulary)
α	The hyperparameter of the Dirichlet prior distribution with respect to $\boldsymbol{\theta}_m$
β	The hyperparameter of the Dirichlet prior distribution with respect to $\boldsymbol{\varphi}_k$
$\boldsymbol{\theta}_m$	The conditional topic distribution given the m-th document
$\boldsymbol{\varphi}_k$	The conditional term distribution given the k-th topic
N_m	The length of m-th document
$z_{m,n}$	The topic assigned to the n-th word position in the m-th document
$w_{m,n}$	The term assigned to the n-th word position in the m-th document
$z_m = \left\{z_{m,n}\right\}_{n=1}^{N_m}$	The topic sequence of the m-th document
$\boldsymbol{w}_m = \left\{w_{m,n}\right\}_{n=1}^{N_m}$	The term sequence of the m-th document
$\boldsymbol{w} = \{\boldsymbol{w}_m\}_{m=1}^{M}$	The term sequence of a set of documents
$z = \{z_m\}_{m=1}^{M}$	The topic sequence of a set of documents

It is worth noting that the original paper introducing LDA (Blei et al. 2003) only presented the Dirichlet prior to the document-conditional topic distribution $\boldsymbol{\theta}_m$, and the topic-conditional term distribution $\boldsymbol{\varphi}_k$ was modeled in the same way as that in PLSA. The following research further introduced the Dirichlet prior for $\boldsymbol{\varphi}_k$. In addition, the original paper used a Poisson distribution to model the document length. For each document, the document length is first chosen according to a Poisson distribution. However, this assumption does not affect the inference of the distribution of terms and topics in documents. Therefore, most of the following studies have removed the modeling of document length.

7.4.2 *Joint Probability*

As defined in Table 7.2, $\boldsymbol{w}$ and $\boldsymbol{z}$ denote the sequence of terms and topics of a set of documents. The joint distribution of $\boldsymbol{w}$ and $\boldsymbol{z}$ can be factorized into two parts:

$$p(\boldsymbol{w}, \boldsymbol{z}) = p(\boldsymbol{w}|\boldsymbol{z})\, p(\boldsymbol{z}) \tag{7.17}$$

where $p(\boldsymbol{z})$ is the probability of the topic sequence and $p(\boldsymbol{w}|\boldsymbol{z})$ is the probability of the term sequence given the topic sequence.

According to the former hypothesis, the topic of the n-th word position in the m-th document $z_{m,n}$ is assigned according to the categorical distribution $\mathrm{Cat}(\boldsymbol{\theta}_m)$, that is $p(z_{m,n} = k|\boldsymbol{\theta}_m) = \theta_{m,k}$. Multiple trials of the categorical distribution correspond to a multinomial distribution; therefore, the probability of the topic sequence $\boldsymbol{z}_m$ given $\boldsymbol{\theta}_m$ is:

$$p(\boldsymbol{z}_m|\boldsymbol{\theta}_m) = \prod_{n=1}^{N_m} p\left(z_{m,n}|\boldsymbol{\theta}_m\right) = \prod_{k=1}^{K} \theta_{m,k}^{n_{m,k,\cdot}} \tag{7.18}$$

where N_m and K have been defined in Table 7.2, and $n_{m,k,\cdot}$ denotes the number of occurrences of topic k in the m-th document.

In LDA, the document-conditional topic distribution $\boldsymbol{\theta}_m$ is drawn from the Dirichlet prior distribution:

$$p(\boldsymbol{\theta}_m) = \frac{1}{\Delta(\boldsymbol{\alpha})} \prod_{k=1}^{K} \theta_{m,k}^{\alpha_k - 1} \tag{7.19}$$

where $\Delta(\boldsymbol{\alpha}) = \frac{\prod_{i=1}^{K} \Gamma(\alpha_i)}{\Gamma\left(\sum_{i=1}^{K} \alpha_i\right)}$.

The joint probability of $\boldsymbol{z}_m$ and $\boldsymbol{\theta}_m$ is:

$$\begin{aligned} p(\boldsymbol{z}_m, \boldsymbol{\theta}_m) &= p(\boldsymbol{z}_m|\boldsymbol{\theta}_m)p(\boldsymbol{\theta}_m) \\ &= \frac{1}{\Delta(\boldsymbol{\alpha})} \prod_{k=1}^{K} \theta_{m,k}^{n_{m,k,\cdot}+\alpha_k - 1} \end{aligned} \tag{7.20}$$

The marginal distribution of z_m is obtained by integrating $\boldsymbol{\theta}_m$ in the joint probability:

$$\begin{aligned} p(z_m) &= \int p(z_m, \boldsymbol{\theta}_m)\, \mathrm{d}\boldsymbol{\theta}_m \\ &= \frac{1}{\Delta(\boldsymbol{\alpha})} \int \prod_{k=1}^{K} \theta_{m,k}^{n_{m,k,\cdot}+\alpha_k-1} \mathrm{d}\boldsymbol{\theta}_m \\ &= \frac{\Delta(\boldsymbol{n}_{m,\cdot,\cdot}+\boldsymbol{\alpha})}{\Delta(\boldsymbol{\alpha})} \end{aligned} \tag{7.21}$$

where $\boldsymbol{n}_{m,\cdot,\cdot} = \{n_{m,k,\cdot}\}_{k=1}^{K}$.

The entire corpus is composed of M independent documents; therefore, the probability of the topic sequence of the entire corpus is:

$$\begin{aligned} p(z) &= \prod_{m=1}^{M} p(z_m) \\ &= \prod_{m=1}^{M} \frac{\Delta(\boldsymbol{n}_{m,\cdot,\cdot}+\boldsymbol{\alpha})}{\Delta(\boldsymbol{\alpha})} \end{aligned} \tag{7.22}$$

Similarly, the probability of the terms sequence given the topic sequence can be obtained. According to the former hypothesis, the conditional probability of term $w^{(i)} = t$ given topic $z^{(i)} = k$ at the i-th word position is:

$$p\left(w^{(i)} = t | z^{(i)} = k\right) = \varphi_{k,t}$$

The generation of each word in a document is independent. Let $\boldsymbol{w}_k$ denote a sequence of terms where the topic at each position is k across the entire corpus, and z_k is the corresponding sequence of topics where each element is topic k. Given parameter $\boldsymbol{\varphi}_k$, the probability of $\boldsymbol{w}_k$ is:

$$\begin{aligned} p(\boldsymbol{w}_k | z_k, \boldsymbol{\varphi}_k) &= \prod_{\{i:z^{(i)}=k\}} p(w^{(i)} | z^{(i)} = k, \boldsymbol{\varphi}_k) \\ &= \prod_{t=1}^{V} \varphi_{k,t}^{n_{\cdot,k,t}} \end{aligned} \tag{7.23}$$

where $n_{\cdot,k,t}$ denotes the frequency of term t under topic k in the corpus.

Similarly, $\boldsymbol{\varphi}_k$ is drawn from the Dirichlet prior distribution:

$$p\left(\boldsymbol{\varphi}_k\right) = \frac{1}{\Delta(\boldsymbol{\beta})} \prod_{t=1}^{V} \varphi_{k,t}^{\beta_t - 1} \tag{7.24}$$

where $\Delta(\boldsymbol{\beta}) = \frac{\prod_{i=1}^{K} \Gamma(\beta_i)}{\Gamma\left(\sum_{i=1}^{K} \beta_i\right)}$.

Given the topic z_k, the joint probability of $\boldsymbol{w}_k$ and $\boldsymbol{\varphi}_k$ is:

$$\begin{aligned} p(\boldsymbol{w}_k, \boldsymbol{\varphi}_k | z_k) &= p(\boldsymbol{\varphi}_k) p(\boldsymbol{w}_k | z_k, \boldsymbol{\varphi}_k) \\ &= \frac{1}{\Delta(\boldsymbol{\beta})} \prod_{t=1}^{V} \varphi_{k,t}^{n_{\cdot,k,t} + \beta_t - 1} \end{aligned} \tag{7.25}$$

By integrating $\boldsymbol{\varphi}_k$ in joint probability, the distribution $p\left(\boldsymbol{w}_k | z_k\right)$ is obtained as follows:

$$\begin{aligned} p\left(\boldsymbol{w}_k | z_k\right) &= \int p\left(\boldsymbol{w}_k, \boldsymbol{\varphi}_k | z_k\right) \mathrm{d}\boldsymbol{\varphi}_k \\ &= \frac{1}{\Delta\left(\boldsymbol{\beta}\right)} \int \prod_{t=1}^{V} \varphi_{k,t}^{n_{\cdot,k,t} + \beta_t - 1} \mathrm{d}\boldsymbol{\varphi}_k \\ &= \frac{\Delta(\boldsymbol{n}_{\cdot,k,\cdot} + \boldsymbol{\beta})}{\Delta(\boldsymbol{\beta})} \end{aligned} \tag{7.26}$$

where $\boldsymbol{n}_{\cdot,k,\cdot} = \left\{n_{\cdot,k,t}\right\}_{t=1}^{V}$.

Because the generation of each word is independent, the probability of the term sequence given the topic sequence $\boldsymbol{z}$ in the whole corpus is:

$$\begin{aligned} p\left(\boldsymbol{w} | \boldsymbol{z}\right) &= \prod_{k=1}^{K} p\left(\boldsymbol{w}_k | z_k\right) \\ &= \prod_{k=1}^{K} \frac{\Delta(\boldsymbol{n}_{\cdot,k,\cdot} + \boldsymbol{\beta})}{\Delta(\boldsymbol{\beta})} \end{aligned} \tag{7.27}$$

By integrating the above two factors (Eqs. 7.22 and 7.27), we can obtain the joint probability of the term and topic sequences as follows:

$$\begin{aligned} p\left(\boldsymbol{w}, \boldsymbol{z}\right) &= p\left(\boldsymbol{w} | \boldsymbol{z}\right) p\left(\boldsymbol{z}\right) \\ &= \prod_{k=1}^{K} \frac{\Delta(\boldsymbol{n}_{\cdot,k,\cdot} + \boldsymbol{\beta})}{\Delta(\boldsymbol{\beta})} \prod_{m=1}^{M} \frac{\Delta(\boldsymbol{n}_{m,\cdot,\cdot} + \boldsymbol{\alpha})}{\Delta(\boldsymbol{\alpha})} \end{aligned} \tag{7.28}$$

7.4.3 Inference in LDA

In the context of the probabilistic graphical model, inference refers to the process of inferring the values of hidden variables according to observed variables. The key inferential problems in LDA are to infer the distribution of latent topics given the observed words $p(z|\boldsymbol{w})$, based on the Bayesian inference framework, and estimate the posterior distribution of $\boldsymbol{\theta}_m$ and $\boldsymbol{\varphi}_k$, i.e., $p(\boldsymbol{\theta}_m|z_m, \boldsymbol{w}_m)$ and $p(\boldsymbol{\varphi}_k|\boldsymbol{w}_k, z_k)$.

However, LDA is somehow difficult to infer exactly, and approximate inference algorithms, such as variational Bayes expectation maximization (VBEM), expectation propagation (EP), and Markov chain Monte Carlo (MCMC), have been proposed for inference in LDA. VBEM was first used in the original paper on LDA (Blei et al. 2003). Griffiths and Steyvers (2004) proposed the LDA approximation inference based on Gibbs sampling, which is a representative method of MCMC.

MCMC is a sampling technique based on the Markov chain. It is often used to solve the challenge of sampling from high-dimensional random variables. By constructing a Markov chain that takes the target distribution as its stationary distribution, a sample of the target distribution can be approximately generated for each transition in the chain after the "burn-in" period, which eliminates the influence of initialization parameters.

Gibbs sampling is a special case of MCMC that is relatively simple to understand and implement. Suppose that the target distribution is $p(\boldsymbol{x})$. In Gibbs sampling, a randomly selected dimension $\boldsymbol{x}^{(i)}$ of the distribution is sampled alternately for each time, given the values of all the other dimensions $\boldsymbol{x}^{(\neg i)}$:

$$p\left(x^{(i)}|\boldsymbol{x}^{(\neg i)}\right) = \frac{p\left(x^{(i)}, \boldsymbol{x}^{(\neg i)}\right)}{p\left(\boldsymbol{x}^{(\neg i)}\right)} \tag{7.29}$$

When applying Gibbs sampling to the inference of LDA, we need to sample the following conditional distribution $p(z^{(i)}|z^{(\neg i)}, \boldsymbol{w})$:

$$\begin{aligned} p\left(z^{(i)}|z^{(\neg i)}, \boldsymbol{w}\right) &= \frac{p\left(\boldsymbol{w}, z\right)}{p\left(\boldsymbol{w}, z^{(\neg i)}\right)} \\ &\propto \frac{n^{(\neg i)}_{\bullet,k',t'} + \beta_{t'}}{\sum_t n^{(\neg i)}_{\bullet,k',t} + \beta_t}\left(n^{(\neg i)}_{m',k',\bullet} + \alpha_{k'}\right) \end{aligned} \tag{7.30}$$

where $\boldsymbol{w}$ is the sequence of terms in the training corpus and z is the corresponding sequence of topics. m', k', and t' are the index for document, term, and topic at the i-th position; $n^{(\neg i)}_{m,k,\cdot}$ denotes the frequency of topic k in document m after the removal for the i-th position; and $n^{(\neg i)}_{\cdot,k,t}$ denotes the frequency of term t under topic k in the sequence after removal of the i-th position. The detailed derivation of Formula (7.30) can be found in (Heinrich 2005).

By sampling $p\left(z^{(i)}|z^{(\neg i)}, \boldsymbol{w}\right)$, the Gibbs sampling algorithm generates a topic each time at one word position in the sequence with alternation and constructs a Markov chain that can be transformed between states. After passing the "burn-in" period, the Markov chain enters into a stable state. The stationary distribution can afterwards be used as an approximation of the target distribution $p(z|\boldsymbol{w})$. The flow chart of LDA inference is as follows.

Algorithm 5: Gibbs sampling for the inference of LDA

Input : The number of documents M, the length of each document N_m, the sequence of terms corresponding to the document set w, the number of topics k, the hyperparameters $\boldsymbol{\alpha}$ and $\boldsymbol{\beta}$. the maximum number of iterations T;

Output: Topic vectors z, estimates of the expectations of the posterior distributions $\hat{\boldsymbol{\Phi}}$ and $\hat{\boldsymbol{\Theta}}$.

1 #Initialization
2 zero all count variables: $n_{m,k,\cdot}$, $n_{\cdot,k,t}$
3 **for** $m = 1, \ldots, M$ **do**
4   **for** $n = 1, \ldots, N_m$ **do**
5   Random Initialization $z_{m,n} \sim \text{Cat}\left(\frac{1}{K}, \frac{1}{K}, \ldots, \frac{1}{K}\right)$
6   **end**
7 **end**
8 #Gibbs Sampling
9 $t = 0$
10 **while** $t < T$ *or until the algorithm converges* **do**
11   **for** $m = 1, \ldots, M$ **do**
12   **for** $n = 1, \ldots, N_m$ **do**
13              $t = w_{m,n}, \quad k = z_{m,n}$
14              $n_{m,k,\cdot} - = 1, \quad n_{\cdot,k,t} - = 1$
15   Sampling based on Formula (7.30) $z_{m,n} = \tilde{k} \sim p\left(z^{(i)}|z^{(\neg i)}, \boldsymbol{w}\right)$
16              $n_{m,\tilde{k},\cdot} + = 1, \quad n_{\cdot,\tilde{k},t} - = 1$
17              $t + = 1$
18   **end**
19   **end**
20 **end**
21 #Parameter Estimation
22 According to Formulas (7.33) and (7.34), we can estimate $\hat{\varphi}_{k,t}$ and $\hat{\theta}_{m,k}$, respectively.

In LDA, the parameters θ and φ are considered to be random rather than deterministic variables. Instead of estimating their exact values, we can infer their posterior distributions and observe their properties based on statistics such as expectation and variance. Based on the conjugate relationship between the Dirichlet distribution and categorical distribution, we can further determine that the posterior distribution of $\boldsymbol{\theta}$ and $\boldsymbol{\varphi}$ still follow the Dirichlet distributions:

$$p\left(\boldsymbol{\theta}_m|\boldsymbol{z}_m, \boldsymbol{w}_m\right) = \frac{1}{\Delta\left(n_{m,\cdot,} \cdots + \alpha\right)} \prod_{k=1}^{K} \theta_{m,k}^{n_{m,k,\cdot}+\alpha_k-1} \tag{7.31}$$

$$p\left(\boldsymbol{\varphi}_k|\boldsymbol{w}_k, z_k\right) = \frac{1}{\Delta(n_{\cdot,k,\cdot} + \beta)} \prod_{t=1}^{V} \varphi_{k,t}^{n_{\cdot,k,t}+\beta_t-1} \tag{7.32}$$

The expectations of the posterior distributions of $\boldsymbol{\varphi}$ and $\boldsymbol{\theta}$ based on the training corpus are as follows:

$$\hat{\varphi}_{k,t} = E(\varphi_{k,t}) = \frac{n_{\cdot,k,t} + \beta_t}{\sum_{t=1}^{V} n_{\cdot,k,t} + \beta_t} \tag{7.33}$$

$$\hat{\theta}_{m,k} = E(\theta_{m,k}) = \frac{n_{m,k,\cdot} + \alpha_k}{\sum_{k=1}^{K} n_{m,k,\cdot} + \alpha_k} \tag{7.34}$$

which contains not only the likelihood information in the training corpus but also the prior information of the parameters.

7.4.4 *Inference for New Documents*

The inference of the conditional topic distribution θ given new document d_m can be achieved by continuously performing Gibbs sampling only on the new document d_m while keeping the topic-conditional term distribution $\boldsymbol{\varphi}_k$ unchanged.

$$\begin{aligned} p\left(\tilde{z}^{(i)} = k|\tilde{w}^{(i)} = t, \tilde{\boldsymbol{w}}^{(\neg i)}, \tilde{z}^{(\neg i)}; \boldsymbol{w}, z\right) &\propto \hat{\varphi}_{k,t}\left(n_{\tilde{m},k,\cdot}^{(-i)} + \alpha_k\right) \\ &= \frac{n_{\cdot,k,t} + \beta_t}{\sum_t n_{\cdot,k,t} + \beta_t} \cdot \left(n_{\tilde{m},k,}^{(\neg i)} + \alpha_k\right) \end{aligned} \tag{7.35}$$

where $\tilde{w}^{(i)}$ and $\tilde{z}^{(i)}$ denote the sequences of terms and topics at the i-th position of the new document, respectively, $\tilde{\boldsymbol{w}}^{(\neg i)}$ and $\tilde{z}^{(\neg i)}$ denote the term and topic in the new document $\tilde{m}$ after the removal of the i-th position, respectively, and $n_{\tilde{m},k,\cdot}^{(\neg i)}$ denotes the frequency of topic k in the new document $\tilde{m}$ after the removal of the i-th position.

After the Markov chain converges to the stationary distribution, the expectation of Eq. (7.35) will be used as the "estimation" of the conditional topic distribution θ given the new document $\tilde{m}$:

$$\hat{\theta}_{\tilde{m},k} = E\left(\theta_{\tilde{m},k}\right) = \frac{n_{\tilde{m},k,\cdot} + \alpha_k}{\sum_k n_{\tilde{m},k,\cdot} + \alpha_k} \tag{7.36}$$

where $n_{\tilde{m},k,\cdot}$ denotes the frequency of topic k in the new document $\tilde{m}$.

7.5 Further Reading

LDA is one of the topic models in the field of text analysis receiving the most attention, and it is widely used in many text data mining tasks. A document can be represented as a distribution of topics based on the LDA model. Such topic-based text representation can be further used in many downstream tasks, such as word or document similarity calculation, text clustering, text classification, text segmentation, collaborative filtering, and so on.

Meanwhile, many LDA variants have been proposed to either improve its topic modeling ability or extend its applications to different scenarios.

For example, Blei and Lafferty (2006) proposed a correlated topic model (CTM), which captured the correlation between potential topics by using a logistic normal distribution instead of a Dirichlet distribution. Griffiths et al. (2004) proposed a hierarchical LDA for modeling tree-level topics. Li and McCallum (2006) proposed a PAM (Pachinko allocation model) that represented the relationship between topics as a directed acyclic graph. Chang and Blei (2009) proposed RTM (relational topic models) for modeling the topics among interlinked documents (i.e., document networks).

To incorporate the annotations into LDA for supervised learning, Mcauliffe and Blei (2008) proposed a supervised latent Dirichlet allocation (SLDA) model, and Ramage et al. (2009) proposed a labeled LDA model, both of which extend the standard LDA from unsupervised learning to supervised learning.

In addition to using LDA to model topics in plain text, a series of LDA variants have been proposed to model text-related external attributes, such as author, user, sentiment, and network structure. For example, Steyvers et al. (2004) proposed an author-topic model (ATM), which established a user model in the process of text generation and built a topic-conditional term distribution for each of the authors. McCallum et al. (2005) proposed an author recipient topic (ART) model, where the generation of topics and terms in a document was determined by both author and recipient. In addition to only considering user information, Zhao et al. (2011) proposed a Twitter-LDA model that introduced the concepts of topic categories and topic types to facilitate the analysis of the topical differences between social media and traditional media. Many LDA extensions have also been proposed to model the additional attributes in social media, including time, place, interest, community, and network structure. Aiming to provide topic modeling for review text containing sentiment and opinion, Mei et al. (2007) proposed a topic–sentiment mixture model (TSM) by introducing sentiment variables on the basis of a traditional topic model. Follow-up work included a multiaspect sentiment analysis (MAS) model (Titov and McDonald 2008) and a joint sentiment–topic (JST) model (Lin and He 2009). We will introduce these models in Sect. 8.5.3 in detail.

Traditional topic models are based on static text data, but the topics in a text data stream are dynamic and change with time. Blei and Lafferty (2006) proposed a dynamic topic model, which segmented text streams according to time and assumed that the parameters of the time series satisfied the first-order Markov hypothesis. In the topic over time (TOT) model proposed by Wang and McCallum (2006), the time information was modeled in another way: they introduced the time labels as observable variables into the topic model and associated topic and term sequence generation with time labels.

Exercises

7.1 How does the topic model handle polysemy and synonym in the text?

7.2 What is the difference between LSA and PLSA? Which is more suitable for modeling large-scale text data?

7.3 Please read the paper (Mei and Zhai 2001) and derive the EM algorithm in PLSA in detail.

7.4 Try to analyze the connection between PLSA and LDA and their differences.

7.5 Why is it we cannot directly estimate the specific value of the parameters of document-conditional topic distribution and topic-conditional term distribution in LDA based on a set of documents but can only estimate their posterior probabilities and expectations?

7.6 Try to learn and derive the LDA inference method based on variational EM.

7.7 As a distributed text representation method, please elaborate the topic models' similarities with and differences from the representation learning methods represented by word2vec.

Chapter 8
Sentiment Analysis and Opinion Mining

8.1 History of Sentiment Analysis and Opinion Mining

The main tasks of sentiment analysis and opinion mining include the extraction, classification, and inference of subjective information in texts (e.g., sentiment, opinion, attitude, emotion, stance, etc.). It is one of the most active research areas in natural language processing and text data mining. A large number of research papers in this field have been published in top academic conferences (including ACL, EMNLP, COLING, IJCAI, AAAI, SIGIR, CIKM, WWW, KDD, and so on). At the same time, evaluations and competitions for sentiment analysis and related tasks, including TREC,[1] NTCIR,[2] COAE, NLPCC, SemEval[3] and so on, have been launched and thereby also effectively promoted the development of sentiment analysis research.

Early sentiment classification studies were primarily rule-based approaches. Turney (2002) proposed a PMI-IR method to identify the semantic orientation (SO) of words (or phrases) in text, accumulated the polarities of these words/phrases, and ultimately obtained the polarity of the whole document. Pang et al. (2002) first introduced machine learning into the sentiment classification of movie reviews and compared three classical classification algorithms (naïve Bayes, maximum entropy, and support vector machines). This work laid the foundation for the study of machine learning-based sentiment classification. However, traditional statistical machine learning algorithms use the bag-of-words model for text representation, which disrupts the original structure of the text and loses word order, syntactic structure, and some semantic information.

[1] https://trec.nist.gov/.

[2] https://research.nii.ac.jp/ntcir/index-en.html.

[3] https://en.m.wikipedia.org/wiki/SemEval.

© Tsinghua University Press 2021

C. Zong et al., *Text Data Mining*, https://doi.org/10.1007/978-981-16-0100-2_8

In subsequent studies, sentiment analysis techniques were naturally divided into two categories, i.e., rules-based methods and machine learning-based methods. The former performs sentiment analysis at different granularities of text based on the sentiment orientation of the words provided by a sentiment lexicon. The latter focuses on more effective feature engineering for text representation and machine learning. The main features include word order and its combination, part of speech (POS), high-order n-gram, syntactic structure information, and so on.

From the perspective of methods, although statistical machine learning methods in sentiment classification follow the framework of traditional topic-based text classification models, there are some special problems, such as sentiment polarity shift and domain adaptation, that need to be dealt with separately. Sentiment classification has emerged based on different machine learning settings, such as semisupervised sentiment classification, class-imbalanced sentiment classification, and cross-linguistic sentiment classification. In recent years, the deep learning method represented by artificial neural networks has been gradually applied to many sentiment analysis tasks and has achieved great success.

From the perspective of tasks, in addition to document- or sentence-level sentiment classification, more sentiment analysis tasks have been proposed at different granularities, including aspect-level sentiment analysis, word or phrase-level sentiment classification, and sentiment dictionary construction. In recent years, a series of generalized sentiment analysis tasks, such as emotion analysis and stance classification, have also emerged in the field of sentiment analysis.

8.2 Categorization of Sentiment Analysis Tasks

In the following, we will categorize the sentiment analysis tasks from the perspectives of task output and analysis granularity.

8.2.1 Categorization According to Task Output

Sentiment classification is one of the core tasks of sentiment analysis and can be regarded as a special type of text classification problem. Traditional text classification mainly refers to the classification of text content according to objective information, while sentiment classification aims to classify the text according to subjective information.

The most studied task of sentiment classification is sentiment polarity classification, or positive-negative classification, the goal of which is to predict whether a document or a sentence contains sentiments that are positive (thumbs up) or negative (thumbs down). Polarity classification requires that the content contained in the text be subjective. It is meaningless to conduct sentiment analysis on text that only contains objective information (such as a person's height and weight or the time and location of an event). In the early research on sentiment analysis, there was a small amount of work dedicated to the subjective–objective classification of sentences (which we call subjectivity detection). Subjective–objective classifications are different from polarity classifications, although they look similar. They all belong to a binary classification problem, but they have different types of classification labels: the former is subjective or objective, while the latter is positive or negative. Wiebe et al. (2004) provided a detailed review of subjectivity detection based on different methods and features.

In addition to the positive and negative categories, the neutral category is often considered in sentiment classification, which results in the three-class (positive–negative–neutral) sentiment classification problem. Neutral texts normally contain two cases: objective text that does not contain sentiments and subjective text that contains mixed positive and negative sentiments. In addition, there are also some fine-grained sentiment analysis outputs, such as sentiment classification according to rating levels (such as 1 to 5 stars), sentiment regression based on sentiment strength (from 0 to 100%), emotion classification based on psychological emotions (such as angry, sad, happy, etc.), or stance classification toward a topic (support, against, or neural).

8.2.2 According to Analysis Granularity

Based on analysis granularity, sentiment analysis tasks can be divided into document-level sentiment analysis, sentence-level sentiment analysis, word-level sentiment analysis, and aspect-level sentiment analysis.

(1) Document-level sentiment analysis

Early research focused on sentiment classification for the entire document; in other words, it aimed to determine the sentiment expressed by the whole document. Document-level sentiment analysis task is defined as follows: given a document d that may contain multiple sentences or even multiple paragraphs, the goal is to determine the sentiment polarity $o(d)$ of the entirety of d. Figure 8.1 shows a document-level book review that contains three paragraphs. The goal of document-level sentiment analysis is then to identify the author's sentiment (e.g., positive or negative) based on the whole document.

Turney (2002) and Pang et al. (2002) are the representative works from the early research on document-level sentiment classification focused on book or movie reviews. In addition to the book and movie domains, there are many document-level reviews on the Internet, including electronic product reviews, hotel reviews, restaurant reviews, etc. The tasks involved in detecting the sentiment of reviews at the document-level are all within the scope of document-level sentiment analysis.

Debbie

☆☆☆☆☆ **Take your time and savor every sentence.**
Reviewed in the United States on September 3, 2018
Verified Purchase

I am extremely stingy with my compliments for good books, but this tale is well-deserving of the praise. Of the last dozen or so books I've read, only two others earned five complete stars by me: She Read to Us in the Late Afternoons: A Life in Novels by Kathleen Hill, and Circe by Madeline Miller.

I have to confess that I have also had magical moments with marsh creatures such as herons, eagles, and mud turtles. Like the main character, Kya, I am a compulsive collector of treasures from those Great Rock Tumblers: the Chesapeake Bay and Atlantic Ocean which makes this book so attractive to me. However, Delia Owens' writing is more than just about the natural world. She spins a good and very well-written tale about murder, courtroom drama, nature, poetry, and even love.

Another reviewer described Owens' writing as lyrical. It is. Take your time and savor every sentence.

1,073 people found this helpful

Fig. 8.1   An example of document-level sentiment classification

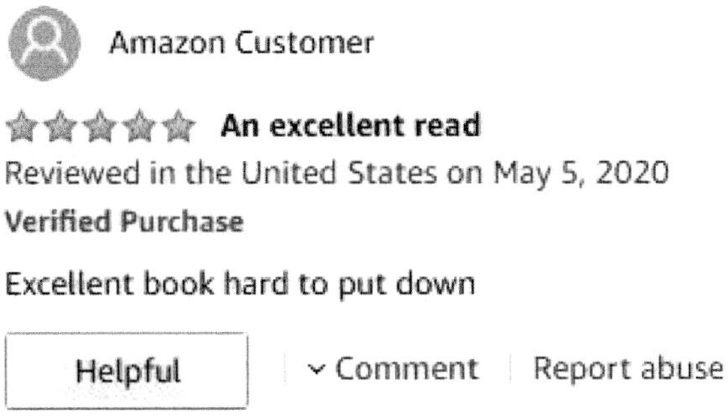

Fig. 8.2   An example of sentence-level sentiment analysis

(2) Sentence-level sentiment analysis

The granularity of document-level sentiment analysis is relatively rough because the entire document usually contains multiple topics, and the sentiment may vary for these different topics. In contrast, the topics involved in a sentence tend to be relatively simple, therefore the sentiment in a sentence is more unique. Many natural language processing techniques, such as syntactic parsing, use sentences as the processing unit, and sentiment analysis at the sentence level is more likely to incorporate such linguistic knowledge.

Sentence-level sentiment analysis can be defined as follows: given a sentence s, the goal is to determine the sentiment polarity $o(s)$ of s. Figure 8.2 gives a sentence-level review, and the goal of the sentence-level sentiment analysis task is to identify the sentiments expressed by this sentence.

Early research on sentence-level sentiment analysis included the subjective–objective classification of sentences. Supervised learning methods used vocabulary,

n-grams, POS, word order, and other features for text representation and naïve Bayes, maximum entropy and other classifiers for classification (Wiebe et al. 1999, 2004). Pang and Lee (2004) used the minimum cut algorithm in graph theory to extract a set of sentences that represents the sentiments of the entire document and then performed sentiment classification on these representative sentences. One disadvantage of sentence-level sentiment classification is that the sentiments need to be manually annotated when building a classifier based on supervised learning algorithms, while the sentiments for a document-level review can often be obtained based on natural annotation (such as the rating of a review provided by the users).

With the development of social media (such as Twitter, Facebook, Weibo, and WeChat) in recent years, there has been a class of message-level sentiment analysis tasks. Such message-level texts are usually short and contain a small number of sentences, often referred to as "short text." This type of sentiment analysis task can actually be handled as a sentence-level sentiment analysis or a short document-level sentiment analysis without considering the structure of the social network.

(3) Word-level sentiment analysis and sentiment lexicon construction

In addition to document- and sentence-level sentiment analysis, some studies have focused on sentiment analysis of more fine-grained units. Words and phrases are often considered to be the smallest unit of sentiment expression. For convenience of description, we refer to the sentiment analysis of both words and phrases as word-level sentiment analysis. Word-level sentiment analysis is defined as follows: given a word or phrase p, the goal is to determine the sentiment polarity $o(p)$ of p.

For a given corpus, word-level sentiment analysis is equivalent to the task of sentiment lexicon construction. Most of the current general-purpose sentiment lexicons are constructed manually, and although they are suitable for many fields, they have difficulty fully covering the sentiment vocabulary from different domains in practical applications. Furthermore, the manual construction of sentiment lexicons is time- and labor-consuming. Therefore, the research community has mostly focused on the study of the automatic construction of sentiment lexicons. These methods are mainly divided into three categories: knowledgebase-based methods, corpus-based methods, and a combination of the two.

(4) Aspect-level sentiment analysis

Aspect-level sentiment analysis is the task of extracting the aspects associated with an opinion target in the review text and identifying the fine-grained sentiment polarity toward a specific aspect. Figure 8.3 displays an example of an opinion summary based on aspect-level sentiment analysis. In comparison, document- and sentence-level sentiment analysis only identify the overall sentiment of a document or a sentence and do not refer to the sentiments of specific aspects of the opinion target.

In early research (Hu and Liu 2004), aspect-level sentiment analysis was called feature-based opinion mining. Later work (Liu 2012, 2015) further defined an opinion as a quadruple (g, s, h, t), where g represents the opinion target, s represents the sentiment toward g, h denotes an opinion holder, and t denotes

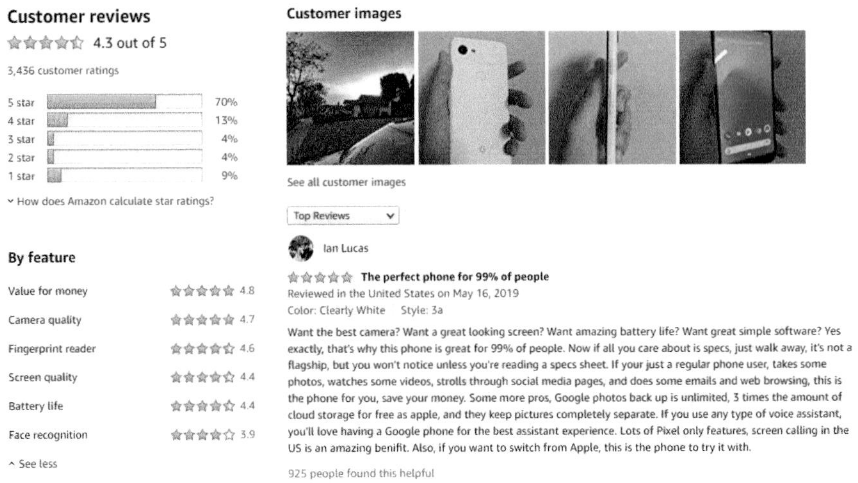

Fig. 8.3 An example of an opinion summary based on aspect-level sentiment analysis

Table 8.1 An example of aspect-level sentiment analysis

Review text	The phone looks good, the speed is very fast, the camera is also good, but the battery capacity is a little small, and the battery life is normal
Analysis result	{(appearance, positive), (speed, positive), (camera, positive),(battery capacity, negative), (battery life, negative)}

the time of the comment. An opinion target usually contains an entity and its attributes. In this case, the above quadruple becomes a quintuple (e, a, s, h, t), and the aspect-level sentiment analysis is defined as the extraction and identification of such quintuples from review texts. For simplicity, in this book, we only focus on the opinion pair (g, s), as shown in Table 8.1.

8.3 Methods for Document/Sentence-Level Sentiment Analysis

Except for slight differences in text representation, the methods for document- and sentence-level sentiment analysis are similar. Therefore, in this section, we introduce the two levels together.

8.3.1 Lexicon- and Rule-Based Methods

The rule-based approach to sentiment classification usually leverages external linguistic knowledge, such as a sentiment lexicon, to determine the sentiment polarity of a given document or sentence.

Das and Chen (2001) identified the sentiment of words (+1 for the positive, −1 for the negative, and 0 for the neutral) based on a manually constructed sentiment lexicon and accumulated the polarities of these words to obtain the polarity of the entire text.

Turney (2002) proposed a PMI-IR (pointwise mutual information-information retrieval) approach to estimate the sentiment orientation (SO) of phrases with predefined patterns and determine the sentiment polarity of the whole document by averaging the SO of each of the candidate phrases that appear in the document. This approach mainly consists of the following three steps:

Step 1: Extracting candidate words and phrases containing sentiment (mostly adjectives and adverbs and their phrases) based on a set of predefined patterns presented in Table 8.2.

Step 2: Estimating the SO of the candidate phrase using the PMI-IR method. By using "excellent" as the positive seed word and "poor" as the negative seed word and estimating the PMI score between each candidate phrase and the seed word, they defined the difference between two PMI scores as the SO value of the candidate phrase:

$$\text{SO (phrase)} = \text{PMI (phrase, "excellent")} - \text{PMI (phrase, "poor")} \tag{8.1}$$

where the PMI score between two words (phrases) is calculated as follows:

$$\text{PMI}\,(w_1, w_2) = \log \frac{p\,(w_1, w_2)}{p\,(w_1)p(w_2)} \tag{8.2}$$

where $p(w_1, w_2)$ is the probability that the words or phrases w_1 and w_2 cooccur in the text. PMI(w_1, w_2) measures the similarity between w_1 and w_2 from the perspective of co-occurrence.

Table 8.2 Part-of-speech patterns for extracting candidate phrases from reviews (Turney 2002)

First word	Second word	Third word (not extracted)
Adjective (JJ)	Noun (NN, NNS)	Any
Adverb (RB, RBR, RBS)	Adjective (JJ)	Nonnoun (NN, NNS)
Adjective (JJ)	Adjective (JJ)	Nonnoun (NN, NNS)
Noun (NN, NNS)	Adjective (JJ)	Nonnoun
Adverb (RB, RBR, RBS)	Verb (VB, VBD, VBN, VBG)	Any

PMI-IR empirically estimates the PMI and SO values based on the information retrieval results from the AltaVista search engine as follows:

$$\text{SO (phrase)} = \log \frac{\text{hits (phrase NEAR "excellent")} \cdot \text{hits ("poor")}}{\text{hits (phrase NEAR "poor")} \cdot \text{hits ("excellent")}} \tag{8.3}$$

where the NEAR operator indicates that two words cooccur in a certain window length, and hits(query) indicate the number of queries returned by the search engine.

Step 3: Accumulating the SO values of all candidate phrases appearing in the text and determining the sentiment polarity according to the average SO value (positive or negative).

In addition to the PMI-IR method, there are still many studies that directly obtain the sentiment polarity or intensity of a candidate word or phrase based on an external sentiment lexicon. We refer to this kind of method as lexicon-based unsupervised sentiment classification. Instead of simply accumulating the SO values of candidate words in the document, Taboada et al. (2011) designed a set of delicate rules to deal with special linguistic phenomena such as negation, contrast, sentiment intensifiers, sentiment diminishers and irrealis in the language of English.

8.3.2 Traditional Machine Learning Methods

The advantage of rule-based methods is that they are easy to use and do not rely on a manually annotated corpus. However, their performance is limited by the quality of rules and lexicons. In the past two decades, statistical machine learning methods have been rapidly developed and widely applied to sentiment analysis and opinion mining.

(1) Early research

The early research followed the framework of traditional text classification, which used a bag-of-words model for text representation followed by traditional statistical learning algorithms for classification.

Pang et al. (2002) first introduced statistical machine learning algorithms into the task of sentiment polarity classification of movie reviews. A supervised learning classifier was trained on manually labeled positive and negative reviews. They compared three kinds of classification algorithms, including naïve Bayes (NB), maximum entropy (ME), and support vector machines (SVM). In terms of feature engineering, they investigated several kinds of features, including n-grams (unigrams, bigrams), POS and position, and compared two feature weighting schemes (term frequency and presence). Table 8.3 reports the classification performance for different settings.

Table 8.3 The sentiment classification performance for different settings on the movie review corpus (Pang et al. 2002)

Features	# of features	Frequency or presence	NB	ME	SVM
Unigrams	16165	Frequency	78.7	N/A	72.8
Unigrams	16165	Presence	81.0	80.4	82.9
Unigrams+bigrams	32330	Presence	80.6	80.8	82.7
Bigrams	16165	Presence	77.3	77.4	77.1
Unigrams+POS	16695	Presence	81.5	80.4	81.9
Adjectives	2633	Presence	77.0	77.7	75.1
Top 2633 unigrams	2633	Presence	80.3	81.0	81.4
Unigrams+position	22430	Presence	81.0	80.1	81.6

Pang et al. (2002) reported that the machine learning methods have a higher classification accuracy than human predicted results. The performance gap in the three classifiers is small, where SVM performs slightly better than ME and NB. However, their performance is not as good as that on traditional topic-based text classification. Meanwhile, regarding feature engineering, the performance of unigrams only is generally the best and slightly better than that of unigrams+bigrams, and the performance of presence term weighting is slightly better than that of term frequency (TF). Subsequent studies have shown that classifier performance is domain dependent in sentiment classification, and there is no permanent winner across different domains (Xia and Zong 2011).

On the basis of (Pang et al. 2002), a large number of studies have been conducted by employing machine learning for sentiment classification. On the one hand, researchers have tried to design new text representation methods that are more suitable for sentiment classification from the perspective of feature engineering. On the other hand, they have also explored the usage of new machine learning algorithms for sentiment classification. In recent years, in particular, with the rapid development of deep learning, a large number of deep neural networks have been proposed in sentiment classification. We will introduce these methods in Sect. 8.3.3.

(2) Deep linguistic features

The BOW text representation used by traditional machine learning methods breaks the inherent structure of the text, ignores the word order information, destroys the syntactic structure, loses a portion of the semantic information, and finally makes sentiment classification less effective.

To address this problem, many researchers have attempted to explore more deep-level linguistic features from text that can effectively express sentiments. These linguistic features include position information (Pang et al. 2002; Kim and Hovy 2004), POS information (Mullen and Collier 2004; Whitelaw et al. 2005), word order information (Dave et al. 2003; Snyder and Barzilay 2007), high-order n-grams (Pang et al. 2002; Dave et al. 2003), and syntactic structure (Dave et al. 2003; Gamon 2004; Ng et al. 2006; Kennedy and Inkpen 2006).

In (Pang et al. 2002; Kim and Hovy 2004), position information was used as an auxiliary feature of words to generate a feature vector, and it potentially complemented the information contained in a simple vocabulary. POS information plays an important role in describing sentiment in texts. In the early subjectivity detection research, adjectives were already used as features (Hatzivassiloglou and McKeown 1997), where the experimental results showed that the subjectivity of a sentence has a higher correlation with the adjectives than with the other POS tags. Mullen and Collier (2004), Whitelaw et al. (2005) argued that although adjectives are an important feature of sentiment classification, this does not mean that other parts of POS tags have no effect on sentiment classification. For example, some nouns and verbs often contain important sentiment information (such as the noun "genius," the verb "recommend," etc.). The comparative experiments in (Pang et al. 2002) showed that the classification results using only adjective features are significantly poorer than those using the same number of high-frequency words.

High-order n-grams are widely used features in many natural language processing tasks, including sentiment classification. Pang et al. (2002) reported that the performance of unigrams alone is higher than that of bigrams. However, Dave et al. (2003) reported that in some cases, bigrams and trigrams achieve better performance than unigrams alone. Generally, in practice, higher-order n-gram features are often used as a complement to unigram features rather than being used separately.

Although higher-order n-grams can reflect some of the word order and dependency relationship (especially in adjacent words), they cannot capture the long-term dependencies of words in a sentence. To capture this kind of information, deeper linguistic analysis tools such as syntactic parsing must be incorporated in feature engineering.

A simple method is to extract interdependent word pairs as features. For example, "recommend" and "movie" are a pair of interdependent words in the dependency tree shown in Fig. 3.3. In this way, the long-term dependency between "recommend" and "movie" can be captured, although they are not neighbors. These interdependent word pairs contain some of the syntactic structure information that may contribute to sentiment classification. However, studies have drawn inconsistent conclusions about whether such features are effective. For example, Dave et al. (2003) argued that the introduction of "adjective–noun" dependency does not provide useful information in addition to the traditional bag-of-words feature template. Ng et al. (2006) made use of not only the "adjective–noun" dependency but also the "subject–predicate" and "verb–object" dependencies as supplemental features to traditional unigram, bigram, and trigram features, ultimately failing to improve sentiment classification performance. Gamon (2004) improved sentiment classification performance by using the syntactic relationship features extracted from the constituent tree, but the performance from using these linguistic features alone is still lower than that from using simple unigram features. In Kennedy and Inkpen (2006), syntactic analysis was also employed to model sentiment contrast, sentiment intensifier, and sentiment diminisher problems.

(3) Term weighting and feature selection

In traditional topic-based text classification, the frequency of words is important information. TF and TF-IDF are frequently used term-weighting schemes. However, in sentiment classification, Pang et al. (2002) found that using Boolean weight (i.e., the presence of words) can achieve even better results than TF weights. One possible explanation for such a finding is that the repetition of words contains more topic information but does not provide more sentiment information. In subsequent studies, presence has become the most commonly used term-weighting scheme for sentiment classification based on traditional machine learning.

On the other hand, as we introduced in Sect. 5.3, feature selection methods, such as mutual information (MI), information gain (IG), and so on, have been widely used in traditional topic-based text classification. Cui et al. (2006), Ng et al. (2006), Li et al. (2009a) applied these feature selection methods (such as IG) to sentiment classification tasks and proved the effectiveness of feature selection in sentiment classification.

(4) Ensemble learning

Aue and Gamon (2005) first applied ensemble learning to sentiment classification by combining training corpora from different source domains based on ensemble learning and obtained improved classification performance. Whitehead and Yaeger (2010) used SVM as the base classifier algorithm and studied four ensemble learning algorithms based on feature subset extraction to test the performance of sentiment classification. Their results showed that ensemble learning can significantly improve system performance. Xia et al. (2011) conducted a comparative study by examining the effects of different classification algorithms, different feature representations, and different ensemble learning strategies for sentiment classification on five document-level reviews and proved the effectiveness of ensemble learning in sentiment classification.

(5) Hierarchical sentiment classification

McDonald et al. (2007) transformed the research perspective from traditional single-grained document-level sentiment classification to multilevel granularity hierarchical sentiment classification. The document–sentence level hierarchical sentiment classification problem was modeled as a joint sequence labeling problem. The respective sentiments of sentences and the overall sentiment of the whole document were learned under a unified conditional random field (CRF) framework; then, the sentiment label of the document was used to correct the sentiment labels of sentences. Mao and Lebanon (2007) considered the sentiment labels of consecutive sentences as a kind of sentiment flow and used the CRF model to capture the change in sentiment intensity in the flow.

8.3.3 Deep Learning Methods

In recent years, deep learning methods represented by artificial neural networks have been widely used in many fields of natural language processing and text data mining because of their deep feature representation ability and joint end-to-end learning architecture. In Sect. 5.5 of this book, we have already introduced the commonly used neural networks for text classification (e.g., CNN and RNN). As mentioned, sentiment classification can be viewed as a kind of special text classification task. Therefore, the neural networks in Sect. 5.5 can still be applied to sentiment classification. In this subsection, we focus on several deep learning models that are used for sentiment classification, in particular.

(1) Recursive neural network based on constituent tree

In the previous section, we mentioned that syntactic features play an important role in sentiment classification. To this end, Socher et al. (2011a) first proposed a recursive neural network by using the syntactic tree as the input for sentence-level sentiment classification. In Fig. 8.4, we display two kinds of RNNs for sentence modeling, where subfigure (a) is a recurrent neural network modeling a sentence described by a sequence of words, and subfigure (b) is a recursive neural network modeling a sentence described by a constituent tree.

Each nonleaf node of the tree has two child nodes. In representation, the vector representations of the two child nodes are concatenated as the input of the parent node. After linear transformation and nonlinear activation are performed, we obtain the vector representation of the parent node. Let $\boldsymbol{c}_1 \in \mathbb{R}^d$ and $\boldsymbol{c}_2 \in \mathbb{R}^d$ denote the vector representations of the two child nodes, respectively, and $\boldsymbol{p} \in \mathbb{R}^d$ denote the vector representation of the parent node. This process can be written as:

$$\boldsymbol{p} = f\left(\boldsymbol{W}\begin{bmatrix} \boldsymbol{c}_1 \\ \boldsymbol{c}_2 \end{bmatrix} + \boldsymbol{b}\right) \tag{8.4}$$

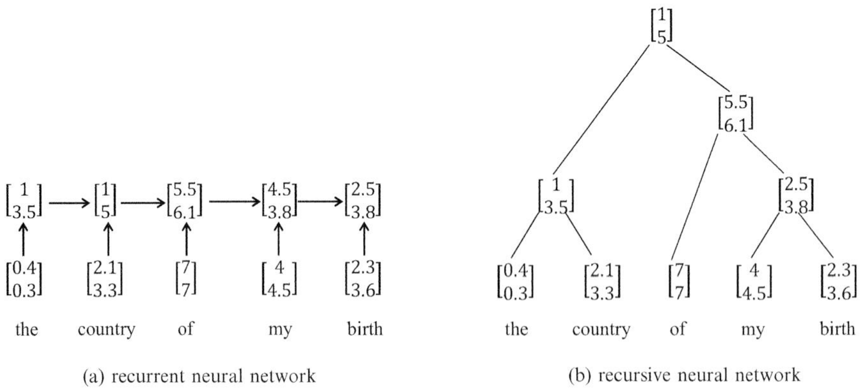

Fig. 8.4 Examples of two neural network modeling methods for sentence classification

where $W \in \mathbb{R}^{d \times 2d}$ is the weight matrix, $b \in \mathbb{R}^{d}$ is the bias, and $f(\cdot)$ is the activation function.

The topology defined by the constituent tree is recursively forwarded from bottom to top until the entire tree is processed, and all nodes share parameters W and b. Finally, the representations of the root note and all intermediate nodes can be obtained. The representation of the root node is used as the representation of the entire sentence and then fed to a softmax layer for classification.

Similar to the BPTT algorithm used for training the recurrent neural network, the back-propagation through structure (BPTS) algorithm is used for training a recursive neural network.

(2) Matrix–vector recursive neural network

On the basis of the abovementioned recursive neural network, Socher et al. (2012) further proposed a matrix–vector recursive neural network (MV-RNN). The structure of MV-RNN is shown in Fig. 8.5.

Each node of the tree in Fig. 8.5 is represented by a vector–matrix pair (a, A), where $a \in \mathbb{R}$ encodes the representation of the current node, and $A \in \mathbb{R}^{d \times d}$ encodes a node's modifying effect upon its adjacent child node. For example, the node "very" may enhance the semantics of "good," while "not" may reverse the semantics of "good." Assuming that the representations of two child nodes are (a, A) and (b, B), the process of integrating the representations of the two child nodes into the representation of the parent node p is described as

$$p = f\left(W\begin{bmatrix} Ba \\ Ab \end{bmatrix}\right) \tag{8.5}$$

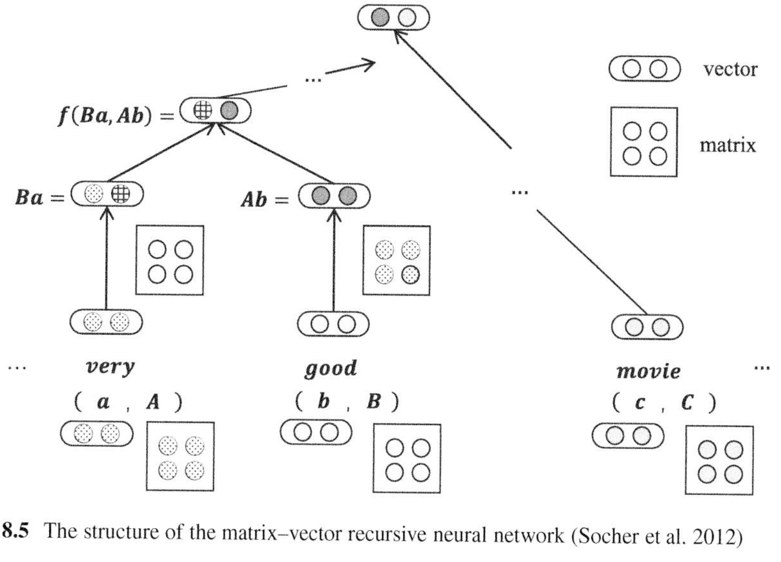

Fig. 8.5 The structure of the matrix–vector recursive neural network (Socher et al. 2012)

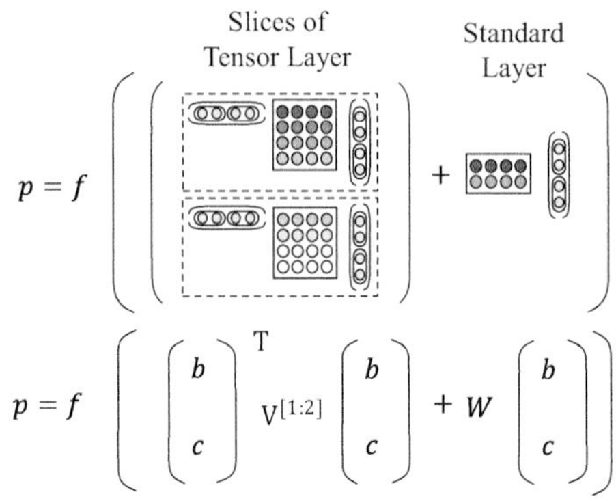

Fig. 8.6 Illustration of a recursive neural tensor network (Socher et al. 2013)

where $\boldsymbol{W} \in \mathbb{R}^{d\times 2d}$ is the weight matrix, and $f(\cdot)$ is the activation function. $\boldsymbol{Ab}$ implements the modification of "very" upon "good" in Fig. 8.5.

Socher et al. (2012) conducted a binary sentiment classification experiment on the sentence-level IMDB movie review dataset and obtained state-of-the-art results. The constituent tree of the sentence was obtained based on the Stanford Parser. One disadvantage of the MV-RNN model is that it requires an additional $\mathbb{R}^{d\times d}$ parameter matrix for each word in the vocabulary, which greatly increases the model's parameter space.

(3) Recurrent neural tensor network

To address MV-RNN's shortcomings, Socher et al. (2013) further proposed a recursive neural tensor network (RNTN), as shown in Fig. 8.6. In the RNTN model, two child nodes $\boldsymbol{a}$ and $\boldsymbol{b}$ are combined into a parent node $\boldsymbol{p}$ in the following way:

$$\boldsymbol{p} = f\left([\boldsymbol{a}, \boldsymbol{b}]\, \boldsymbol{V}^{[1:d]} \begin{bmatrix} \boldsymbol{a} \\ \boldsymbol{b} \end{bmatrix} + \boldsymbol{W} \begin{bmatrix} \boldsymbol{a} \\ \boldsymbol{b} \end{bmatrix}\right) \tag{8.6}$$

where $\boldsymbol{V}^{[1:d]} \in \mathbb{R}^{2d\times 2d\times d}$ is a tensor, and $\boldsymbol{h} = \boldsymbol{x}^T \boldsymbol{V}^{[1:d]}\boldsymbol{x} \in \mathbb{R}^d$ is called the tensor product of $\boldsymbol{x}$. h_i can be calculated from each channel $\boldsymbol{V}^{[i]}$ of the tensor $\boldsymbol{V}$ as follows: $\boldsymbol{h} = \boldsymbol{x}^T \boldsymbol{V}^{[i]}\boldsymbol{x}$.

As mentioned above, MV-RNN models the interaction of two child nodes by introducing a matrix representation for each word. In contrast, in RNTN, such interaction was modeled by a tensor $\boldsymbol{V}^{[1:d]}$ shared by different nodes, which significantly reduced the scale of the parameter space.

Meanwhile, they released the Stanford Sentiment Treebank (SST) dataset, which has become the benchmark dataset for subsequent sentence-level sentiment analysis studies. Other studies using recursive neural networks include (Irsoy and Cardie 2014), etc.

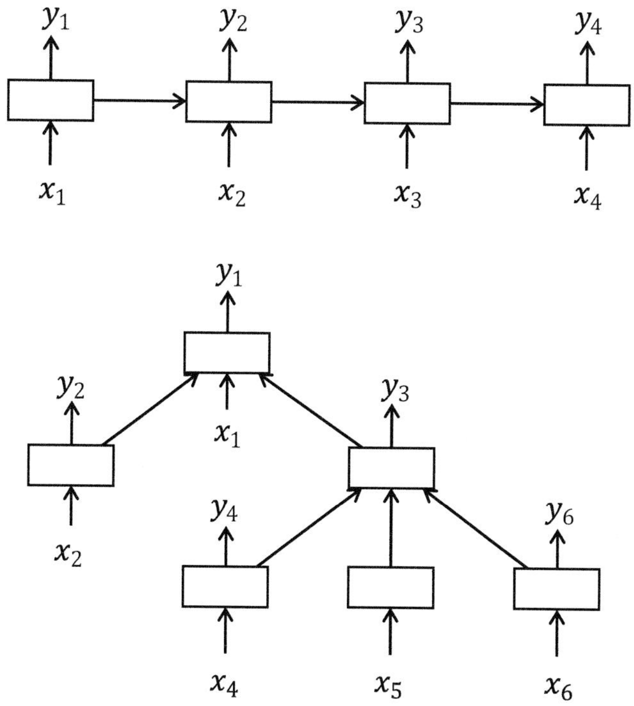

Fig. 8.7 The architecture of LSTMs and Tree-LSTMs

(4) Tree-structured long short-term memory networks

A recurrent neural network is established according to time steps, and a recursive neural network is established according to the tree structure. To combine the advantages of the two networks, Tai et al. (2015) proposed a tree-structured long short-term memory network (Tree-LSTM), which gives the recurrent neural network the ability to model tree structures. The basic idea of Tree-LSTM is illustrated in Fig. 8.7. It consists of two variants:

(a) Child-Sum Tree-LSTM: The hidden state of a node in Child-Sum Tree-LSTMs is determined by the sum of the hidden states of all its child nodes. It is suitable for a dependency tree when the number of child nodes is uncertain and the child nodes are order-independent (e.g., the dependency tree). In this case, the model is called a Dependency Tree-LSTM.

(b) N-ary Tree-LSTM: This variant is suitable for tree structures where the child nodes are ordered but the number of child nodes is N at most. In comparison with the Child-Sum Tree-LSTM, the N-ary Tree-LSTM introduces a separate parameter matrix for each child node. Both models define separate forget gates for each child node, but in N-ary Tree-LSTMs, one forget gate contains the interaction between all child nodes. When the N-ary Tree-LSTM is applied to constituent trees, it is called Constituent Tree-LSTM.

The binary constituent tree was used in (Tai et al. 2015), where each intermediate node contains only two child nodes: the left child and the right child. They conducted experiments on the SST dataset, and the results proved that Tree-LSTMs can achieve significant improvements over the standard LSTM and its existing variants.

(5) Hierarchical modeling for document-level sentiment analysis

A document can be viewed as a sentence sequence, where each sentence is a word sequence. To capture the word–sentence–document hierarchy, Tang et al. (2015b) proposed two hierarchical document classification models named Conv-GRNN and LSTM-GRNN.

In contrast to the standard RNN that encodes a document by a word sequence, Tang et al. (2015b) modeled a document in two levels. In the first level, a set of sentence-level LSTMs (or CNNs) was used to model each sentence and produce the representation of each sentence from the representations of the words comprising it, based on average pooling. In the second level, a simplified GRU module called GRNN was exploited to obtain the document representation by using the sentence representations as the input. The document representation is ultimately fed to a softmax layer for document-level sentiment classification. Experimental results on the restaurant datasets of Yelp 2013 through 2015 and the IMDB movie review dataset proved the effectiveness of their hierarchical approach in comparison with existing nonhierarchical approaches.

8.4 Word-Level Sentiment Analysis and Sentiment Lexicon Construction

The sentiment lexicon is an important resource in rule-based sentiment classification. The automatic construction of sentiment lexicons is an important research direction in the field of sentiment analysis and opinion mining. The method applied to this task can be mainly divided into three categories: knowledgebase-based methods, corpus-based methods, and a combination of the two. Wang and Xia (2016) present a detailed review of the automatic methods used to construct English and Chinese sentiment lexicons.

8.4.1 *Knowledgebase-Based Methods*

Some languages have open and complete semantic knowledgebases, such as WordNet in English. A sentiment lexicon can then be constructed by mining the relationships between words (such as synonym, antonym, hypernym, and hyponym).

On the basis of some already-known positive and negative seed words, Hu and Liu (2004) made use of synonym, antonym, and other relationships defined in WordNet to expand the seed word set and ultimately obtained a general-purpose sentiment lexicon.

The above sentiment lexicon was constructed based on adjectives only. However, sentiment words include not only adjectives but also some nouns, verbs, and adverbs. The lexicon also only provided sentiment polarity without sentiment intensity or neutral words. Some research work has been proposed to address these problems (Strapparava et al. 2004; Kim and Hovy 2004; Blair-Goldensohn et al. 2008). For example, Blair-Goldensohn et al. (2008) added a set of neutral words in the process of lexicon expansion. In addition to the relationship between words, some research work, such as (Kamps et al. 2004; Andreevskaia and Bergler 2006; Baccianella et al. 2010; Esuli and Sebastiani 2007), made use of the path of relationships between two words as well as the interpretation of words provided by the knowledgebase to construct the sentiment lexicons.

The knowledgebase-based method can quickly build a general-purpose sentiment lexicon. However, it has obvious disadvantages, such as a strong dependence on the quality of the knowledgebase and low coverage of domain-specific sentiment words.

8.4.2 Corpus-Based Methods

As we have mentioned, sentiment analysis is a domain-related task. There are large gaps between the usages of sentiment words in different domains. Even the same sentiment word used for different domains or different targets may express different sentiment polarities. For example, for "fast" in the two clauses "the computer runs fast" and "the battery powers off fast," the sentiment polarities are completely opposite.

When general-purpose sentiment lexicons are used for sentiment analysis in a specific domain, the recall rate is usually very low. To solve this problem, it is necessary to construct a domain-specific sentiment lexicon. The corpus-based sentiment lexicon construction method exactly meets this need. It can automatically extract sentiment words from the corpus and identify their sentiment polarities and intensities, and it has the characteristics of better domain adaptability and higher sentiment classification accuracy. Its methods can be further divided into conjunction methods, co-occurrence methods, and representation learning methods.

(1) Conjunction methods

The essence of conjunction methods is to infer the sentiment polarity changes of neighboring words based on conjunctions in texts. For example, the sentiment polarity before and after certain coordinate conjunctions (e.g., "also," "and") usually does not change, but the sentiment polarity before and after the adversative conjunctions (e.g., "but," "however") usually reverses. Take a look at the following review text: "*Overall good, although it is a bit expensive, and delivery is not fast. But it is still very satisfying online shopping.*"

There are four clauses in this review, separated by three conjunctions. The first is an adversative conjunction ("although") that reverses the sentiment from positive to negative ("good" → "expensive"); the second is a coordinate conjunction ("and") that holds the sentiment ("expensive" → "not fast"); and the third is another adversative conjunction ("but") that reverses the sentiment from negative to positive ("not fast" → "satisfied").

Hatzivassiloglou and McKeown (1997) summarized conjunction patterns between words in the English language and investigated the polarity relations of the words before and after the conjunction on a large-scale corpus. On this basis, they proposed a rule-based approach to identify the sentiment polarities of candidate words (adjectives in this work). Based on the corpus and a sentiment seed word set, they first collected the adjectives connected by the conjunctions, marked the polarity of the high-frequency words in them, and used a logistic regression model to determine whether the two words connected by conjunction have the same sentiment polarity or opposite sentiment polarity. Then, they made use of the clustering algorithm to generate two clusters of words and annotated sentiment polarities of the two clusters. Some subsequent research (Kanayama and Nasukawa 2006) has further improved the algorithm. Wang and Xia (2015) applied a similar approach to the automatic construction of the Chinese sentiment lexicon.

The disadvantage of conjunction methods is that they largely depend on linguistic rules, and their coverage is relatively low because they normally use adjectives as candidates.

(2) Co-occurrence methods

The principle of the co-occurrence methods is that words that have similar contexts in the text have similar semantics, including sentiment.

As described in Sect. 8.3.1, Turney (2002) estimated the PMI scores between the candidate word and the positive/negative seed words. The difference between two PMI scores is finally used to measure the SO of the candidate word:

$$\mathrm{SO}\,(t) = \mathrm{PMI}\left(w, w^{+}\right) - \mathrm{PMI}(w, w^{-}) \tag{8.7}$$

where w represents the candidate word, and w^{+} and w^{-} represent the positive and negative seed words, respectively. If the SO score is larger than a preset threshold (normally 0), the word will be categorized as positive, and vice versa.

Apart from PMI, co-occurrence can also be measured by the other metrics. For example, Turney and Littman (2003) made use of latent semantic analysis (LSA) to calculate the SO as follows:

$$\mathrm{SO\_LSA}\,(w) = \sum_{w^{+} \in \mathrm{Pwords}} \mathrm{LSA}\left(w, w^{+}\right) - \sum_{w^{-} \in \mathrm{Nwords}} \mathrm{LSA}(w, w^{-}) \tag{8.8}$$

where Pwords and Nwords represent a set of positive seed words and a set of negative seed words, respectively.

In addition to considering the co-occurrence between the candidate word and the seed word, another approach is to directly calculate the co-occurrence between the candidate word and the sentiment category of the text (usually short text such as a sentence, message, or short review). In Wang and Xia (2017), the PMI between the candidate word and the naturally labeled sentiment of the review that contains the word is used instead:

$$\mathrm{PMI}\,(t,+) = \log\frac{p\,(+|t)}{p\,(+)} \tag{8.9}$$

$$\mathrm{PMI}\,(t,-) = \log\frac{p(-|t)}{p(-)} \tag{8.10}$$

Then, the SO of the candidate word can be calculated as follows:

$$\mathrm{SO}\,(t) = \mathrm{PMI}\,(t,+) - \mathrm{PMI}(t,-) \tag{8.11}$$

In comparison, the co-occurrence methods are simple to implement and can obtain not only the sentiment polarity but also the sentiment intensity. Therefore, they are widely used in practice. However, they wholly rely on the assumption that words in similar contexts have similar sentiments, which does not hold well due to the polarity shift problem in texts (such as negation and contrast). If two words have high co-occurrence in a contrast structure (e.g., "it is good, but a bit expensive"), although they may have similar contexts, they actually have opposite sentiments.

(3) Representation learning methods

As we have introduced in Chaps. 3 and 4, many word representation learning techniques (such as NNLM, log-bilinear, word2vec, GloVe, etc.) have been proposed in the literature and successfully applied to text data mining. The existing representation learning methods assume that words with similar contexts have similar semantics. However, such methods only consider the semantic similarity while ignoring the sentiment similarity of words in a context.

To solve this problem, Tang et al. (2014b) proposed a sentiment-aware representation learning method that incorporated both semantic and sentiment information. It added sentence-level sentiment supervision based on the traditional skip-gram model in addition to language model supervision for word representation learning. Based on such sentiment-aware word embedding, a softmax regression classifier achieved better results on the SemEval 2013 sentiment classification task than were achieved with traditional features. To further construct a sentiment lexicon based on sentiment-aware word embedding, Tang et al. (2014a) collected a large set of sentiment seed words as the training dataset and then trained a word-level sentiment prediction model by using softmax regression as the classifier and sentiment-aware representation as the features. The classifier was finally used to predict the sentiment polarity for unknown words.

Vo and Zhang (2016) proposed a document-level sentiment-aware representation learning method. It established a neural network to learn a two-dimensional word embedding for each word. The two dimensions represent the word's positive and negative probabilities, and the sentiment lexicon is constructed by treating the difference between the two scores as the sentiment score of each word.

Wang and Xia (2017) proposed sentiment-aware representation learning as well as the sentiment lexicon construction method by incorporating both document and word-level sentiment supervision. In addition to document-level sentiment labels, the PMI-SO method is used to generate word-level pseudo sentiment labels. Sentiment supervision at both the document and word levels together was used to better learn sentiment-aware word representation. Based on such a representation, they examined two kinds of lexicon construction methods similar to (Tang et al. 2014b; Vo and Zhang 2016).

8.4.3 Evaluation of Sentiment Lexicons

There are two main kinds of evaluation methods for sentiment lexicons. Direct evaluation is performed by comparing the constructed lexicon with a ground truth (e.g., a general-purpose lexicon); the indirect evaluation applies the sentiment lexicon to a sentiment analysis task and evaluates its performance by using the sentiment lexicon as features.

For direct evaluation, one simple approach is to randomly extract a certain number of sentiment words in the lexicon and calculate the precision, recall, and F_1 scores in comparison with a ground truth lexicon.

The indirect evaluation needs to be performed based on a downstream sentiment analysis task such as document-level sentiment classification, which can be further divided into supervised and unsupervised sentiment classification.

In the case of supervised sentiment classification, the sentiment lexicon is typically used to build a feature template to train supervised classifiers (e.g., softmax regression, SVM, etc.) based on a sentiment classification corpus and evaluate the performance of sentiment classification. As shown in Table 8.4, Mohammad et al.

Table 8.4 Sentiment classification feature template based on sentiment lexicon (Mohammad et al. 2013)

Feature ID	Meaning
1	Total count of tokens in the text with sentiment score greater than 0
2	The sum of the sentiment scores for all tokens in the text
3	The maximal sentiment score
4	The nonzero sentiment score of the last token in the text

(2013) designed a sentiment lexicon feature template for each sentiment category (positive and negative). Tang et al. (2014a), Wang and Xia (2017) also made use of this template for sentiment lexicon evaluation.

In the case of unsupervised sentiment classification, rule-based sentiment classification methods (introduced in Sect. 8.3.1) are commonly adopted to determine the sentiments in a given corpus. The sentiment classification metrics such as accuracy or F_1 score are then used to evaluate the quality of the sentiment lexicon.

8.5 Aspect-Level Sentiment Analysis

As mentioned earlier, sentiment analysis can be carried out at multiple levels, such as the document, sentence, word, and aspect levels. The purpose of word/phrase-level sentiment analysis is to identify the sentiment polarity of a word or phrase. The purpose of document/sentence-level sentiment analysis is to identify the sentiment of a document or sentence without involving the specific opinion target. In contrast, the goal of aspect-level sentiment analysis is to extract the opinion target (also called the aspect) in the review and identify the user's sentiment toward this aspect.

For simplicity, in this section, we only focus on the aspect-level sentiment analysis task, which extracts and recognizes the aspect–sentiment pair (g, s), where g is the aspect, and s is the sentiment category, rather than trying to capture the quadruple or quintuple mentioned in Sect. 8.2.2. The aspect-level sentiment analysis mainly includes two basic tasks: aspect term extraction and aspect-based sentiment classification. We next review the methods for conducting these two tasks.

8.5.1 Aspect Term Extraction

Aspect term extraction can be viewed as a special type of information extraction problem. The aspect and sentiment often appear in pairs in a review, which is a unique characteristic of aspect term extraction in comparison with the other information extraction tasks. The methods can be mainly divided into three categories:

(1) Unsupervised learning methods

Early aspect term extraction methods were mainly based on heuristic rules. In general, domain-specific aspects are concentrated on certain nouns or noun phrases. Therefore, high-frequency nouns or noun phrases are usually explicit aspect expressions. The pioneering work of (Hu and Liu 2004) involved POS tagging to select high-frequency nouns and noun phrases as candidate aspect terms. Although this method was simple and easy to use, the extracted aspect terms usually contained considerable noise. To improve the extraction performance, Popescu and Etzioni (2007) tried to filter out nonopinion aspects in high-frequency nouns and noun phrases by calculating the PMI between the candidate aspects (e.g., "Epson 1200") and the automatically generated discriminator phrases (e.g., "is a scanner"). Ku et al. (2006) calculated the TF-IDF value of words at the document and paragraph granularity level and then judged whether the candidate word was a valid aspect by comparing its frequencies across the documents/paragraphs and inside the document/paragraph. Yu et al. (2011) used a shallow parser to extract suitable noun phrases as candidate aspects, based upon which an aspect ranking algorithm is employed to extract important aspect terms.

In addition to using the aspect's noun characteristics, some other studies have also attempted to exploit the relationship between aspect term and opinion term to assist in aspect term extraction because aspects and their corresponding sentiments usually appear in pairs in reviews. To utilize this relationship, Hu and Liu (2004) supposed that if there is no high-frequency aspect term, but there is an opinion term in a review, the noun or noun phrase closest to the opinion term will be extracted as an aspect term. Similar methods are also applied in (Blair-Goldensohn et al. 2008). Zhuang et al. (2006) used a dependency parser to identify the relationship between the aspect and the opinion terms to extract aspect terms. Qiu et al. (2011) further proposed a double-propagation algorithm based on dependency trees to extract aspect terms and opinion terms simultaneously.

(2) Traditional supervised learning methods

Kobayashi et al. (2007) made use of a dependency tree to find candidate pairs of aspect terms and opinion terms and then used the tree-structure-based classifier to classify the aspect–opinion pairs.

Because aspect term extraction is a special case of information extraction, sequence labeling models, such as hidden Markov models (HMMs) and CRFs, can be used for aspect term extraction. Jin et al. (2009) made use of an HMM framework to extract aspects and their sentiments.

Based on the linear-chain CRF model, Li et al. (2010a) proposed Skip-chain CRF, Tree CRF, and Skip-tree CRF for aspect term extraction. Jakob and Gurevych (2010) studied aspect term extraction under both single-domain and cross-domain settings based on CRF. They developed a feature template including token, POS, dependency relations, word distance, and opinion features. The feature template is summarized in Table 8.5. In the cross-domain setting, they found that the same sentiment words may have different polarities in different domains; for example, "unpredictable" in the movie review is positive, but in the automobile domain, it is negative. Moreover, the vocabulary of aspects in different domains are substantially different from each other; that is, aspect terms are closely related to their domains. This is also the main difficulty in cross-domain aspect term extraction.

In the aspect term extraction task of the SemEval 2014 competition, Chernyshevich (2014) proposed a new tagging scheme to replace the previous BIO scheme in CRF. In their scheme, FA denotes the aspect word before the head word of a noun phrase; FPA refers to the aspect words after the head word; FH denotes the head word of a noun phrase; FI denotes other nouns in the noun phrase; and "O" represents nonaspect words or symbols. The new scheme can force the head word aspect to always be labeled with the same tag FH, which helps provide more

Table 8.5 Feature templates used for aspect extraction with CRF (Jakob and Gurevych 2010)

Feature	Description	Example
Token	Current token	"food"
Part of speech	The part-of-speech tag of the current token	Noun
Dependency path	The direct dependency relationship between the current token and the opinion expression in a sentence	In the sentence "I like the food," suppose "food" is the current token and "like" is an opinion expression; there is a direct dependency relation DOBJ, between "food" and "like"
Word distance	Whether the current token is in the phrase closest to the opinion expression	Yes
Opinion sentence	Whether the current token contains an opinion expression	No

accurate aspect extraction. They also defined a rich feature template, including 15 types of features in three categories (lexical level, semantic level, and sentiment level). Inspired by the named entity recognition (NER) task, Toh and Wang (2014) introduced the head word feature, POS of the head word, and the index feature, in addition to the traditional features such as token, POS, and dependency, into the CRF feature template. In addition to these, they also added features generated from external sources including the token's syntactic categories (e.g., "noun.food") defined in WordNet, word cluster information trained using Yelp and the Amazon corpora, etc., and ultimately obtained state-of-the-art performance, ranking 1st and 2nd for the restaurant and laptop domains, respectively, at SemEval 2014.

(3) Deep learning methods

Liu et al. (2015b) proposed the first deep learning architecture for aspect term extraction based on RNN. They compared the effects of a variety of settings in RNN (Elman-type RNN, Jordan-type RNN, LSTM), bidirectionality, training for word embedding from different corpora, and fine-tuning during training. The results proved that word embedding improves the performance of both CRF and RNN models for aspect extraction, and without using any hand-crafted features, RNN outperforms feature-rich CRF-based models.

Wang et al. (2016a) proposed a recursive neural conditional random field (RNCRF) model for the coextraction of aspect and opinion terms and in reviews. As shown in Fig. 8.8, a recursive neural network is first employed by using the dependency tree of a given sentence as the input to recursively obtain the representations of each word in the tree and the inter-word dependencies. These representations are fed to the softmax layer to predict the probability of each word. A linear-chain CRF is arranged at the top to obtain the optimal labels over the entire sequence. They manually annotated the opinion terms for each review sentence based on the SemEval 2014 aspect term extraction dataset. The experiments showed

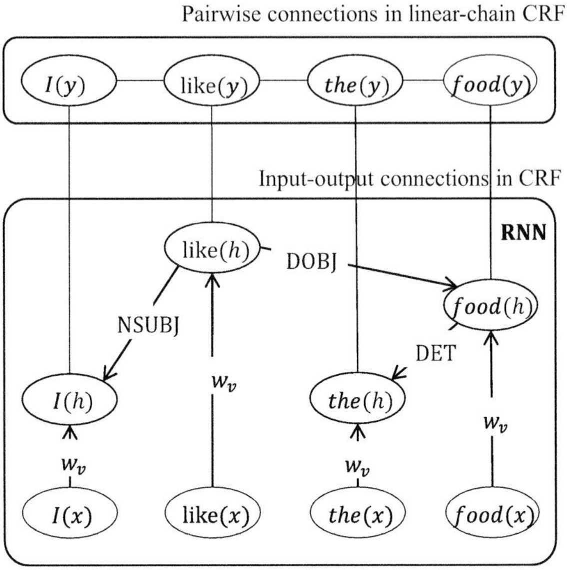

Fig. 8.8 Coextraction of aspect and opinion terms based on RNCRF (Wang et al. 2016a)

that RNCRF significantly outperforms traditional CRF methods that used many feature engineering approaches.

Li and Lam (2017) proposed a deep multitask learning framework for aspect and opinion term extraction with memory interaction. Since the aspect term and the opinion term often appear in pairs, they defined two modules, namely, Aspect-LSTM and Opinion-LSTM, for the extraction of aspect and opinion terms, respectively. The two LSTM modules exchange information through a memory interaction mechanism. Finally, they concatenated the sentence representation based on a Sentence-LSTM and the hidden states of Aspect-LSTM for aspect term extraction. The model structure is shown in Fig. 8.9.

8.5.2 Aspect-Level Sentiment Classification

The aspect-level sentiment classification refers to the task of identifying the sentiment polarity toward a given aspect in a review. The main methods of aspect-level sentiment classification include lexicon-based methods, traditional machine learning methods, and deep learning methods.

(1) Lexicon-based approach

The basic idea of the lexicon-based approach is to design opinion-oriented rules to determine the sentiment of each aspect in a sentence by taking compound expressions, syntactic trees, and the phenomenon of sentiment polarity shift, which all may affect sentiments, into consideration.

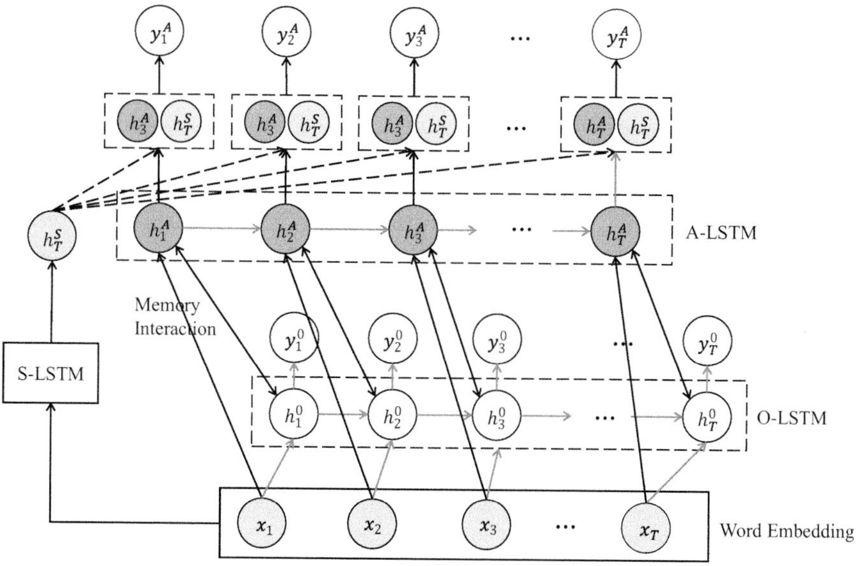

Fig. 8.9 Deep multitask learning with memory interaction for aspect and opinion term extraction (Li and Lam 2017)

Hu and Liu (2004) simply used the sum of the scores of all sentiment words in a sentence as the sentiment score of the aspect in the sentence based on a sentiment lexicon. Kim and Hovy (2004) considered not only the polarity of sentiment words but also their sentiment intensity in the opinion region and used the sum or the product of the sentiment scores in the region as the sentiment score of the aspect. Ding et al. (2008) designed detailed sentiment calculation rules for different aspects and considered the distance between sentiment words and aspect words when integrating the scores of sentiment words in the sentence.

$$\text{score}\,(f) = \sum_{w_i \in s} \frac{\text{SO}(w_i)}{\text{dist}(w_i, f)} \tag{8.12}$$

where $\text{SO}(w_i)$ represents the sentiment of word w_i, and $\text{dist}(w_i, f)$ represents the distance between w_i and the aspect word f. According to this rule, a sentiment word that is closer to the aspect word will contribute more to the sentiment score of the aspect. They also considered complex linguistic phenomena such as negation, contrast, synonym, antonym, and sentiment dependence in the context.

Although the lexicon-based approach is simple and straightforward, it relies heavily on rules and lexicons. To improve its performance, Blair-Goldensohn et al. (2008) enhanced this approach in conjunction with a supervised learning approach. Thet et al. (2010) used the sentiment-based lexical resource SentiWordNet to determine the sentiment polarity and intensity of each aspect in a review.

(2) Traditional classification methods

Jiang et al. (2011) analyzed the dependency of aspect words and other words in a review sentence and found the importance of using aspects as features for aspect-based sentiment classification. They designed a series of aspect-related features, added them to the traditional sentiment classification feature template, and significantly improved the performance of aspect-based sentiment classification.

Kiritchenko et al. (2014) designed a complex feature template that includes three types of features—shallow features, lexicon features, and syntactic features—each of which incorporates the aspect information. Based on this feature template, they employed an SVM classifier and achieved the best performance in the SemEval 2014 aspect-based sentiment classification task. To reduce the above method's reliance on syntactic analysis, Vo and Zhang (2015) divided a review into three parts: aspect term, left context, and right context. Based on the three component parts, a feature template including traditional word embedding, sentiment-specific word embedding, and lexicon features was extracted, and finally, an SVM classifier was used for aspect-based sentiment classification. Although this work used word embedding features, it was still based on a traditional machine learning framework.

(3) Deep learning method

With the development of deep learning methods for application in the field of natural language processing, some end-to-end deep learning frameworks have also been developed for aspect-based sentiment classification.

Dong et al. (2014) proposed an adaptive recursive neural network (AdaRNN) for aspect-based sentiment classification on Twitter. They made use of a dependency parser to parse tweets, performed semantic compositions in a bottom-up manner, propagated the sentiment information to the target node based on recursive neural networks, and finally fed the representation to a softmax layer for sentiment classification. They also established an aspect-based sentiment classification corpus from Twitter. They used the official API to collect tweets according to preset keywords, which were later considered aspects, and manually annotated their sentiment labels. The corpus contained 6,248 training examples and 692 test examples with 25%, 50%, and 25% positive, neutral, and negative examples, respectively. This Twitter corpus, together with the restaurant and laptop datasets from SemEval 2014, was widely used in subsequent research on aspect-based sentiment classification.

Tang et al. (2015a) proposed three LSTM-based aspect-level sentiment classification neural networks, as shown in Fig. 8.10: (1) A standard LSTM, which encodes each sentence and uses the last hidden state to represent the sentence, without special treatment of aspect terms; (2) A TD-LSTM, in which a sentence is divided into left and right parts according to the position of the aspect term, and the two subsentences are encoded by two LSTMs along opposite directions. Finally, the last hidden states of the two LSTMs are used for classification and achieve better results than the standard LSTM. (3) Based on the TD-LSTM, a TC-LSTM that appends the embedding of the aspect term to the embedding of each word and obtains further improvement.

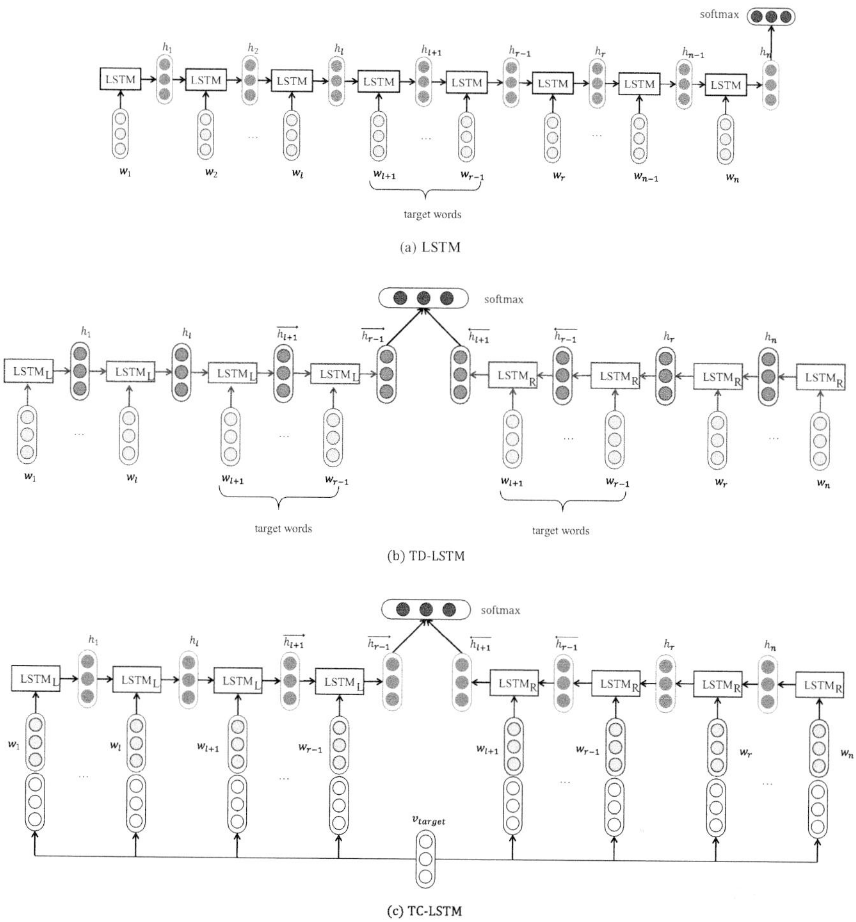

Fig. 8.10 LSTM-based aspect-level sentiment classification models (Tang et al. 2015a)

Wang et al. (2016c) proposed an attention-based LSTM model based on aspect embedding (ATAE-LSTM) for aspect-based sentiment classification. The main structure is shown in Fig. 8.11. Their model first concatenates the embeddings of each word with the aspect term for use as the input of the LSTM. The hidden state is then appended with the aspect term embedding again and operated with a standard attention mechanism. Finally, the sentence is represented by the weighted average of the hidden states and fed to a softmax layer for classification.

Tang et al. (2016) proposed a deep memory network (DMN) model. They designed a context- and position-based attention mechanism to capture the influences of words at different positions on the sentiment of the aspect, and they enhanced the model's representation ability through a multihop neural network. Chen et al. (2017a) further proposed a recurrent attention network on memory

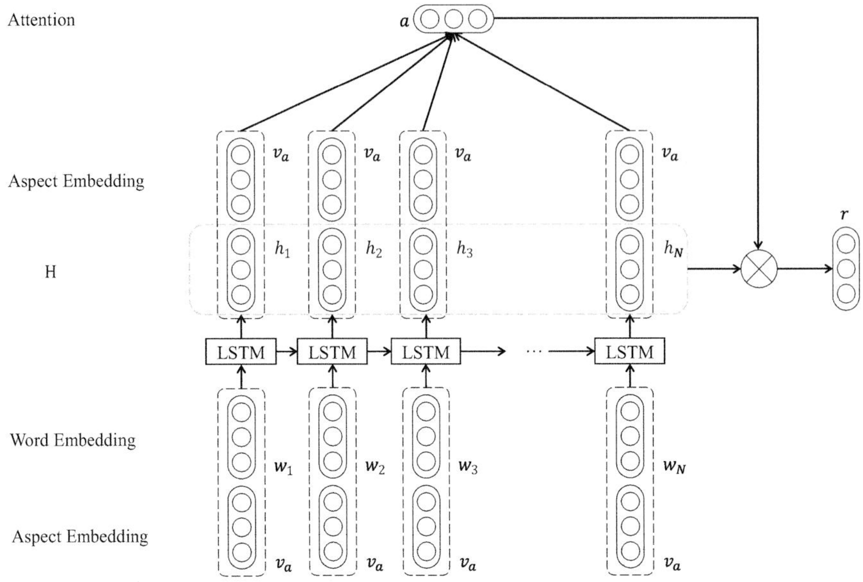

Fig. 8.11 Aspect-level sentiment classification model based on the attention mechanism (Wang et al. 2016c)

(RAM) based on DMN. In contrast to modeling the embedding of context as memory in DMN, they employed a bidirectional LSTM to encode the review sentence, used the hidden state matrix as memory, and replaced the ordinary linear transformation in DMN with RNN to construct the multilayer network.

As we introduced earlier, a review can be divided into three parts: left context, aspect term, and right context. The early work of (Vo and Zhang 2015), discussed above, learned the embeddings for each of the three parts and employed an SVM for classification. On this basis, Zhang et al. (2016b) further proposed a three-way gated neural network that first uses a bidirectional gated neural network to encode the sentence to obtain the hidden states and then divides them into three parts based on the aforementioned approach. The three parts are pooled separately, resulting in three representations. Finally, a three-way gated structure is employed to trade off the three representations to better obtain the aspect-related sentence representation. Liu and Zhang (2017) used a similar method to obtain the representations of the above three parts based on LSTM, coupled with contextualized attention. Ma et al. (2017) also used LSTM to model the aspect term and its context separately (but without distinguishing left and right contexts) and designed an interactive attention network (IAN) to obtain better aspect and context representations.

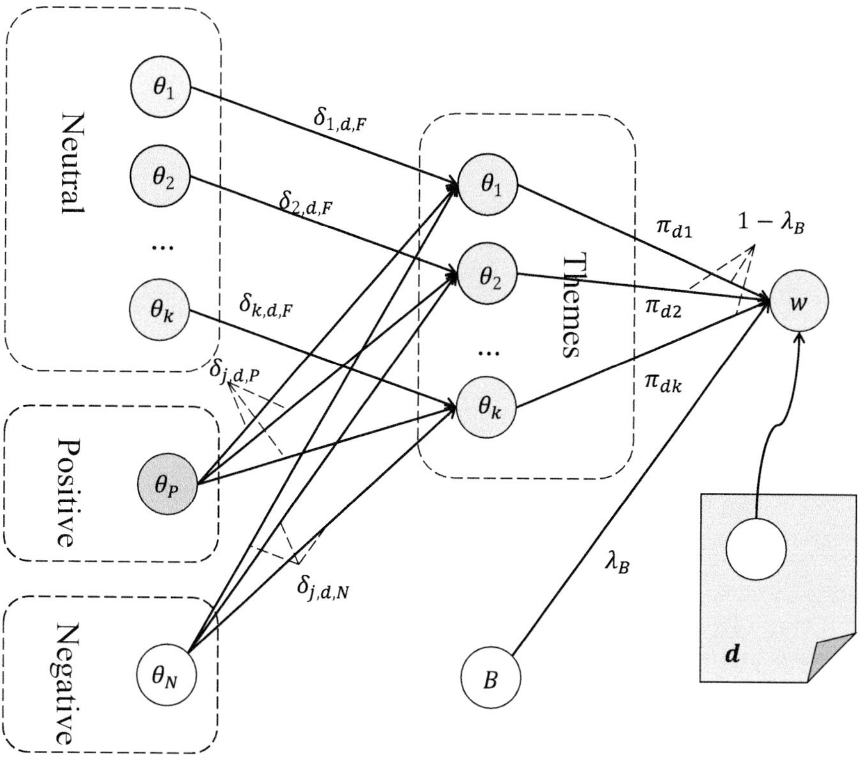

Fig. 8.12 Topic–sentiment mixture (TSM) model (Mei et al. 2007)

8.5.3 Generative Modeling of Topics and Sentiments

Because the aspects in the review are often strongly related to topics, generative topic models such as PLSA and LDA have been used by researchers to model the topics and sentiments in reviews. Mei et al. (2007) first defined the topic–sentiment analysis (TSA) problem and proposed a probabilistic mixture model called the topic–sentiment mixture (TSM) to model and extract multiple topics and sentiments in a collection of blog articles. A blog article was assumed to be generated by sampling words from a mixed model involving a background language model, a set of topic language models, and two (positive and negative) sentiment language models, as shown in Fig. 8.12. TSM can extract the topic/subtopics from blog articles, reveal the correlation of these topics and different sentiments, and further model the dynamics of each topic and its associated sentiments.

Titov and McDonald (2008) proposed a model for multiaspect sentiment analysis (MAS) that first employs the multigrain LDA model to discover the topics in reviews, then identifies the aspects associated with the topics, and finally performs sentiment analysis directed toward the aspects. The model uses aspect ratings to

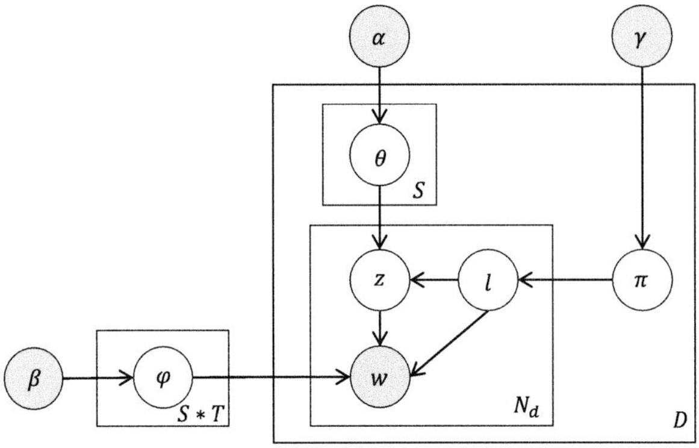

Fig. 8.13 Joint sentiment–topic (JST) model (Lin and He 2009)

discover the corresponding topics and can thus extract fragments of text discussing these aspects without the need for annotated data. It has been demonstrated that the model can discover corresponding coherent topics and achieve aspect rating accuracy comparable to that of a standard supervised model.

Lin and He (2009) also introduced sentiment information into traditional LDA by proposing a joint sentiment–topic (JST) model, as shown in Fig. 8.13. For each document, a parameter $\boldsymbol{\pi}_d$ is sampled according to $\boldsymbol{\pi}_d \sim \text{Dir}(\boldsymbol{\gamma})$; for each sentiment category of each document, a parameter $\boldsymbol{\theta}_{d,l}$ is sampled according to $\boldsymbol{\theta}_{d,l} \sim \text{Dir}(\boldsymbol{\alpha})$; for each topic under each sentiment category, a parameter $\boldsymbol{\varphi}_{l,k}$ is sampled according to $\boldsymbol{\varphi}_{l,k} \sim \text{Dir}(\boldsymbol{\beta})$. On this basis, for each word position i in a document, the sentiment label l_i, the topic z_i, and the term w_i are generated according to three categorical distributions: $l_i \sim \text{Cat}(\boldsymbol{\pi}_d)$, $z_i \sim \text{Cat}(\boldsymbol{\theta}_{d,l_i})$, and $w_i \sim \text{Cat}(\boldsymbol{\varphi}_{l_i,z_i})$, respectively. A sentiment lexicon was used as a priori information to guide topic and sentiment detection.

Jo and Oh (2011) proposed a sentence-level LDA (Sentence-LDA) model based on traditional LDA. This model added sentence-level topic modeling between the granularity of documents and words. On this basis, they proposed an aspect–sentence unification model (ASUM) similar to JST. The difference is that ASUM assumes that different words of the same sentence have the same topic and sentiment. Figure 8.14 compares the structures of standard LDA, Sentence-LDA, and ASUM.

Brody and Elhadad (2010) introduced a local topic model (Local LDA), which works at the sentence level and employs a small number of topics to automatically infer the aspects. For sentiment detection, they presented a method for automatically deriving an unsupervised seed set of positive and negative adjectives that can replace manually constructed sets.

Zhao et al. (2010) proposed a MaxEnt-LDA hybrid model to jointly discover both the aspects and the aspect-specific opinion words. The novelty of their model is the integration of a discriminative maximum entropy (MaxEnt) component with the standard generative component (LDA). The MaxEnt component allows the model

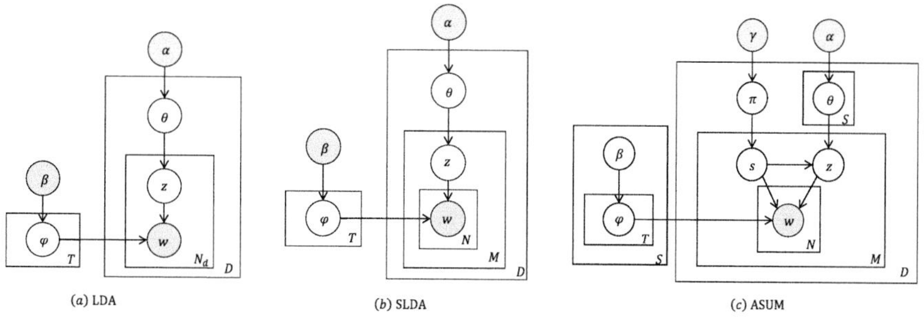

Fig. 8.14 Comparison of standard LDA, SLDA, and ASUM (Jo and Oh 2011)

to leverage arbitrary features such as POS tags to help separate aspect and opinion words.

Mukherjee and Liu (2012) proposed a semisupervised aspect–sentiment joint extraction model that allows users to provide some seed aspect words for a few aspect categories to guide the topic model so that the extracted aspects and sentiments are more in line with the user's needs.

8.6 Special Issues in Sentiment Analysis

8.6.1 Sentiment Polarity Shift

Sentiment polarity shift refers to a linguistic phenomenon wherein the sentiment in a text changes due to special linguistic structures such as negations, contrasts, intensifiers, and diminishers. These structures are also called "sentiment shifters" in (Liu 2012) and "contextual valence shifters" in (Polanyi and Zaenen 2006).

The sentiment-shifted text is often similar to the original text in text representation. For example, "I don't like this book" and "I like this book" have a high similarity when using the bag-of-words model for text representation, but their sentiment polarities are completely opposite. According to the statistics in (Li et al. 2010b), more than 60% of the sentences in product reviews contain explicit polarity shifts, which makes research on polarity shifts in sentiment analysis necessary. Liu (2015) performed a detailed analysis of different types of polarity shifts. In practical applications, negations and contrasts will reverse the polarity of sentiment, while intensifiers and diminishers will only change the intensity of sentiment. Therefore, negations and contrast were more discussed in the literature on polarity shifts in sentiment classification.

Focusing on word/phrase-level sentiment analysis, Wilson et al. (2005) began with a sentiment lexicon whose sentiment polarity was determined in advance as prior knowledge and built a supervised classifier to learn the contextual polarity of words. Based on syntactic patterns, Choi and Cardie (2008) designed a series of hand-written rules motivated by compositional semantics to address the polarity shift problem for phrase-level sentiment analysis. Nakagawa et al. (2010) proposed a semisupervised clause-level sentiment analysis model that captured the negation structures based on the interactions between nodes in a dependency tree.

In aspect-level sentiment classification, Hu and Liu (2004), Ding and Liu (2007), Ding et al. (2008) designed a set of linguistic rules, including the negation rule, the but-clause rule, the intrasentence conjunction rule, and the pseudo intrasentence conjunction rule, to match various types of sentiment shifts, such as negations, contrasts, intensifiers, and diminishers.

In document-level and sentence-level sentiment classification, the methods for addressing the polarity shift problem vary with the type of sentiment classification method. Generally, in lexicon- and rule-based sentiment classification approaches, it is relatively easy to address polarity shifts by designing a set of rules to first match the patterns of different polarity shifts and then adjust the sentiment score accordingly. For example, in the case of negation, the sentiment score of the polarity-shifted word will be reversed; in the case of intensifiers and diminishers, the sentiment score will be increased and decreased, respectively. Finally, the scores of the respective parts are accumulated to obtain the sentiment score of the entire sentence or document. Taboada et al. (2011) is a representative work on these approaches and presented detailed rules for different types of polarity shifts (negations, contrasts, intensifiers, diminishers, irrealis, etc.) in the English language.

In traditional machine learning-based approaches, the bag-of-words is a widely used model for text representation. However, it is relatively difficult to integrate polarity shift information into the BOW model.

A simple way to deal with this is simply attaching "NOT" to words in the scope of negation, so that in the text "I don't like book," the word "like" becomes a new word "like-NOT" (Das and Chen 2001, 2007). However, Pang et al. (2002) reported that this only had negligible effects on improving the sentiment classification accuracy.

There were also some attempts to model polarity shift phenomena by using more linguistic features and lexicon resources. For example, Na et al. (2004) proposed modeling negation by looking for specific POS tag patterns. Kennedy and Inkpen (2006) made use of a syntactic tree to model three types of sentiment shift (negations, intensifiers, and diminishers). Their experimental results showed that handling polarity shifts can significantly improve the performance of rule-based systems, but the improvements were very slight compared to the baselines of machine learning systems. Ikeda et al. (2008) proposed a machine learning method based on a sentiment lexicon extracted from General Inquirer to model polarity shifts for both word-wise and sentence-wise sentiment classification.

Li and Huang (2009), Li et al. (2010a) proposed a method that first classified each sentence in a text into a polarity-unshifted part and a polarity-shifted part based on certain rules and on a trained binary detector, respectively. The two components were then represented as two bags of words and finally combined for sentiment classification. Orimaye et al. (2012) proposed an inter-sentence polarity shift detection algorithm to identify consistent sentiment polarity patterns and used only the sentiment-consistent sentences for document-level sentiment classification.

Xia et al. (2016) divided polarity shifts into two cases: explicit polarity shifts and implicit polarity shifts. The former includes explicit linguistic structures such as negation and contrast; the latter mainly refers to implicit sentiment incoherence between sentences. On this basis, they proposed a rule-based method to detect explicit polarity shifts and a statistics-based method to detect implicit polarity shifts. Different processing methods were adopted to eliminate the polarity shift of the different parts, and then the different parts were integrated to obtain the sentiment of the entire document.

Xia et al. (2013b, 2015b) proposed a framework called dual sentiment analysis (DSA) to address the polarity shift problem. By making use of the sentiment classification characteristic of having two opposite class labels (i.e., positive and negative), they propose a data expansion technique by creating sentiment-reversed (antonymous) reviews for each training and testing review. The original and reversed reviews were constructed in a one-to-one correspondence and were modeled by a pair of bags-of-words (Dual BOW) models. On this basis, a dual training algorithm and a dual prediction algorithm were proposed, respectively, to use both the original and the reversed samples in pairs for training a statistical classifier and then to make predictions by considering the two sides of one review. Because of the removal of polarity shift structures during the reversed review construction process and the usage of Dual BOW text representation, the DSA framework can potentially alleviate the polarity shift problem. Xia et al. (2015a) further extended the DSA framework from the supervised learning to the semisupervised learning setting by proposing a dual-view cotraining method for semisupervised sentiment classification.

Qian et al. (2017) proposed a solution to address the sentiment polarity shift problem under the deep learning framework. A linguistically regularized LSTM was developed that encoded a sentence from right to left based on LSTM and predicted the sentiment at each position. They performed sentiment prediction at each word position by modeling the relationship between neighboring sentiment words, negation words, and intensity words as constraints, so as to learn linguistically regularized representation and ultimately improve the performance of sentiment classification.

8.6.2 Domain Adaptation

In statistical machine learning, the learning process for a specific domain typically requires a large number of annotated samples in that domain, and the training and test data are assumed to have identical distributions. Therefore, statistical machine learning usually suffers from the "domain-dependence" problem: classifiers trained on annotated samples in a certain domain (we refer to it as the source domain) usually perform well on test samples in the same domain but perform poorly on test samples in a different domain (which we refer to as the target domain), especially when the distributions of the source and target domains are significantly different.

Since the distribution of review data is highly dependent on the type of product, this "domain-dependence" problem is very common in sentiment analysis. To address this problem, "domain adaptation" (also known as "transfer learning" in the field of machine learning) has become a popular research topic in the fields related to sentiment analysis.

The goal of domain adaptation is to train an adaptive classifier for the target domain, conditioned on labeled data from the source domain, with the help of a large amount of unlabeled data (or a small amount of labeled data) from the target domain. Since it is next to impossible to annotate sufficient labeled data for each product type, domain adaptation is very important and truly beneficial for sentiment classification. For example, given some labeled reviews from the restaurant domain and unlabeled reviews from the electronics domain, the goal of domain adaptation is to directly utilize these labeled and unlabeled data to train a robust model that can effectively classify product reviews in the electronic domain.

In sentiment analysis, Aue and Gamon (2005) first explored the domain adaptation problem; they proposed using the EM algorithm to train a classifier with both the labeled samples of the source domain and the unlabeled samples of the target domain, but the results were not satisfactory. Later, Jiang and Zhai (2007) analyzed the domain adaptation problem and proposed two kinds of methods: instance adaptation and labeling adaptation. Furthermore, Pan and Yang (2009) divided existing transfer learning methods into three categories: the instance-based transfer, feature-based transfer, and parameter-based transfer.

The feature-based approaches aim at finding "good" feature representations for the target domain based on the labeled data of the source domain and a large amount of unlabeled data (or a small amount of labeled data) of the target domain and then using the new feature representations to train a domain adaptive classifier. Structure correspondence learning (SCL), proposed by Blitzer et al. (2007), is a representative work in this field. Blitzer et al. (2007) first defined a set of pivot features, usually domain-invariant sentiment words, and used the remaining words as nonpivot features. They then learned the mapping matrix between the two feature spaces, subsequently using singular value decomposition (SVD) to obtain the principal component subspace of the mapping matrix. Finally, they projected the nonpivot features to the mapping matrix subspace to obtain new feature representations for sentiment classification; this approach has been demonstrated to achieve good performance. Inspired by SCL, a series of similar approaches have been proposed. In these, the basic idea is to use the domain-invariant features of the source and target domains as a bridge to associate each domain's domain-specific features based on their co-occurrence with the domain-invariant features. Based on the co-occurrence among features, these approaches typically utilize different subspace methods to map the features of the source domain and the target domain to the same subspace and finally perform classification in the new subspace. For example, Pan et al. (2010b) and Pan et al. (2010a) proposed two transfer learning algorithms: transfer component analysis (TCA) and spectral feature alignment (SFA). These two methods employ the concepts of principal component analysis and spectral clustering to construct the association between the source domain and the target domain and learn the effective feature representations for the target domain. Xia and Zong (2011) believed that features with different types of POS tags have different domain independence properties. For example, adjectives and adverbs in two different domains tend to be similar, while nouns tend to differ greatly between two different domains. According to this characteristic, they divided the original feature space into several subsets and trained base classifiers with each feature subset in the source domain. Finally, they adopted ensemble learning to learn appropriate weights for each component, which derives a new labeling function for the target domain.

With the recent trend of deep learning in NLP, a series of neural transfer learning algorithms have been proposed. Similar to SCL, the basic approach underlying these algorithms is to first construct some domain-independent auxiliary tasks as a bridge to associate the domain-specific features of each domain and then optimize the auxiliary tasks via neural networks to map all the domain-specific features to the same subspace for sentiment classification. For example, Yu and Jiang (2016) designed two auxiliary tasks that used nonpivot features to predict the occurrence of the positive pivot feature and the negative pivot feature. With the two auxiliary tasks, they proposed using the nonpivot features and the original features as an input pair and employed a bichannel convolutional neural network (Bichannel CNN) to jointly train the auxiliary task and the main sentiment classification task. In this way, the nonpivot features and the original features are mapped to two target-domain-aware subspaces, which are combined for the final sentiment classification on the target domain. Based on this work, Ding et al. (2017) and Li et al. (2017c) respectively used traditional rule-based predictions and adversarial domain discriminators as auxiliary tasks and then used LSTM to jointly train the auxiliary tasks and the main task, respectively. Subsequently, Li et al. (2018) proposed combining the auxiliary tasks in (Yu and Jiang 2016) and (Li et al. 2017c) and used the hierarchical attention network for multitask learning, achieving the best results among the current feature-based approaches on several benchmark datasets.

The parameter-based transfer assumes that the model parameters of the source domain and the target domain have the same prior distribution, and the shared priors can be used as constraints for model optimization to transfer the classification knowledge from the source domain to the target domain. Xue et al. (2008) proposed a PLSA-based domain adaptation method, where they extended the traditional PLSA to exploit the shared topics between the two domains with topic bridges (referred to as topic-bridged PLSA). This method assumes that the source and target domains share the prior distribution $p(z|w)$ in the graphical model, and it uses the EM algorithm to optimize the model based on the shared prior distribution, which was shown to be effective in cross-domain text classification. Li et al. (2009b) extended the traditional nonnegative matrix decomposition to domain adaptation. Specifically, they proposed performing nonnegative matrix decomposition on both the source and target domains under the constraint that the two domains share the same $p(z|c)$ matrix; this transfers the shared sentiment knowledge from the source domain to the target domain. Their key idea is similar to the topic-bridged PLSA model.

The instance-based transfer approaches aim to assign appropriate weights to each source-labeled sample based on its similarity to the distribution of the target domain and to train an adaptive classifier by importance sampling on the training samples in the source domain. This problem belongs to the sample selection bias problem (Zadrozny 2004) in machine learning, where the key challenge is estimating the density ratio, i.e., the ratio of the probabilities of the samples occurring in the source domain and target domain. The density ratio is used to measure the probability of each source sample occurring in the target domain (also known as the weight for importance sampling). However, the estimation of the density ratio is challenging. Although several theoretical approaches have been proposed in the field of machine learning (Shimodaira 2000; Huang et al. 2007; Sugiyama et al. 2008; Bickel et al. 2009), most of them failed to achieve satisfactory performance in many NLP tasks, including sentiment analysis. To address this problem, Xia et al. (2013a) proposed a method based on positive-unlabeled learning (PU Learning) to calculate the similarity between each source sample and the target distribution. They identified the samples in the source and the target domains as the U-set and the P-set, respectively, and identified some reliable nontarget domain samples from the source domain as the N-set. Based on the EM algorithm, they further established a semisupervised classifier to predict the probability of each source sample belonging to the target domain and used this probability as the similarity between the source sample and the target distribution. The probabilities after calibration were then used for importance sampling and achieved good performance in cross-domain sentiment classification. On this basis, Xia et al. (2014) further proposed a logistic approximation approach for jointly calculating the similarity between each source sample and the target distribution and estimating the density ratio for importance sampling. More recently, Xia et al. (2018) analyzed the bias-variance dilemma in instance-transfer approaches and proposed the idea of controlling sample weight variance while overcoming sample selection bias, which largely improved the stability of the instance-based approaches.

8.7 Further Reading

Pang and Lee (2008), Liu (2012), Liu (2015) provided comprehensive reviews of sentiment analysis and opinion mining.

In addition to the traditional sentiment analysis and opinion mining tasks, some generalized sentiment analysis tasks, such as emotion classification and stance classification, have been proposed in this field.

Emotion classification can be viewed as an extension of sentiment classification. It aims to identify people's emotions in multiple dimensions from the perspective of human psychology (Ekman et al. 1972; Plutchik and Kellerman 1986). In text analysis, emotions are usually divided into six categories: love, joy, surprise, anger, sadness, and fear. The main techniques in emotion classification include lexicon-based methods, traditional machine learning methods, and deep learning methods.

Based on emotion classification, a new task called emotion cause analysis has been recently proposed, with the goal of identifying and extracting the corresponding causes of emotions in texts (Gui et al. 2016; Ding et al. 2019; Xia et al. 2019; Xia and Ding 2019; Ding et al. 2020).

Stance classification is another new area of sentiment analysis that has emerged in recent years. Sentiment classification aims to identify the sentiment polarity (positive, negative, or neutral) in texts, while stance classification focuses on detecting people's stance (support, deny, query, comment) toward a given target (e.g., "homosexual love"). In aspect-level sentiment classification, the opinion target is usually a fine-grained explicit aspect in reviews, while in stance classification, the target is normally a topic or an event. SemEval 2016 launched an evaluation of stance detection in tweets (Mohammad et al. 2016). Similar evaluations based on a Chinese microblog were organized by NLPCC 2016.

Exercises

8.1 What is the difference between sentiment classification and traditional topic-based text classification tasks?

8.2 What is the difference between document-level sentiment classification and sentence-level sentiment classification? What are their respective advantages and disadvantages?

8.3 Please analyze the similarities and differences between the PMI method shown in Formulas (8.12) and (8.13) and the mutual information feature selection method introduced in Sect. 4.3.1 of this book.

8.4 Please analyze the differences between the two hierarchical sentiment classification methods mentioned in Sect. 8.3.2 and Sect. 8.3.3, respectively.

8.5 What is the difference between sentence sentiment classification using a recurrent neural network and that using a recursive neural network? What are the advantages and disadvantages of each method?

8.6 Please point out the advantages and disadvantages of the three corpus-based sentiment dictionary construction methods described in Sect. 8.4.2.

8.7 What are the advantages and disadvantages of treating aspect extraction and aspect-based sentiment classification as two separate tasks in aspect-level sentiment analysis? Can you design a joint end-to-end model of aspect extraction and aspect-based sentiment classification?

Chapter 9
Topic Detection and Tracking

9.1 History of Topic Detection and Tracking

Traditional TDT technology was established and developed in an evaluation-driven manner. The original motivation for TDT research was proposed by the Defense Advanced Research Projects Agency (DARPA) in 1996. Their aim was to explore a new technology to automatically detect and track topics in news data streams without human intervention.

In 1997, researchers from DARPA, Carnegie Mellon University (CMU), and University of Massachusetts (UMass) initiated preliminary studies of TDT, later called TDT1997 or TDT Pilot. They focused on how to find topic-related information from data streams (text or voice) and included two parts: enabling the system to automatically locate the boundaries of two events by searching for fragments consistent with the intrinsic theme and detecting the emergence of new events and the reproduction of old events. They carried out basic research (Allan et al. 1998a) and established a TDT pilot corpus.[1] This corpus includes nearly 16,000 stories, from July 1, 1994, to June 30, 1995, taken half from Reuters newswire and half from CNN broadcast news transcripts. For the evaluation of TDT performance, they proposed the metrics of miss and false alarm rates and used a detection error tradeoff (DET) plot to visually display the errors in the TDT system.

Starting in 1998, the National Institute of Standards and Technology (NIST), sponsored by DARPA, hosted the annual TDT evaluation conference, which was one of two conferences in the Translingual Information Detection, Extraction and Summarization (TIDES) project (the other is the Text REtrieval Conference, TREC). Many famous universities, companies, and research institutes, such as IBM Watson Research Center, BBN Technologies Company, CMU, UMass, University of Pennsylvania, University of Maryland, and Dragon Systems Company, actively participated in the conference. TDT1998 held the first public TDT evaluation, with

[1] https://catalog.ldc.upenn.edu/LDC98T25.

© Tsinghua University Press 2021

C. Zong et al., *Text Data Mining*, https://doi.org/10.1007/978-981-16-0100-2_9

the evaluation tasks including news story segmentation, topic detection, and topic tracking, and for the first time, it introduced the Chinese corpus. TDT1999 added two new tasks: first story detection (FSD) and link detection (LD).

TDT2002, the 5th TDT conference, held in autumn 2002, further enriched the corpus by incorporating the Arabic corpus. At the same time, new technologies such as text filtering, speech recognition, machine translation, and text segmentation were added to the research content of TDT.

TDT2004 canceled the task of news story segmentation because most instances of practical application were easily separable. Meanwhile, two new tasks, supervised adaptive topic tracking and hierarchical topic detection, were added. The TDT conference was held for 7 consecutive years, with the last one being TDT2004. The TDT corpus is still open to the public, and researchers can obtain the TDT tasks and corpus through the website of the Linguistic Data Consortium[2] (LDC).

In recent years, social media platforms, such as Twitter, Facebook, Weibo, and WeChat, have developed into important channels for discussions on current affairs, information exchanges, and the expression of opinions. A large number of users participate in discussions of events, persons, products, and other content on these platforms and generate a large amount of text data, which becomes a mirror reflecting society. The research on TDT in social media has even more important practical significance. However, at the same time, texts on social media platforms raise new problems and challenges to TDT research because of their properties, such as short contexts, rich forms, dynamic topics, massive volumes, and a large number of nonstandard language phenomena.

Allan (2012) and Yu et al. (2007) summarized the studies of traditional TDT. In the following sections, we first introduce the terminologies and tasks in TDT and then review the traditional technologies considering four aspects: text representation, text similarity, topic detection, and topic tracking. Finally, following the extension from traditional media to social media, we introduce the research of TDT in social media.

9.2 Terminology and Task Definition

9.2.1 Terminology

The goal of TDT is to automatically discover topics from text data streams and link topic-related content together. It involves concepts such as event, topic, story, and subject.

Event: In a TDT study, an event refers to an activity or a phenomenon that occurs at a specific time and place and is associated with certain

[2] https://www.ldc.upenn.edu/.

actions or conditions. Usually, an event is a story or a series of stories, which consist of detailed descriptions of the cause, time, place, process, and result of the event. For instance, "Trump defeated Hillary in the November 8, 2016, presidential election and became the 45th President of the United States" is an event in TDT. It has specific attributes such as a time, place, and person.

Topic: Topic was defined as an event in the original TDT Pilot study, but since TDT1998, it has been given a broader meaning that includes not only the initial event but also the subsequent events and other events directly related to it. In other words, a topic can be viewed as a core event together with its direct-relevant events, making it a collection of related stories about one event. Assume that the "512 Wenchuan earthquake" is a topic and "a strong earthquake of magnitude 8.2 occurred in Wenchuan, China, on May 12, 2008" is the core event of this topic. Subsequent events, such as earthquake rescue and post-earthquake reconstruction, are also part of the topic because they are directly related to the core event. The research on TDT originated from early event detection and tracking (EDT). However, compared with EDT, the object of TDT extends from events occurring at specific times and places to topics with more relevant extensions.

Subject: The subject in TDT is a summary of one kind of event or topic; it covers a group of similar events but does not involve any specific events. It therefore has a wider meaning than a topic in TDT. For instance, "earthquake disaster" is a subject, and the "512 Wenchuan earthquake" is a specific topic under that subject. Note that the concept of a "topic" in the topic model is different from that in TDT. Specifically, "topic" and "subject" in TDT are concepts describing events, representing a series for a specific event and a group of similar events, respectively, while "topic" in a topic model represents the underlying semantics of words in the text.

Story: A story in TDT denotes an article in a newswire or a piece of broadcast news that is composed of two or more statements of independent events.

9.2.2 *Task*

NIST divides TDT into the following five basic tasks.

(1) Story Segmentation

The purpose of story segmentation (SS) is to discover all topics and their boundaries in a news story and divide the story into multiple substories with a complete structure and independent topics, as shown in Fig. 9.1. For example, given a piece of broadcast news that includes multiple topics such as politics, sports events, finance, and economics, an SS system needs to divide the broadcast into

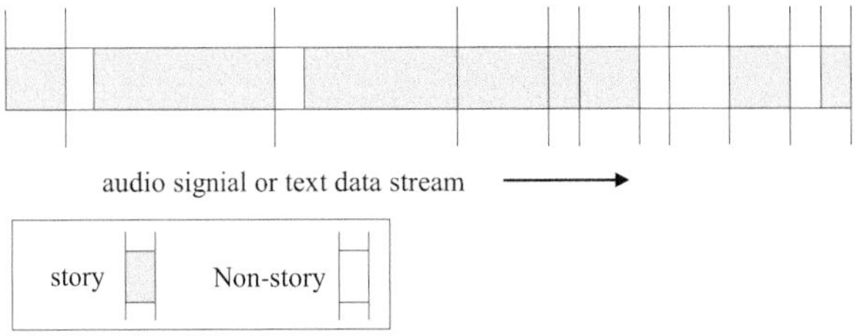

Fig. 9.1 Story segmentation task in TDT

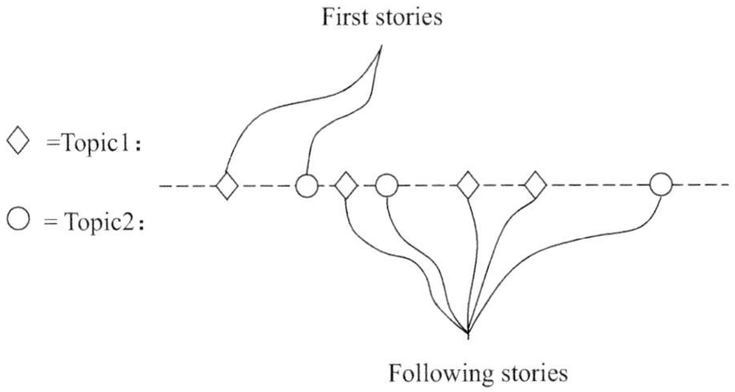

Fig. 9.2 First story detection task in TDT

segments of different topics. SS is designed mainly for news broadcasting, which contains two kinds of data streams: the audio signal and the text data stream transcribed from the audio signal. TDT2004 removed this task because most of the instances can be easily segmented in practice.

(2) First Story Detection

The FSD task aims to automatically detect the first discussion of a given topic from a chronological stream of news, as shown in Fig. 9.2. This task requires judging whether a new topic is discussed in each story. It is therefore considered to be the basis for topic detection, and it is called transparent testing of topic detection. TDT2004 renamed FSD to new event detection (NED).

(3) Topic Detection

The goal of the TD task is to detect topics in the news data streams without providing prior knowledge about any topics, as shown in Fig. 9.3. The output of FSD is one story, while the output of TD is a collection of stories that discuss the same topic. The difficulty of TD is the absence of prior knowledge of the topic; it means that the TD system must be independent of a certain topic but apply to any topic.

Although most of the stories refer to only one topic, there are also some stories that involve multiple topics organized in a hierarchical structure. In response to this problem, TDT2004 defined a hierarchical topic detection (HTD) task that

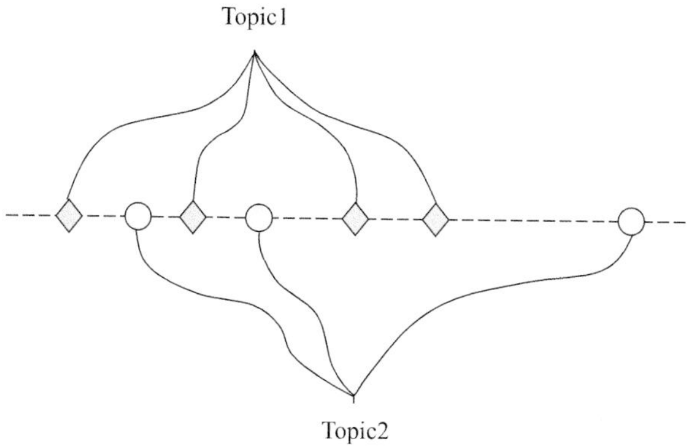

Fig. 9.3 Topic detection task in TDT

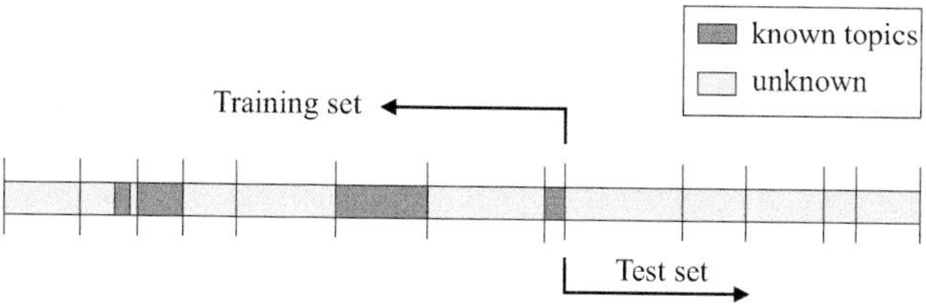

Fig. 9.4 Topic tracking task in TDT

transformed the organization of topics from a parallel relationship in FSD and TD to a hierarchical structure.

(4) Topic Tracking

The goal of topic tracking (TT) is to track subsequent stories of known topics, that is, to detect the related follow-up stories in the data streams given one or more stories associated with a topic, as shown in Fig. 9.4. The topic is denoted by several related stories rather than by a query (the NIST evaluation usually provided one to four stories for each topic).

(5) Link Detection

The goal of link detection (LD) is to judge whether two stories belong to the same topic, as shown in Fig. 9.5. Similar to TD, no prior knowledge is provided; the LD system establishes a topic relevance detection model that does not depend on stories from one topic as the reference. LD is often considered a core module in other TDT tasks (e.g., topic detection and topic tracking) rather than an independent task. A good link detection system can improve the performance of other TDT tasks.

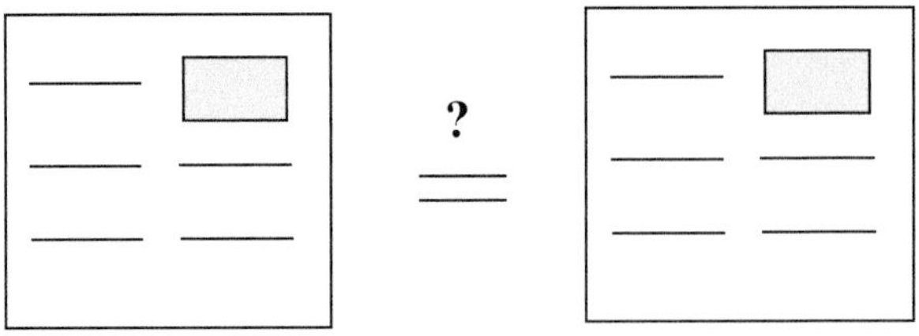

Fig. 9.5 Link detection in TDT

Table 9.1 Basic text mining techniques involved in TDT

Main task	Basic technique
Representation of topic/story	Text representation
Link detection	Text similarity computation
Topic detection	Text clustering
Topic tracking	Text classification

In general, TDT studies the relationship between stories and topics. It mainly solves the following technical problems: (a) representation of topics and stories; (b) similarity between topics and stories; (c) clustering of topics and stories; and (d) classification of topics and stories. Table 9.1 shows the text mining techniques involved in different TDT tasks.

9.3 Story/Topic Representation and Similarity Computation

In Chaps. 3, 5, and 6, we introduced the standard text representation and text similarity computation methods in detail. This section will briefly introduce the methods of text representation and similarity computation in TDT. Text preprocessing techniques, including stemming, lemmatization, and stop word filtering, are usually employed first. Then, a vector space model (VSM) or language model (LM) is often applied for text representation.

VSM is one of the most commonly used text representation models in TDT. It regards a story as a document, ignores the order of terms, and uses a vector to represent this document. TF-IDF and its variants are usually used as the term-weighting scheme. Allan et al. (2000) pointed out the limit of text similarity computation in VSM. Many researchers have proposed improving the representation ability of VSMs based on information extraction and feature engineering. For example, the information of the name entity (Yang and Liu 1999; Kumaran and Allan 2004, 2005), the 4Ws (who, what, when, where) (Kumaran and Allan 2004), and semantic concepts (Kumaran and Allan 2004) have been added to vector space to improve the performance.

There are three kinds of similarity measures between stories or topics: the similarity between two stories, the similarity between a story and a topic, and the similarity between two topics; these correspond to the content in Sect. 6.2.

Identifying the similarity between two stories is also called link detection in TDT. The goal of link detection is to determine whether two randomly selected stories discuss the same topic. A basic approach is as follows: first, each story is represented as a vector based on VSM, and then the similarity is calculated by the cosine distance of the two vectors. Finally, a preset threshold is used to determine whether the corresponding stories are relevant to each other. If the cosine similarity is greater than the threshold, the two stories are relevant; otherwise, they are irrelevant. The similarities between stories can also be measured by traditional Euclidean distance, the Pierson correlation coefficient, and other similarity measures.

The correlation between a story and a topic can be transformed into the problem of computing the similarities between the story and all the stories that constitute the topic, in which the key problem is link detection for a pair of stories. In some work, a topic is represented as a single model (e.g., using the centroid vector of all stories under the topic to represent the topic), thereby converting the similarity between a story and a topic into that between the story and the centroid vector, where the key technique is still link detection.

Researchers at the University of Massachusetts (UMass) studied a variety of similarity computation methods, including cosine distance, weighted sum, language models, and Kullback-Leibler divergence. Experiments on the TDT3 corpus showed that cosine distance performs best in link detection (Allan et al. 2000).

Another line of methods applies language models to story representation and link detection. The language model has been widely used in text mining as a generative probability model for representing natural language. Let the random variables C and S represent a topic and a story, respectively. According to Bayes' theorem, the posterior probability $p(C|S)$ of the topic C conditioned on the story S is proportional to the product of the prior probability $p(C)$ and the conditional probability $p(S|C)$, that is

$$p\left(C|S\right) = \frac{p(C)p(S|C)}{p(S)} \propto p\left(C\right) p\left(S|C\right) \tag{9.1}$$

Assuming that the terms in story S are independent of each other given the topic, we obtain

$$p\left(C|S\right) \propto p\left(C\right) \prod_i p\left(t_i|C\right) \tag{9.2}$$

where $p(t_i|C)$ is the probability that term t_i appears in topic C.

Language modeling furthermore provides a method for computing the similarity between a story and a topic (or two stories). In the unigram language model, a subset

C_j with respect to the j-th topic can be represented as a multinomial distribution as follows:

$$p\left(S|C_j\right) = \prod_i p(t_i|C_j) \tag{9.3}$$

where t_i denotes the ith term in the vocabulary. Based on maximum likelihood estimation, we can estimate $p(t_i|C_j)$ as the term frequency of t_i in C_j divided by the total number of terms in C_j.

In practice, the data sparsity problem may cause $p(t_i|C_j)$ to equal zero. To avoid this problem, we can use the smoothing technique to estimate $p(t_i|C_j)$

$$p_{\text{smooth}}\left(t_i|C_j\right) = \lambda p\left(t_i|C_j\right) + (1-\lambda)\, p\left(t_i|G\right) \tag{9.4}$$

where $p(t_i|G)$ is an estimated probability of word t_i in a general corpus G. Since the texts in TDT appear in a time series, and new texts may have words that did not appear in previous documents, as a kind of prior knowledge, estimation based on a general corpus is reasonable.

The problem determining which topic is most likely to generate the given story S can be described as

$$\arg\max_j \frac{p(S|C_j)}{p(S)} = \arg\max_j \prod_i \frac{p(t_i|C_j)}{p(t_i)} = \arg\max_j \log \prod_i \frac{p(t_i|C_j)}{p(t_i)} \tag{9.5}$$

Therefore, $D\left(S, C_j\right) = \sum_i \log \frac{p(t_i|C_j)}{p(t_i)}$ can be defined as the similarity between story S and topic C_j.

If a story is regarded as a distribution of words, then the similarity between a story S and a topic C can be measured by the similarity between two distributions, e.g., Kullback-Leibler divergence:

$$D_{\text{KL}}\left(C \| S\right) = -\sum_i p\left(t_i|C\right) \log \frac{p\left(t_i|S\right)}{p\left(t_i|C\right)} \tag{9.6}$$

Moreover, if two stories S_a and S_b to be compared are regarded as two multinomial distributions of terms, the Kullback-Leibler (KL) divergence can also be used for LD. Similarly, KL divergence can also measure the similarity between two topics. These techniques have been applied in Lavrenko and Croft (2001) and Leek et al. (2002).

On the basis of story/topic representation and similarity computation, most TDT tasks, such as topic detection and topic tracking, can be formulized as clustering or classification problems.

9.4 Topic Detection

The purpose of topic detection is to capture new (i.e., previously undefined) topics from a continuous stream of stories. The topic information, such as time, content, and number of stories, is unknown in advance, and there are also no annotated data for supervised learning. Therefore, topic detection is an unsupervised learning task and usually considered to be a clustering problem. Therefore, most topic detection algorithms can be regarded as a kind of modification or extension to standard text clustering algorithms. The standard clustering algorithms take the whole dataset as the input, while the input of topic detection is a continuous data stream of stories with a clear temporal relationship. The topics in the data stream also tend to change dynamically. These issues need to be addressed when using traditional clustering methods for topic detection.

Topic detection can be divided into two main types: online topic detection and retrospective topic detection. The input of online topic detection is a real-time story data stream, and thus subsequent stories do not yet exist. When a new story appears, the system is required to make a real-time decision on whether the story is a new topic. The input of retrospective topic detection is the whole corpus, containing all stories over time. Retrospective topic detection requires the system to decide for each story which topic it belongs to in an offline manner and to divide the whole corpus into several topic clusters accordingly. In comparison, the focus of online topic detection is to detect new topics from real-time data streams, while the purpose of retrospective topic detection is to discover previously unmarked news topics from existing stories.

In the following, we will describe the two topic detection tasks separately.

9.4.1 Online Topic Detection

Online topic detection aims to detect new topics from real-time stories. Since the information for new topics is unknown beforehand, it cannot be retrieved by a certain query. In addition, the task requires that the system make real-time decisions as soon as each story appears. For these reasons, incremental clustering algorithms are usually employed for online topic detection.

One simple method is based on single-pass clustering. The algorithm processes the input stories sequentially and represents each story based on a VSM. The model uses words (or phrases) as terms and TF-IDF (or its variants) as the term-weighting scheme to represent each story. Then, the similarities between the new story and all existing topics are computed. The similarity between a story and a topic is usually transformed into the similarity between the story and the centroid vector of the topic. If the similarity is higher than a preset merge-split threshold, the story will be classified into the most similar cluster (a cluster represents a topic); otherwise, the story will establish a new cluster. The above process is repeated until all the stories in the data stream have been processed. This algorithm ultimately forms a set of flat clusters, where the number of clusters depends on the merge-split threshold. More details of the single-pass clustering algorithm can be found in Sect. 6.3.2.

In the early research into TDT, researchers at UMass and CMU adopted the single-pass clustering method (Allan et al. 1998b; Yang et al. 1998). To make the algorithm better suited to real-time data streams, they made some modifications to the text representation and similarity computations.

Specifically, Allan et al. (1998b) represented the content of a story as a query and compared it to all previous queries. If a new story triggers an existing query, it is assumed that the story discusses the topic corresponding to the triggered query. Otherwise, the story is considered to contain a new topic.

Assume that q is a query and denoted as a vector over a set of terms. Based on the term set, a document is represented as a representation vector $\boldsymbol{d}$. The correlation between a query $\boldsymbol{q}$ and a story $\boldsymbol{d}$ is defined as

$$\mathrm{eval}\left(\boldsymbol{q}, \boldsymbol{d}\right) = \frac{\sum_{i=1}^{N} w_i \cdot d_i}{\sum\limits_{i=1}^{N} w_i} \tag{9.7}$$

where w_i represents the relative weight of a query term q_i and d_i is the appearance of term q_i in the story.

Because the future documents (i.e., stories) are unknown, the inverse document frequency (IDF) is estimated based on an auxiliary corpus c (which should belong to the same domain):

$$\mathrm{idf}_i = \frac{\log \frac{|c|+0.5}{\mathrm{df}_i}}{|c|+1} \tag{9.8}$$

where df_i represents the document frequency of q_i in corpus c and $|c|$ is the number of documents contained in corpus c. Meanwhile, the average term frequency is calculated as

$$\mathrm{tf}_i = \frac{t_i}{t_i + 0.5 + 1.5 \cdot \frac{\mathrm{dl}}{\mathrm{avg\_dl}}} \tag{9.9}$$

where t_i denotes the frequency of q_i in $\boldsymbol{d}$, dl is the length of $\boldsymbol{d}$, and avg_dl is the average length of all documents in c. On this basis, they set the weight of q_i as

$$\mathrm{tw}_i = 0.4 + 0.6 \cdot \mathrm{tf}_i \cdot \mathrm{idf}_i \tag{9.10}$$

In addition, the features in query $\boldsymbol{q}$ are dynamic. Each time a new story appears, the top n high-frequency words of all existing documents in the data stream are selected to construct the new term set. Thus, all query representations in the past need to be updated. The corresponding weight of q_i is the average value of tf_i in all existing stories.

Many studies have observed that documents that appear more closely in time in the data stream are more likely to discuss the same topic; therefore, using the timing of news stories may improve NED performance. Based on this idea, a time penalty was added to the threshold model. When the jth document in the data stream is compared with the ith query ($i < j$), $j - i$ is introduced to the threshold as a time penalty:

$$\theta\left(\boldsymbol{q}^{(i)}, \boldsymbol{d}^{(j)}\right) = 0.4 + p \cdot \left(\mathrm{eval}\left(\boldsymbol{q}^{(i)}, \boldsymbol{d}^{(j)}\right) - 0.4\right) + \mathrm{tp} \cdot (j - i) \tag{9.11}$$

where $\mathrm{eval}(\boldsymbol{q}^{(i)}, \boldsymbol{d}^{(i)})$ is the initial threshold of query $\boldsymbol{q}^{(i)}$, p is the weight of the initial threshold, and tp is the weight of the time penalty.

As mentioned in Sect. 6.3.2, the single-pass clustering algorithm is very sensitive to the order of the input sequence. Once the order changes, the clustering results may vary greatly. However, in TDT, the order of the input stories is fixed, which makes single-pass clustering highly suitable for TDT. Meanwhile, single-pass clustering has its advantage for real-time large-scale topic detection because it is simple and fast and supports online operations. The aforementioned work mainly involves three aspects of improvement upon standard single-pass clustering: (1) establish a better story representation, (2) find a more reasonable similarity computation method, and (3) make full use of the time information in the data stream.

9.4.2 Retrospective Topic Detection

The main goal of retrospective topic detection (GTD) is to review all news stories that have happened in the past and detect topics from them.

To address this task, the researchers at CMU proposed a hierarchical clustering algorithm based on group average clustering (Allan et al. 1998a; Yang et al. 1998), which has since been widely used in retrospective detection. This method adopts a divide-and-conquer strategy to hierarchical clustering: it divides the ordered story stream into several averaged buckets, adopts a bottom-up hierarchical clustering in each bucket, and then aggregates the more proximate clusters into a new cluster. Through repeated iterations, a topic cluster structure with a hierarchical relationship can ultimately be obtained.

Subsection 6.2.3 has already introduced bottom-up hierarchical clustering in detail. The basic idea is to initially treat each example as a separate cluster and then repeatedly merge the two most similar clusters until all examples have been merged into one cluster.

Finally, the algorithm constructs a hierarchical clustering dendrogram. The top level of the dendrogram represents a coarse-grained topic partition, and the lower level represents a more fine-grained topic partition. The time complexity of the algorithm is $O(mn)$, where n is the number of stories in the corpus and m is the size of the bucket. The disadvantage of the algorithm is that it is only suitable for retrospective topic detection and cannot be applied to online topic detection.

9.5 Topic Tracking

The goal of topic tracking is to detect follow-up related stories from the news data stream given a small number of stories related to the topic as a priori knowledge.

On the one hand, topic tracking is related to information filtering in the information retrieval field. We can thereby perform topic tracking based on the information filtering techniques. The basic approach in topic tracking is to establish a query filter that takes a small number of stories to be tracked as positive examples, where the other stories are the negative examples. We then compute the similarity between the query and each subsequent story and finally determine whether the story matches the tracking topic by comparing the similarity score to a preset threshold. There are normally two ways to build a query filter in practice. The first focuses on how to better represent the topics to be tracked based on VSM, including establishing queries based on relevance feedback, extracting features based on shallow parsing, and attempting different feature weighting methods. The other is based on language modeling, which usually requires a large-scale background corpus.

On the other hand, topic tracking can also be viewed as two kinds of text classification tasks. Stories are categorized into two classes: the positive class denotes the relevance to the topic, and the negative class denotes irrelevance to the topic. A training set is constructed based on a small number of positive stories and a large number of negative stories, and a linear classifier is trained to predict the category of new stories.

CMU is the representative for research institutes using the k-NN classifier for topic tracking. Their algorithm incrementally builds a training set comprising positive and negative stories. When a new story appears, the similarities between it and each example in the training set are calculated. After comparing the similarity with a preset threshold, the new story is first classified as positive or negative. Then, the nearest k training examples are assessed to determine which topic the story belongs to. Although the k-NN method is simple and straightforward, the class imbalance problem (i.e., the number of negative samples is much higher than that of positive samples) makes it difficult to find a reasonable threshold for the algorithm. One improvement is a k-NN model based on positive and negative examples: the former k-NN is used to compute the similarity between a new story and positive examples S^+, while the latter is used to compute the similarity between a new story and the negative examples S^-. Last, a linear weighted combination of the two K-NN predictions is used for the final prediction.

Researchers from UMass used the Rocchio algorithm for topic tracking. They used three different term-weighting schemes for story representation and similarity calculation. They also tried to dynamically adjust the topic vector during the tracking process.

Some researchers have employed decision trees for topic tracking. The major drawback of this method is that it can only give prediction results such as "yes" or "no" and cannot output a continuous prediction score, which is needed to produce a valid DET curve. Subsequent research includes introducing more information on news stories (such as "when," "where," and "who") into story and topic representation and constructing a strong topic tracker with an ensemble of multiple weak trackers.

Since the initial training data used to construct a topic model is normally very sparse, and there is also insufficient prior knowledge about the tracking topics, a topic tracking model that is trained based only on initial training data is often insufficient and inaccurate. Furthermore, because the topics are dynamic in topic tracking, the model cannot always track effectively after a period of time. To address this problem, some researchers proposed a new subtask called adaptive topic tracking (ATT), with the goal of adjusting the topic tracking model dynamically during the tracking process.

The work on ATT mainly focused on modifying the topic tracking model based on the system's pseudolabels. Most approaches established a dynamic term vector, adjusted the weight of terms dynamically, and trained the model in an incremental learning manner. The systems developed by the Dragon company (Yamron et al. 2000) and UMass (Connell et al. 2004) were the first to attempt unsupervised learning for ATT. The former added relevant stories into the training corpus and learned a new language model for topic tracking. The latter took the centroid of all prior stories as the representation of a topic and used the average correlation between prior stories and the centroid topic as the threshold. Each time a relevant story is detected during the follow-up process, it is added into the corpus, and the centroid and threshold are re-estimated correspondingly. By self-learning, ATT gradually adds pseudolabeled examples for model learning and modification, which reduces the limitation created by training only on the initial training corpus. However, the self-learning module in ATT is totally based on pseudolabeled examples. When the pseudolabels are not correct, this method can easily lead to the incorporation of irrelevant information, subsequently cause concept drift, and ultimately affect the performance of follow-up topic tracking.

9.6 Evaluation

TDT is an evaluation-driven technology. The TDT conferences have released five TDT corpora, including the TDT pilot corpus, TDT2, TDT3, TDT4, and TDT5. These corpora are provided by the Linguistic Data Consortium (LDC).

The corpora contain both broadcasting and text data except TDT5. The initial TDT corpus contained only English languages and subsequently added the Chinese and Arabic languages. Three annotations ("yes," "brief," and "no") were employed in TDT2 and TDT3, and two annotations ("yes" and "no") were employed in TDT4 and TDT5, where "yes" means that the story and the topic are highly correlated, "brief" means that the correlation score is less than 10%, and "no" means that the two evaluate as uncorrelated. The broadcasting corpus includes not only news stories but also non-news stories such as commercial trade and financial stories, for which LDC provided three additional annotations: "news," "miscellaneous," and "untranscribed."

The TDT task can be essentially considered as a binary classification problem. Similar to the method of evaluating text classification described in Sect. 5.6, we can categorize the prediction results for TDT into four different cases, as shown in Table 9.2. By using the missed detection rate (MDR) and false alarm rate (FAR) as the basis, a DET curve can be plotted to observe the mistakes of a TDT system. Figure 9.6 is an example of the DET curve, where the x-axis is FAR and the y-axis is MDR. The closer the DET curve is to the lower-left corner of the coordinate, the better the TDT system performance is.

The performance of a TDT system can be quantified by a C_{Det} indicator defined as

$$C_{\mathrm{Det}} = C_{\mathrm{MD}} \cdot p_{\mathrm{MD}} \cdot p_{\mathrm{target}} + C_{\mathrm{FA}} \cdot p_{\mathrm{FA}} \cdot p_{\mathrm{non\_target}} \tag{9.12}$$

where p_{MD} and p_{FA} are the conditional probabilities of missed detections (MD) and false alarms (FA), respectively, C_{MD} and C_{FA} are preset coefficients of MD and FA, p_{target} represents the prior probability of a target topic, and $p_{\mathrm{non\_target}} = 1 - p_{\mathrm{target}}$. C_{MD}, C_{FA}, and p_{target} are all preset parameters. The formulations of p_{MD} and p_{FA} are as follows:

$$p_{\mathrm{MD}} = \frac{\#\mathrm{missed\_detections}}{\#\mathrm{targets}} \times 100\% \tag{9.13}$$

$$p_{\mathrm{FA}} = \frac{\#\mathrm{false\_alarms}}{\#\mathrm{non\_targets}} \times 100\% \tag{9.14}$$

Generally, the normalized C_{Det} is used as the final performance of a TDT system:

$$C_{\mathrm{Det-Norm}} = \frac{C_{\mathrm{Det}}}{\min\left\{C_{\mathrm{MD}} \cdot p_{\mathrm{target}}, C_{\mathrm{FA}} \cdot p_{\mathrm{non\_target}}\right\}} \tag{9.15}$$

Table 9.2 Four kinds of prediction results from the TDT tasks

		Reference	
		Target	Non-target
Prediction	Yes	Correct	False alarm
	No	Missed detections	Correct

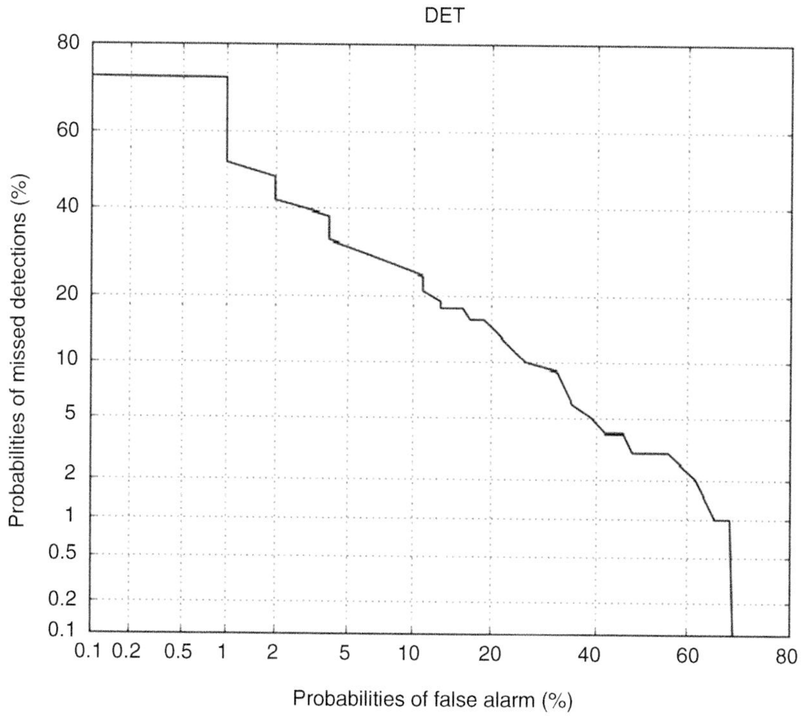

Fig. 9.6 TDT detection error tradeoff (DET) diagram

9.7 Social Media Topic Detection and Tracking

In recent years, the ways in which information is shared and disseminated on the Internet have gradually moved from the Web 1.0 era, which is represented by mainstream media websites, to the Web 2.0 era, which is represented by social media websites and applications. Traditional TDT mainly focuses on the content of traditional media, while social media TDT confront the following challenges: (1) the characteristics of user-generated context (UGC) in social media (e.g., short text, dynamic topic, irregular gramma, and diversified modals) increase the difficulty of text representation and TDT modeling; (2) the huge amount of data shared and propagated through social media brings great difficulty to real-time TDT; and (3) due to wide participation and openness, social media platforms are often the first site people use to report many emergencies. Therefore, bursty/breaking topic detection has attracted much attention in social media TDT.

In the following, we first introduce the differences between TDT in social media and traditional TDT and then introduce the main tasks and approaches of social media TDT. Lastly, we emphasize bursty topic detection in social media.

9.7.1 Social Media Topic Detection

The main goal of social media topic detection is to detect hot topics in the social media data stream. Similar to traditional topic detection, the social media topic detection task can also be divided into online topic detection and retrospective topic detection. However, due to the real-time nature of social media, more attention is being paid to online topic detection.

From the perspective of event types, social media topic detection can be categorized into specific and nonspecific topic detection. Specific topic detection aims at discovering historical topics that have already happened or detecting planned topics such as upcoming meetings or festival celebrations. Related information, such as the time, place, and main content of the known events, can be used to construct a topic detection model. Nonspecific topic detection focuses on detecting new topics from real-time data streams without any knowledge of the topic in advance (e.g., earthquakes) and collecting relevant follow-up stories. Nonspecific topic detection is the emphasis of social media topic detection.

(1) Specific Topic Detection

Specific topic detection methods can be divided into unsupervised and supervised machine learning methods. Similar to traditional topic detection, unsupervised topic detection methods in social media are mainly based on clustering or dynamic query expansion. The difference between traditional and unsupervised topic detection is that in addition to text content, the latter normally incorporates more social media-related information for topic representation and similarity calculation. For example, Lee and Sumiya (2010) proposed a local festival detection task from Twitter data streams. They found that the number of users and tweets will significantly increase when there are local festivals. They first collected Twitter data with geographical tags and then used the *k*-means algorithm to cluster these data and find topics in specific areas to detect local festivals. Massoudi et al. (2011) proposed a topic detection model for microblogs based on dynamic query expansion, in which they integrated text content and social media attributes such as emoji, hyperlink, number of fans, and number of retweets and replies for topic representation.

When the topic information is known in advance, such information can be used to compare with a labeled dataset. Then, supervised machine learning algorithms can be applied for topic detection. For example, Popescu and Pennacchiotti (2010) first collected a Twitter corpus and labeled it manually according to known topics. A supervised gradient boosted decision tree was then trained to detect controversial topics. They emphasized the importance of a rich and diverse feature set including hashtags, linguistic structure, and emotion features. Popescu et al. (2011) subsequently tried more features such as location and the number of replies. Supervised topic detection performs more effectively than unsupervised methods.

(2) Nonspecific Topic Detection

Information on nonspecific topics is unknown in advance. Traditional methods mainly use clustering to detect nonspecific topics, but the character of social media content makes these methods less effective.

On the one hand, some studies added social media-related features as new features for topic representation. For example, based on the classical incremental clustering algorithm (Allan et al. 1998a; Becker et al. 2011) explored the usage of retweets, replies, and mentions as features to detect social media topics. Feng et al. (2015) aggregated Twitter data in two dimensions (time and space) and designed a hashtag-based single-pass clustering method. Phuvipadawat and Murata (2010) concluded that accurate recognition of the proper name of an entity could help in the accurate calculation of text similarity and ultimately improve topic detection performance. The topics were then sorted by the number of fans and retweets to identify breaking news in the Twitter data stream.

On the other hand, some research tried to modify existing clustering algorithms or design new clustering algorithms to meet the requirements of social media applications. For example, Petrović et al. (2010) attempted to improve performance when applying traditional topic detection approaches to large-scale real-time data streams from social media. They further proposed an online NED method with constant time and space based on locality sensitive hashing. This method can effectively reduce the search space and significantly improve the efficiency of the system without decreasing the topic detection performance.

9.7.2 Social Media Topic Tracking

The main task of social media topic tracking is detecting microblogs related to existing topics from the social media data stream.

Similar to social media topic detection, existing studies mainly focus on how to use the special attributes of social media for topic representation and how to improve the sparseness of that representation. Phuvipadawat and Murata (2010) used rich social attributes such as URLs, hashtags, number of retweets, and user portraits to calculate the popularity of tweets and successfully tracked unexpected topics in social media. Lin et al. (2011) viewed the hashtag as a kind of topic indicator and used them to train a pretopic language model. Perplexity-based classifiers were then applied to filter the tweet stream to detect topics of interest.

9.8 Bursty Topic Detection

Bursty topic detection, also known as bursty/breaking event detection, refers to the detection of unexpected topics that develop rapidly in microblog data streams.

Bursty topic detection is different from traditional topic detection. Traditional topic detection emphasizes the detection of new topics without judging whether the detected topic is bursty or not. However, bursty topic detection focuses on the detection of topics' bursty features and bursty periods. Fung et al. (2005) divided bursty topic detection methods into document-pivot methods and feature-pivot methods. The former first detects topics through document clustering and then evaluates the burst of topics; the latter first extracts bursty features and then clusters these features to generate bursty topics.

The traditional topic detection approaches are mainly document-pivot methods. However, because the number of topics and stories is huge and hot topics change rapidly in social media, traditional document-pivot detection methods are often inefficient for social media bursty topic detection, and feature-pivot methods have attracted more attention.

Both document-pivot and feature-pivot methods need to identify the bursty status. The former usually recognizes burst states based on clustered topics, while the latter usually recognizes burst states based on feature discovery. In the following, we first introduce the classical burst status recognition algorithms and then review the representative document-pivot and feature-pivot bursty topic detection methods.

9.8.1 Burst State Detection

Kleinberg (2003) proposed a burst state detection algorithm for text data streams. It was later called the Kleinberg algorithm, and it has been widely used in bursty topic detection. The core idea of the algorithm is to simulate the time intervals between adjacent texts or sets of features in a data stream with an automation model to discover the optimal hidden state of the text at different time points. The states consist of a normal state and a burst state, which are denoted by different distributions of features, and the transition between states indicates the emergence or disappearance of a "burst."

In the Kleinberg algorithm, a text stream is organized into a sequence of messages, where each message has a corresponding arrival time. For a given term w, the algorithm records the arrival time of w and accordingly obtains a sequence of arrival times $\boldsymbol{t}^w = (t_0, t_1, \ldots, t_n)$. This determines a sequence of time intervals (called interarrival gaps) $\boldsymbol{x}^w = (x_1, \ldots, x_n)$ where $x_i = t_i - t_{i-1}$. If $\boldsymbol{x}^w$ is assumed to be generated by a binary state automaton, the problem will be transformed into a hidden Markov problem with the goal of solving the hidden state sequence with a known observation sequence. Finally, the bursty period is determined based on the obtained dynamic hidden state of the feature from the bursty and normal periods.

In detail, an exponential distribution is used to simulate the interarrival gaps. Suppose the interval x is distributed according to the density function as follows:

$$f(x) = \alpha \mathrm{e}^{-\alpha x}, \quad \alpha > 0, x > 0 \tag{9.16}$$

and the corresponding cumulative distribution function is

$$F(x) = 1 - \mathrm{e}^{-\alpha x}, \quad \alpha > 0, x > 0 \tag{9.17}$$

The expectation of x is α^{-1}, where α represents the arrival rate of the documents.

For a two-state model, a normal state q_0 (low state) and a burst state q_1 (high state) are defined. At each arrival time for w, the automaton must be in one of the states, which potentially affects the next arrival time of the w. The state will switch to another state or remain unchanged with a certain probability. Bursty topics are recognized as transitioning from a low state to a high state in a period of time.

As shown in Fig. 9.7, when a term is in a low state q_0, the interval x has a density function $f_0(x) = \alpha_0 \mathrm{e}^{-\alpha_0 x}$. When a term is in a high state q_1, the interval x has a different density function $f_1(x) = \alpha_1 \mathrm{e}^{-\alpha_1 x}$. Obviously, the arrival rate $\alpha_1 > \alpha_0$.

Suppose that the corresponding state sequence of $\boldsymbol{x}$ is $\boldsymbol{q} = (q_{i_1}, q_{i_2}, \ldots, q_{i_n})$, where $i_n \in \{0, 1\}$, the probability of state transition is p, and the number of state transitions in the sequence is b. Then, the density function for interval sequence $\boldsymbol{x}$ is

$$f_q(\boldsymbol{x}) = \prod_{t=1}^{n} f_{i_t}(x_t) \tag{9.18}$$

and the prior probability of $\boldsymbol{q}$ is

$$p(\boldsymbol{q}) = p^b (1-p)^{n-b} \tag{9.19}$$

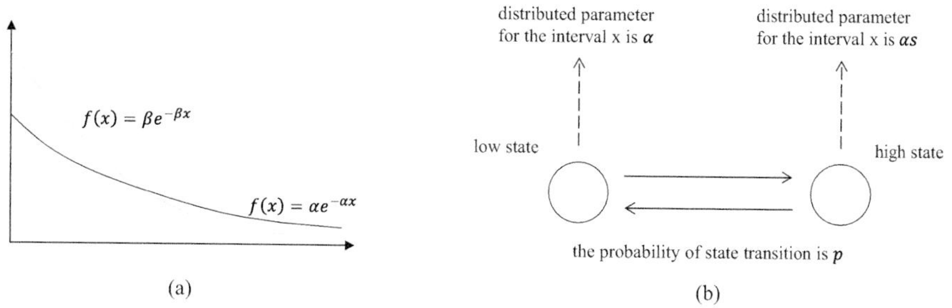

Fig. 9.7 (**a**) The distribution of the interval time for a normal and a burst state. (**b**) State transition model

According to Bayes' theorem, the posterior probability of $\boldsymbol{q}$ under $\boldsymbol{x}$ can be written as

$$p\left(\boldsymbol{q}|\boldsymbol{x}\right) = \frac{p(\boldsymbol{q})f_q(\boldsymbol{x})}{\sum\limits_{q'} p(\boldsymbol{q}')f_{q'}(\boldsymbol{x})}$$
$$= \frac{1}{Z}\left(\frac{p}{1-p}\right)^b (1-p)^n \prod_{t=1}^{n} f_{i_t}(x_t) \tag{9.20}$$

where $Z = \frac{p(\boldsymbol{q})f_q(\boldsymbol{x})}{\sum_{q'} p(\boldsymbol{q}')f_{q'}(\boldsymbol{x})}$.

The negative log-likelihood of the posterior distribution is

$$-\ln p\left(\boldsymbol{q}|\boldsymbol{x}\right) = b\ln\left(\frac{1-p}{p}\right) + \left(\sum_{t=1}^{n} -\ln f_{i_t}\left(x_t\right)\right) - n\ln\left(1-p\right) + \ln Z \tag{9.21}$$

where the third and fourth terms in the above formula are independent of $\boldsymbol{q}$. According to the maximum likelihood estimation, the following loss function can be defined:

$$c\left(\boldsymbol{q}|\boldsymbol{x}\right) = b\ln\left(\frac{1-p}{p}\right) + \left(\sum_{t=1}^{n} -\ln f_{i_t}(x_t)\right) \tag{9.22}$$

Determining the optimal hidden state sequence is equivalent to finding a state sequence that minimizes $c\left(\boldsymbol{q}|\boldsymbol{x}\right)$. The first term in $c\left(\boldsymbol{q}|\boldsymbol{x}\right)$ favors a sequence with a small number of state transitions, while the second term favors state sequences that conform well to the sequence $\boldsymbol{x}$ (i.e., making the value of the density function corresponding to each x_t as large as possible).

If each state in the state sequence $\boldsymbol{q}$ belongs to several continuous state levels $(q_0, q_1, \ldots, q_i, \ldots)$, the Kleinberg algorithm can be further extended from two states to an infinite number. The function $\tau\left(i, j\right)$ is defined to capture the loss of the transition from state s_i to state s_j. The transition loss from the low state to the high state is proportional to the number of intervening states, and the transition loss from the high state to the low state is 0:

$$\tau\left(i, j\right) = \begin{cases} \left(j-i\right)\gamma \ln n, & j > i \\ 0, & j \leq i \end{cases} \tag{9.23}$$

where γ is the state transition control parameter (usually set to 1). Given the parameters s and γ, this automaton can be represented by $A^*_{s,\gamma}$ (asterisk denotes the infinite states). For a given interval sequence $\boldsymbol{x} = (x_1, x_2, \ldots, x_n)$, the algorithm's goal is to solve a state sequence $\boldsymbol{q} = (q_{i_1}, q_{i_2}, \ldots, q_{i_n})$ to minimize the cost function. Let $\delta\left(\boldsymbol{x}\right) = \min\limits_{i=1,\ldots,n} \{x_i\}$, and the maximum state level can be obtained by $k = \lceil 1 + \log_s T + \log_s \delta(\boldsymbol{x})^{-1} \rceil$, where $\lceil \cdot \rceil$ is the ceiling function. It can be proven that if $\boldsymbol{q}^*$ is the optimal state sequence of automaton $A^k_{s,\gamma}$, it is also the optimal sequence of $A^*_{s,\gamma}$. Thus, the infinite state sequence optimization problem is transformed into the finite state optimization problem.

In the last step, a standard forward dynamic programming algorithm (such as the Viterbi algorithm) can be used to solve the above problem. Given an interval sequence $\boldsymbol{x} = (x_1, x_2, \ldots, x_t)$, the minimum loss sequence $C_j(t)$ can be expressed as follows:

$$C_j(t) = -\ln f_j(x_t) + \min_l (C_l(t-1) + \tau(l, j)) \tag{9.24}$$

$C_j(t)$ can be solved iteratively according to time t, where the initial state value is $C_0(0) = 0, C_j(0) = +\infty$. Finally, the optimal state sequence corresponding to $\boldsymbol{x}$ is obtained.

It is worth mentioning that the Kleinberg algorithm can detect bursts at the feature level (detecting the burst state of features/terms), as well as at the topic level (detecting the burst state of clustered topics). Therefore, it can be applied not only to feature-pivot bursty topic detection but also to document-pivot topic detection.

9.8.2 Document-Pivot Methods

Document-pivot methods first detect new topics from text data streams and then determine their burstiness. A traditional method is to first divide the text data stream into different windows according to the time of their appearance and perform clustering on the text in each window. Each cluster represents one topic, and features are extracted from the cluster to represent that topic. Finally, a bursty state recognition algorithm is applied to determine whether the topic is bursty or not.

Chen et al. (2013) first designed a strategy to obtain a real-time microblog data stream related to a given entity (such as a person or company name). For each time step t, a single-pass clustering algorithm is applied to the messages within the time window $[t - T, t]$ (T is the length of a unit window). The similarity between each message and the clustering centers is calculated. If the similarity is larger than the preset threshold, the message will be merged into the existing cluster; otherwise, the message constitutes a new cluster. Finally, each cluster is treated as a topic. The algorithm runs continuously to detect new topics in real-time data streams. They further established a semi-supervised classifier based on cotraining to detect whether the topics are burst or not. Figure 9.8 denotes a bursty topic evolution curve, where t_s denotes the time of one topic's occurrence, t_{hot} denotes the time the topic becomes hot, and the period $[t_s, t_{\text{hot}}]$ was defined as the bursty period. They labeled t_s and t_{hot} for each bursty topic in an offline training dataset. An SVM classifier was trained based on six features, including user growth rate, message growth rate, and response

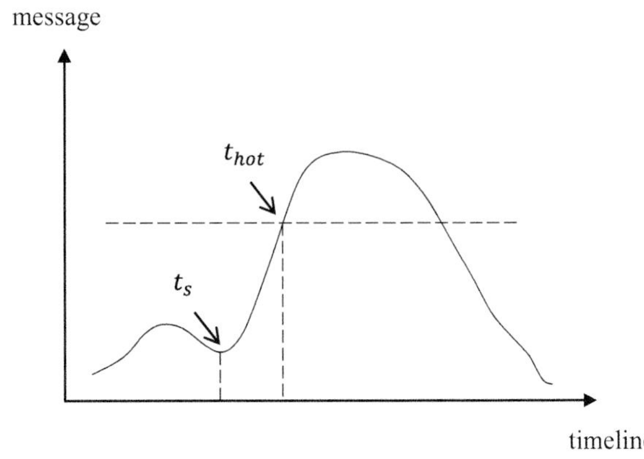

Fig. 9.8 The bursty period of one bursty topic

growth rate, and then it offered predictions for new topics detected from an online data stream as to whether they were bursty topics.

Diao et al. (2012) proposed a topic model called TimeUserLDA to detect bursty topics in social media data streams. The model was motivated by the finding that messages published at the same time are more likely to have the same topic and that messages published by the same author are also more likely to describe the same topic. Based on this, they incorporated the time and author information into a traditional LDA to model the messages. They mined a set of potential concepts C from a large-scale Twitter dataset, with each concept representing a topic in social media. For each topic $c \in C$, they calculated its occurrence frequency $(m_1^c, m_2^c, \ldots, m_T^c)$ along the time axis. Finally, they used an automaton similar to Kleinberg (2003) to identify the bursty topics.

Document-pivot methods are more suitable for topic detection in traditional media. As we have mentioned, the characteristics of social media, such as the short length, high volume, and broad topics, make document-pivot methods less efficient. Therefore, most of the applications for the bursty topic detection of social media are based on feature-pivot detection methods.

9.8.3 Feature-Pivot Methods

As shown in Fig. 9.9, the feature-pivot methods first discover a set of bursty features and then generate bursty topics. Here, "features" usually denote words or terms in texts. Text data streams are generally divided into equal-length and nonoverlapping time windows (such as "hours" or "days") in advance. Then, different kinds of methods, including feature selection methods, probabilistic methods, and the Kleinberg algorithm, are used to identify the bursty features.

One type of simple method uses the absolute or relative number of features and their changing speed as indicators for bursty feature selection. For example, for each

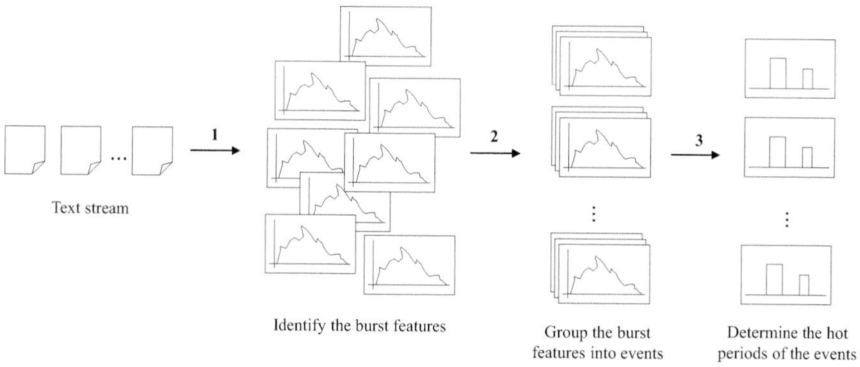

Fig. 9.9 Feature-pivot bursty topic detection methods

feature (e.g., words) in each time window, the relative word frequency $A_{ij} = \frac{F_{ij}}{F_{\max}}$ and the word frequency growth rate $B_{ij} = \frac{F_{ij} - F_{i(j-1)}}{1 + F_{i(j-1)}}$ are calculated. The features are ordered according to these indices, and a set of burst features is finally selected.

Fung et al. (2005) proposed a feature-pivot emergency detection method based on a probabilistic model. They first divided the text stream into $D = \{d_1, d_2, \ldots\}$, where d_i represents the text published on day i. Based on a binomial distribution, they identified a set of bursty features by comparing a feature's daily occurrence probability with its global occurrence probability. Subsequently, these bursty features were grouped into several bursty events, each of which comprised a subset of bursty features. Finally, the probability of the hot bursty event was calculated by computing the expected probability of the bursty event based on the subset of bursty features and comparing it with the expected value to determine the hot periods of each bursty event.

Other studies used spectrum analysis to detect bursty features and topics. For example, He et al. (2007a) employed discrete Fourier transform (DFT) to transform the time-series text data stream from the time domain to the frequency domain. The bursty topics in the time domain were supposed to correspond to the peaks in the frequency domain. The frequency domain attributes were then used to identify bursty features and their related periods.

He et al. (2007b) employed the Kleinberg algorithm to first recognize bursty features. For each bursty feature $f_j(t)$ in time window t, they calculated a bursty weight, which they combined with the static weight (e.g., TF-IDF weight) as a dynamic term-weighting scheme for bursty feature representation tf-idf$_{ij} + \delta w_j(t)$, where i is the index of documents and $\delta > 0$ is the bursty coefficient. Based on such a dynamic weight, topic clustering and classification experiments carried out on the TDT3 corpus achieved better performance than the traditional methods.

Cataldi et al. (2010) proposed a bursty feature extraction and bursty topic detection method based on content aging theory. First, nutrition is defined for each feature k under each time window TW^t taking into account the factors of word frequency and user authority. The energy of feature k in time window TW^t is then defined as the mean square difference between the nutrient value of feature k in current time window TW^t and that in the previous s time windows. Energy is also used as an indicator to measure the burstiness of feature k in time window TW^t: the feature with a larger energy value has higher burstiness. By using all the features in window TW^t as candidates, the bursty feature set EK^t is obtained by sorting the candidates according to their energy value. Finally, a feature relation graph TG^t is constructed with features in a window as nodes and correlation coefficients between features as edge weights. The burst topics are then sorted and annotated based on strongly connected subgraphs containing bursty features.

9.9 Further Reading

At the beginning of this century, TDT was an active research direction in text mining. Recent development in this direction, on the one hand, is reflected in the change in its application (i.e., from traditional media to social media), which we have described in Sects. 9.7 and 9.8. In addition, several studies have attempted to apply the latest machine learning theory to this development. For example, Fang et al. (2016) improved the traditional feature space via word embedding to improve the performance of story and topic representation and similarity computation. However, there are few works of this type. Moreover, since clustering is the most frequent task in TDT and there are few deep learning-based clustering algorithms, research on TDT based on deep learning is also scarce.

Meanwhile, TDT is closely related to several hot areas of text mining, such as information retrieval, sentiment analysis, and event extraction. In comparison with information retrieval, extraction, and summarization, TDT emphasizes the abilities to detect, track, and integrate information. In addition, TDT usually addresses text data streams with temporal relationships rather than static texts. TDT can be used to monitor several kinds of information sources to capture new topics in time and to carry out historical research on the origin and development of topics. It has broad application prospects in many fields, such as information security, public opinion mining, and social media analysis. The joint technique of TDT and sentiment analysis can effectively detect not only hot topics but also people's views and opinions about that topic.

TDT also has a strong correlation with event extraction. The former emphasizes the automatic organization of macrolevel events in text data, while the latter emphasizes fine-grained event recognition and element extraction in a piece of text. The studies of TDT were driven by the TDT conferences, while research into event extraction focused on the evaluations of ACE (automatic content extraction) and KBP (knowledge base population).

Exercises

9.1 Please point out the similarities and differences between topic detection and topic tracking.

9.2 Compared with traditional news texts, what are the characteristics of text representation under social media?

9.3 What are the disadvantages of applying the standard single-pass clustering algorithm to the online topic detection of news stories? Do you have any solutions to these issues?

9.4 What is the difference between feature-pivot methods and document-pivot methods for bursty event detection?

9.5 Please derive the Kleinberg algorithm if there are five states (representing five-level burstiness).

9.6 How can the Kleinberg algorithm be used in the case of feature-pivot methods and document-pivot methods for bursty event detection, respectively?

Chapter 10
Information Extraction

10.1 Concepts and History

IE refers to a text data mining technology that extracts factual information such as entities, entity attributes, relationships between entities and events from unstructured or semistructured natural language texts (such as web news, academic literature, and social media) and generates structured outputs (Sarawagi 2008). Unlike information retrieval technology, which searches for related documents or webpages from document sets or the open Internet based on specific queries, IE technology aims to generate machine-readable structured data and directly provide users with answers to questions instead of letting users find answers from numerous related candidate documents. IE technology also provides technical support for downstream tasks such as intelligent question answering and automatic decision-making. For example, we may want to extract information about natural disaster events from relevant news reports, including the name, time, place, and consequences of natural disasters. We may aim to extract information about a disease from medical records, including etiology, symptoms, drugs, and effects. In addition, we may also attempt to extract information about an acquisition event from a report of one company's acquisition of another, including acquirer, acquiree, time, and amount.

Typical IE tasks include named entity recognition (NER), entity disambiguation, relation extraction, and event extraction. Taking the news report of *Google's acquisition of DeepMind* as an example (as shown in Fig. 10.1), IE will identify entities such as the time, person, location, and organization of the event, analyze the relationship between these entities (for example, *Larry Page* is *CEO* of *Google*), and finally extract all the specific information about the company's acquisition event.

Note that different event types correspond to different event structures. For example, in terrorist attacks, as shown in Fig. 10.2, casualties are very important and should be accurately extracted in addition to time and location information.

IE technology can be traced back to the late 1970s, and since the late 1980s, the US government has sponsored a series of evaluation activities on IE, spurring

© Tsinghua University Press 2021

C. Zong et al., *Text Data Mining*, https://doi.org/10.1007/978-981-16-0100-2_10

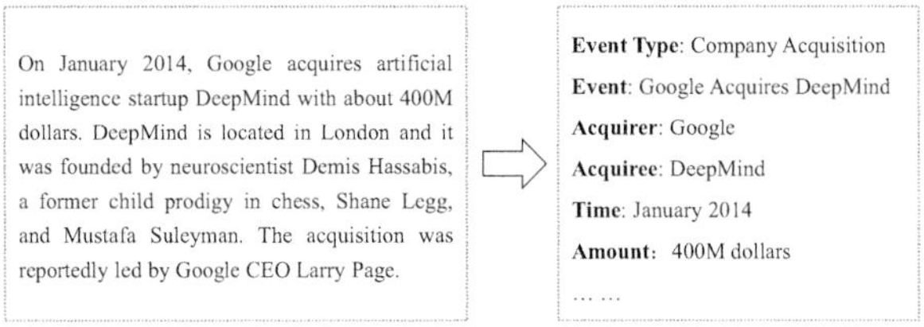

Fig. 10.1 An example of a company acquisition

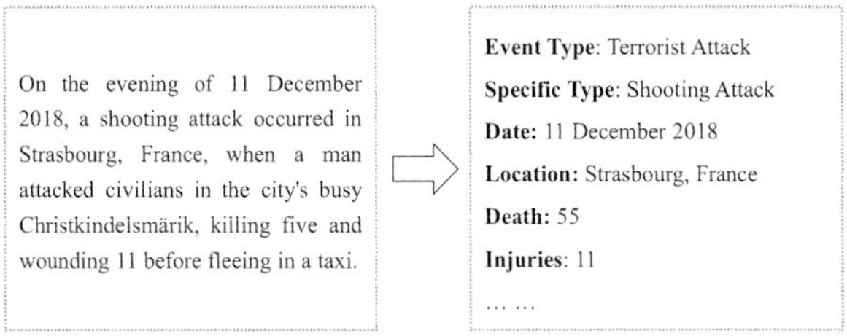

Fig. 10.2 An example of a terrorist attack event

the rapid development of this technology. In 1987, the Defense Advanced Research Projects Agency (DARPA) launched the first Message Understanding Conference (MUC1[1]) to evaluate the performance of IE technology, calling for international research institutions to compete on the same standard datasets provided by DARPA. For example, given ten intelligence texts in the naval military area, the IE system is required to output the coreference relationship between the entities contained in the text. From 1987 to 1997, the MUC evaluation was held seven times, mainly focusing on text in limited fields, such as naval military intelligence, terrorist attacks, personnel position changes, and aircraft crashes. The evaluation tasks included NER, coreference resolution, relation extraction, and slot filling.

In 1999, the Automatic Content Extraction (ACE) meeting began to replace the MUC; it focused on extracting finer-grained entity types (e.g., facility names and geopolitical entities), entity relations, and events from the dialog corpus and broader news data, such as political and international events. ACE was carried out until 2008. In later evaluation activities, IE tasks have been extended by increasing the difficulty accordingly, such as multilingual (English, Chinese, and Arabic) IE, entity detection, and tracking.

The series of MUC and ACE evaluation conferences provide a number of standard test data that play an important role in the development of IE technology. Since 2009, ACE has become a subtask of the Text Analysis Conference (TAC[2]), namely, the knowledge base population (KBP). The KBP has been held every year

[1] https://www.ldc.upenn.edu/collaborations/past-projects/ace.

[2] https://www.idc.upenn.edu/collaborations/past-projects/ace.

since 2000. The KBP focuses more on open domain data (such as webpages), and the extraction task mainly focuses on entity attribute extraction and entity linking, such as mining all relevant information (entity attributes) of a given entity (such as "Steven Jobs") from two million news pages and populating the mined information into the knowledge base (entity linking).

Several other conferences evaluate IE. For example, the Conference on Computational Natural Language Learning (CoNLL[3]) organized a language-independent NER task in 2003. SIGHAN[4] (Special Interest Group on Chinese Language Processing, ACL) held two evaluations on named entity identification in 2006 and 2007. They also effectively promoted the development of IE technology. In China, the CCKS (China Conference on Knowledge Graph and Semantic Computing) and NLPCC have organized several evaluation tasks for Chinese language-oriented entity recognition and entity linking in recent years, actively promoting the development of IE technology in China.

In summary, IE technology can be classified from different perspectives. Considering the domain aspect of the input data, it can be divided into two categories: limited domain and open domain. According to the type of extracted information, it can be divided into entity recognition, relation extraction, and event extraction. From the view of implementation methods, it can be classified into rule-based, statistical learning-based, and deep learning-based approaches.

10.2 Named Entity Recognition

NER is a fundamental task in natural language processing. In IE, NER aims at identifying entities belonging to specified categories in text. These entities mainly represent seven categories: person, location, organization, time, date, currency or quantity number and percentage. Because the constitution of time, date, quantity number, and percentages follows obvious rules, regular expressions are commonly used to accurately recognize them, while the identification of person, location, and organization faces larger challenges. Therefore, NER research mainly focuses on the recognition of these three types of entities.

The NER task can be further divided into two subtasks: entity detection and entity classification. The entity detection task aims to detect whether a word string in a given text is an entity. That is, it determines the beginning and ending boundaries of the entity. The entity classification task aims to judge the specified category of the detected entities.

In the last sentence of the left part of Fig. 10.1 *The acquisition was reportedly led by Google CEO Larry Page*, the detection task first identifies two entities *Google*

---

[3]https://www.sinall.org/conll.

[4]https://www.signll.org/conll.

and *Larry Page*. Subsequently, the classification task predicts that *Google* and *Larry Page* are organization and person, respectively.

As NER is a basic and key technology in natural language processing and text data mining, we will provide a detailed introduction of different NER methods, such as rule-based methods, supervised machine learning methods, and semisupervised machine learning methods.

10.2.1 Rule-based Named Entity Recognition

Since both the internal structure and external context of person, location, and organization names have certain rules to follow, early research on NER mostly focused on rule-based methods, among which regular expression is commonly used.

Compared to location and organization names, person names are much easier to identify for many languages (e.g., English, Chinese, and Japanese). For example, in English, person names usually begin with capital letters and follow some kind of titles such as *Mr.* , *Dr.*, or *Prof.*. Therefore, a regular expression *title [capitalized-token+]* can be designed to recognize such person names efficiently. The regular expression indicates that if the predecessor of the current word is a title and the current word begins with a capital letter, the word will be recognized as a person name. For instance, we can find that *John* in the sentence *Prof. John leaves school* is a person name according to the above regular expression.

In Chinese, the regularity of person names is even stronger. Most person names contain two or three Chinese characters, and these Chinese characters used for person names are very limited. For example, there are approximately 300 popular surnames. According to statistics, the top ten most frequent surnames (*Li, Wang, Zhang, Liu, Chen, Yang, Zhao, Huang, Zhou, Wu*) account for approximately 40% of the population. Chinese characters for given names are also relatively limited. Statistics show that there are approximately 1000 commonly used Chinese characters for given names. In terms of context, Chinese names also have obvious patterns. For example, the titles before and after a person's name usually contain *Mr.*, *Ms.*, *Director*, *Professor*, and so on. Verbs such as *say*, *point out*, and *express* often appear after the person's name. These contexts are key clues for identifying names or excluding impossible candidates. Therefore, in addition to collecting the names of as many famous people as possible and putting them into the predefined vocabulary for retrieval, candidate names can be selected with the help of a dictionary of surnames and given names. Then, a large number of names can be identified more accurately through rules combined with clues such as titles and salient context. Liu (2000) used limited Chinese surnames as a trigger to determine the left boundary of the candidate name. Then, by calculating the probability of the presence of two or three words on the right side, the candidate names with the highest possibility are selected. Finally, these rules are employed to exclude the impossible candidates and determine the most appropriate names.

Organization names and location names also follow some composition rules. Taking Chinese as an example, many organization names end with the words *university*, *company*, *group*, and *center*, while location names usually end with the words *city*, *county*, *town*, and *street*. Chen and Zong (2008) conducted a detailed analysis of the composition patterns of organization names. However, the above clues can only determine the right boundaries of some specific entities, while the left boundaries are very difficult to fix. In addition, the contexts around these entities provide less information to help determine the left and right boundaries. Therefore, a more practical solution is to construct a database of location and organization names.

Nevertheless, even with a large-scale database of person, location, and organization names, rule-based NER methods still face many challenges. On the one hand, a phrase in different contexts may lead to different types of entities. For example, *Washington* can be either a person name or a location name. In addition, common words may also be a type of entity. For example, *Bill* at the beginning of a sentence could be either an ordinary word or a person name. Furthermore, many entities are frequently abbreviated in the text. For example, *United States of America* is usually written as *USA*. This abbreviation can cause ambiguity in many cases. *IFA* can be either *International Franchise Association* or *International Factoring Association*. More importantly, new named entities, especially for person and organization names, are constantly emerging, and the consistency becomes less reliable. These problems mean that rule-based methods enhanced with entity databases are difficult to handle and cannot obtain high recognition accuracy. Furthermore, the rule-based approach also faces the problem of system maintenance, as it requires the constant modification or addition of new rules that may conflict with existing ones. Therefore, the development of a NER model using an automatic learning framework is desirable.

10.2.2 Supervised Named Entity Recognition Method

Given a collection of text data, suppose all person, location, and organization entities are manually labeled, as shown in the examples below. Supervised NER systems attempt to design machine learning methods that learn automatic prediction models based on these correctly labeled training data.

- He graduated last year and worked in [New York]/LOC
- The dog attacked [Louis Booy]/PER in the front garden
- [Akashi]/PER is the head of the [Department of Humanitarian affairs]/ORG

Researchers usually regard this task of supervised NER as a sequence labeling problem. The sequence labeling model first needs to determine the label set and the language granularity for labeling. BIO is a widely used label set in which "B" denotes the beginning of an entity, "I" indicates the middle or end of an entity, and "O" denotes the outside of any entity. For person, location, and organization names, seven tags can be employed: PER-B, PER-I, LOC-B, LOC-I, ORG-B, ORG-1, and O. Among them, PER, LOC, and ORG denote person, location, and organization names, respectively. PER-B indicates the starting point of a person name, and PER-I represents the middle or end part of a person name. LOC-B, LOC-I, ORG-B, and ORG-I have similar meanings.

Words and characters are usually the basic language units for label annotation. For example, the location name *New York* can be annotated as *New*/LOC-B *York*/LOC-I on the word level and *N*/LOC-B *e*/LOC-I *w*/LOC-I *Y*/LOC-I *o*/LOC-I *r*/LOC-I *k*/LOC-I on the character level. In this chapter, we introduce supervised NER methods based on word-level annotations. Accordingly, the above three annotated instances can be converted into word-level annotations as follows:

- He/O graduated/O last/O year/O and/O worked/O in/O New/LOC-B York/LOC-I
- The/O dog/O attacked/O Louis/PER-B Booy/PER-I in/O the/O front/O garden/O
- Akashi/PER-B is/O the/O head/O of/O the/O Department/ORG-B of/ORG-I Humanitarian/ORG-I affairs/ORG-I

Formally, given training data of M annotated sentences, $D = \{(X_m, Y_m)\}_{m=1}^{M}$, X_m is the word sequence (sentence) and Y_m is the corresponding label sequence that shares the same length as X_m. $Y_{mi} \in$ {ORG-B, ORG-I, LOC-B, LOC-I, PER-B, PER-I, O} denotes the gold label corresponding to the i-th word X_{mi}. Sequential labeling-based NER aims to design a parameter model $f(\theta)$ and learns reasonable model parameters θ^* from D. $f(\theta^*)$ will be employed to predict the label sequence of the test sentence. As shown in Fig. 10.3, a reasonable label sequence (the upper row) is correctly predicted for the input sentence (the English sentence in the bottom row).

There are many supervised machine learning models for sequence labeling. We introduce three typical methods for NER.

(1) Named entity recognition method based on hidden Markov model

Given a sentence of word sequence $\boldsymbol{x} = x_0 x_1 \ldots x_T$ (observation sequence), the sequence labeling model aims to search for a hidden label sequence $\boldsymbol{y} = y_0 y_1 \ldots y_T$ (state sequence) to maximize the posteriori probability $p(\boldsymbol{y}|\boldsymbol{x})$. The hidden Markov model (HMM) decomposes $p(\boldsymbol{y}|\boldsymbol{x})$ using Bayesian rules:

$$p(\boldsymbol{y}|\boldsymbol{x}) = \frac{p(\boldsymbol{x}, \boldsymbol{y})}{p(\boldsymbol{x})} = \frac{p(\boldsymbol{y}) \times p(\boldsymbol{x}|\boldsymbol{y})}{p(\boldsymbol{x})} \tag{10.1}$$

Since probability $p(\boldsymbol{x})$ is fixed for a given sentence and has no influence on any label sequence, maximizing conditional probability $p(\boldsymbol{y}|\boldsymbol{x})$ is equivalent

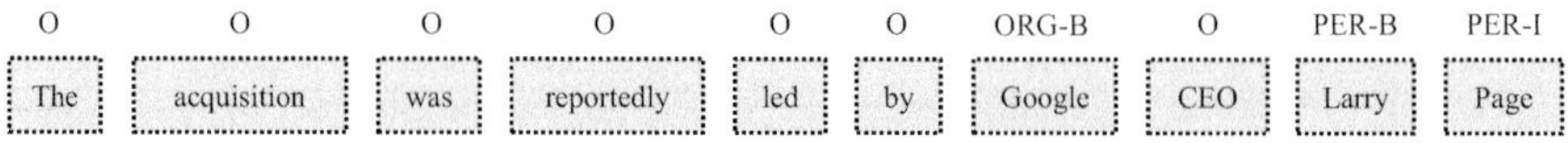

Fig. 10.3 An example of sequence labeling for named entity recognition

to maximizing the joint probability $p(\boldsymbol{x}, \boldsymbol{y})$, namely, the product of the prior probability $p(\boldsymbol{y})$ and the likelihood $p(\boldsymbol{x}|\boldsymbol{y})$. To efficiently calculate $p(\boldsymbol{y})$ and $p(\boldsymbol{x}|\boldsymbol{y})$, HMM assumes that the label sequence satisfies the first order Markov, the label value of the hidden state y_t relies only on y_{t-1}, and the observation value x_t is only generated by y_t. Therefore, the joint probability $p(\boldsymbol{x}, \boldsymbol{y})$ is decomposed into the following formula:

$$p(\boldsymbol{x}, \boldsymbol{y}) = p(\boldsymbol{y}) \times p(\boldsymbol{x}|\boldsymbol{y}) = \prod_{t=0}^{T} p(y_t|y_{t-1}) \times p(x_t|y_t) \tag{10.2}$$

We can see from the above formula that HMM simulates the process of generating the observation sequence. More details about the HMM can be found in Rabiner and Juang (1986), Zong (2013). Then, the HMM only needs to calculate $p(y_t|y_{t-1})$ and $p(x_t|y_t)$. In the HMM, $p(y_t|y_{t-1})$ is the state transition probability from the previous state to the current state, and $p(x_t|y_t)$ is the emission probability from current state to the current observation. Given the training data $D = \{(\boldsymbol{x}^{(m)}, \boldsymbol{y}^{(m)})\}_{m=1}^{M}$, state transition probability $p(y_t|y_{t-1})$ and emission probability $p(x_t|y_t)$ can be calculated by using maximum likelihood estimation:

$$p(y_t|y_{t-1}) = \frac{\text{count}(y_{t-1}, y_t)}{\text{count}(y_{t-1})} \tag{10.3}$$

$$p(x_t|y_t) = \frac{\text{count}(x_t, y_t)}{\text{count}(y_t)} \tag{10.4}$$

$\text{count}(y_{t-1}, y_t)$ denotes the co-occurrence count of y_t and y_{t-1}. $\text{count}(x_t, y_t)$ is the co-occurrence count of y_t and x_t.

Concerning the task of NER, not all the label pairs have transition probabilities. For example, the tags PER-I, LOC-I, and ORG-I cannot appear after tag O, and PER-B cannot be followed by LOC-I or ORG-I. The probability of these transitions should be all 0.

Since some words and labels may never co-occur in the training data, a data sparsity problem will arise. To handle this issue, a smoothing algorithm (e.g., Laplace smoothing and interpolation smoothing) is usually employed to assign a small probability to the unknown emission pairs during probability estimation.

Given an input sentence $\boldsymbol{x} = x_0 x_1 \ldots x_T$, the posterior probability of any label sequence $\boldsymbol{y} = y_0 y_1 \ldots y_T$ can be obtained by using the above HMM formula. The optimal label sequence can be found by naively exhausting all the possible tag sequences and choosing the one with the highest probability. However, the exhaustive search is too inefficient and impossible in practice. Thus, dynamic programming algorithms are commonly applied to solve such problems. The Viterbi decoding algorithm is used in the HMM model.

The Viterbi algorithm needs to maintain two sets of variables, $\delta_t(y)$ and $\varphi_t(y)$, where $\delta_t(y)$ records the maximum probability of the path ending with label y up to time t, and $\varphi_t(y)$ records the label of the previous time-step $t-1$ that leads to $\delta_t(y)$:

$$\delta_t(y) = \max_{y'}\{\delta_{t-1}(y')p(y|y')p(x_t|y)\} \tag{10.5}$$

$$\varphi_t(y) = \operatorname*{argmax}_{y'}\{\delta_{t-1}(y')p(y|y')p(x_t|y)\} \tag{10.6}$$

We can use the following formula to predict the label of the T-th word at the end of the sentence:

$$y_T = \operatorname*{argmax}_{y}\{\delta_T(y)\} \tag{10.7}$$

Then, the following formula is employed to retrospectively find the optimal label sequence:

$$y_t = \varphi_{t+1}(y_{t+1}) \tag{10.8}$$

Figure 9.4 shows an example of NER using the HMM model. The NER method based on HMM is simple and effective. It is a popular generative machine learning method from the early studies of NER (Zhou and Su 2002).

However, the assumption of HMM is too strict to capture more and richer contextual features. For example, in Fig. 10.4, the label of y_8 not only relies on the label of y_7 but also depends on the label of y_6. Moreover, the observed word *Larry* would also provide some clues to predict y_9 in addition to *Page* and y_8. Under the framework of HMM, these contexts cannot be modeled and exploited due to rigorous assumptions, and the absence of richer contexts limits its performance in NER. Accordingly, discriminant models that excel at exploring all kinds of contexts are gradually becoming popular, among which the conditional random field model (CRF) is one of the most widely used sequence labeling models.

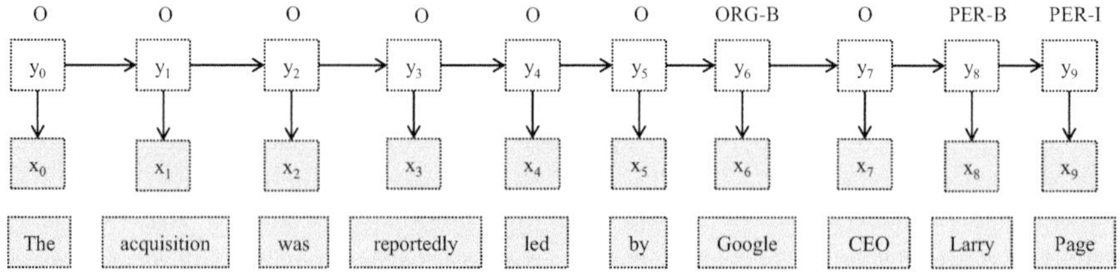

Fig. 10.4 HMM-based named entity recognition

Table 10.1 Feature templates commonly used in NER

Lexical features	Current word x_t, previous word x_{t-1}, next word x_{t+1}, word combinations $x_{t-1}x_t$, $x_t x_{t+1}$, $x_{t-1}x_t x_{t-1}$, and so on
Label features	Current label y_t, previous label y_{t-1}, label combinations $y_{t-1}y_t$, and so on
Label–word combination features	$x_t y_t$, $y_{t-1}x_t$, $y_{t-1}x_t y_t$, and so on
Dictionary features	Whether the string $x_{t-1}x_t$, $x_t x_{t+1}$, $x_{t-1}x_t x_{t-1}$ is in a given dictionary

(2) Named entity recognition method based on the conditional random field model

The CRF model (Lafferty et al. 2001) is a kind of undirected graph model, and the model for sequential labeling tasks is called the linear-chain conditional random field (linear-chain CRF). The linear-chain CRF is a discriminative model that calculates the conditional probability of the label sequence $\boldsymbol{y} = y_0 y_1 \ldots y_T$ given an input sequence $\boldsymbol{x} = x_0 x_1 \ldots x_T$ as follows:

$$p(\boldsymbol{y}|\boldsymbol{x}) = \frac{1}{Z_x} \exp\left\{\sum_{t=1}^{T}\sum_{k} \lambda_k f_k(y_{t-1}, y_t, \boldsymbol{x}, t)\right\} \tag{10.9}$$

where $f_k(y_{t-1}, y_t, \boldsymbol{x}, t)$ denotes a feature function that acts on the labels and inputs. $\lambda_k \geq 0$ is the weight of $f_k(y_{t-1}, y_t, \boldsymbol{x}, t)$, indicating the importance or contribution of this feature. The feature weights are optimized on the training data $D = \{(\boldsymbol{x}^{(m)}, \boldsymbol{y}^{(m)})\}_{m=1}^{M}$. The design of feature functions $f_k(y_{t-1}, y_t, \boldsymbol{x}, t)$ and the learning of parameter weights λ_k are the two key issues of the CRF model (McCallum and Li 2003).

Formally, feature function $f_k(y_{t-1}, y_t, \boldsymbol{x}, t)$ maps discrete feature combinations to Boolean variables as follows:

$$f_k(y_{t-1}, y_t, \boldsymbol{x}, t) = \begin{cases} 1, \text{ if } y_{t-1} = \text{ORG-B}, y_t = \text{ORG-I}, x_t = \text{Page} \\ 0, \text{ Otherwise} \end{cases} \tag{10.10}$$

The above equation tells us that if the labels of the previous and current time steps are ORG-B and ORG-I and the current word input is Page, then the feature function obtains a value of 1. Otherwise, its value is 0. Feature functions can be designed by considering the combinations of y_{t-1}, y_t, and x_t, and usually, each feature function f is called a feature template. For NER, there are many feature templates available. Table 10.1 lists some common feature templates.

Based on the templates, hundreds of thousands or even millions of features can be extracted from training data $D = \{(\boldsymbol{x}^{(m)}, \boldsymbol{y}^{(m)})\}_{m=1}^{M}$. Each feature corresponds to a weight parameter λ_k that needs to be learned. The learning of λ_k is task independent and can be optimized by the conventional CRF training algorithm; one can obtain the parameters by using CRF open source tools (such as CRF++[5]).

Z_x is a normalization factor, which needs to be solved by forward and backward algorithms in the process of model training. Details can be found in Sutton and McCallum (2012). The objective of parameter optimization is to maximize the conditional likelihood over the labeled data:

$$\mathbb{L}(\Lambda) = \sum_{m=1}^{M} \log(p(\boldsymbol{y}^{(m)}|\boldsymbol{x}^{(m)}, \Lambda)) + \log p(\Lambda) \tag{10.11}$$

where $p(\Lambda)$ is the prior probability of the parameters. During testing, the normalization factor Z_x can be ignored since we only aim to find the label sequence with maximum probability, but we are not concerned with the probability value.

$$\begin{aligned} \underset{y}{\operatorname{argmax}}\, p(\boldsymbol{y}|\boldsymbol{x}) &= \underset{y}{\operatorname{argmax}} \frac{1}{Z_x} \exp\left\{\sum_{t=1}^{T}\sum_{k} f_k(y_{t-1}, y_t, \boldsymbol{x}, t)\right\} \\ &= \underset{y}{\operatorname{argmax}} \left\{\exp\left(\sum_{t=1}^{T}\sum_{k} f_k(y_{t-1}, y_t, \boldsymbol{x}, t)\right)\right\} \end{aligned} \tag{10.12}$$

Similar to the HMM model, the optimal label sequence can be efficiently obtained by searching with the Viterbi dynamic programming algorithm. The calculation formula for the two variables is as follows:

$$\delta_t(y) = \max_{y'} \left\{\delta_{t-1}(y') \times \exp\left(\sum_{k} \lambda_k f_k(y', y_t, \boldsymbol{x}, t)\right)\right\} \tag{10.13}$$

$$\varphi_t(y) = \underset{y'}{\operatorname{argmax}} \left\{\delta_{t-1}(y') \times \exp\left(\sum_{k} \lambda_k f_k(y', y_t, \boldsymbol{x}, t)\right)\right\} \tag{10.14}$$

Figure 10.5 gives an illustration of the NER method based on CRF. Compared with HMM, CRF is a global optimization model with no independence assumption, and more contextual features can be exploited in label sequence prediction. As a result, the ultimate NER performance can improve. The famous NER tool developed by Stanford University (Stanford NER) is implemented with CRF as the core model. However, both generative HMM and discriminative CRF are based on discrete symbols such as words or characters, which will lead to two problems. First, data sparsity is a major issue. If a word has not been seen in the training sample, it is impossible to predict its label. Second, these methods cannot capture the semantic similarity between any two strings, such as *speak* and *say*. The semantic

[5]https://taku910.github.io/crfpp.

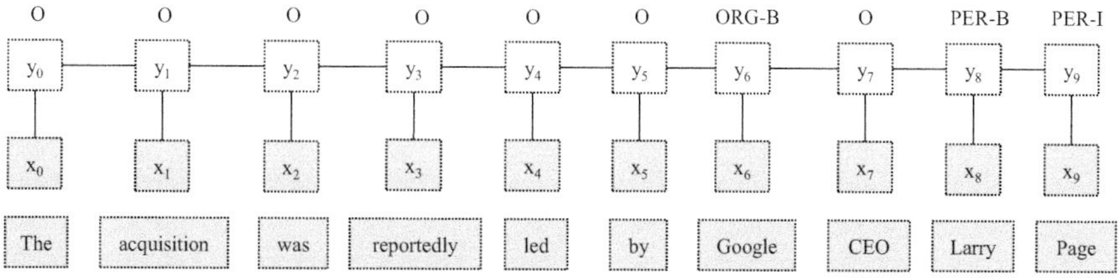

Fig. 10.5 CRF-based named entity recognition

similarity between these two words is close, but the symbolic representations of these two words prevent us from exploiting this similarity. Fortunately, neural network models based on distributed representations are good at abstracting deep semantic information, capturing semantic similarity between words, and thus have become a new popular model in NER research.

(3) Named entity recognition based on neural networks

Neural network models are good at automatically learning global and deep semantic distributed features for the input sentence. The NER model directly utilizes the learned distributed features rather than requiring a manual design that extracts discrete features to feed the classifier.

First, each word is mapped into a low-dimensional real-valued vector. Then, a multilayer network structure is adopted to learn the abstract distributed representation of the word sequence, and finally, the label of each input word is predicted based on the deep abstract representation.

Many neural network models, such as feed-forward neural networks, recurrent neural networks, convolutional neural networks, and recursive neural networks, have been introduced in the previous chapters. This chapter will take the recurrent neural network augmented with a CRF model as an example to introduce the application of neural network-based NER (Huang et al. 2015).

First, let us see how the recurrent neural network learns the deep abstract representation of a sentence. Here, the recurrent neural network adopts bidirectional long short-term memory (Bi-LSTM). As shown in Fig. 10.6, given the word sequence $\boldsymbol{x} = x_0 x_1 \ldots x_T$, Bi-LSTM maps each word x_i into a low-dimensional real-valued vector representation $\boldsymbol{e}_i \in \mathbb{R}^{d_1}$ (see the bottom in Fig. 10.6), where d_1 indicates the vector dimension, and $\boldsymbol{e}_i$ is generally randomly initialized and optimized during training. The forward LSTM obtains the distributed representation $\overrightarrow{\boldsymbol{h}}_i \in \mathbb{R}^{d_2}$ corresponding to each word (d_2 denotes the neuron number in the hidden layer). Similarly, the backward LSTM can obtain another distributed representation $\overleftarrow{\boldsymbol{h}}_i \in \mathbb{R}^{d_2}$ (please refer to the introduction of distributed representation in Chap. 3 for the calculation process). $\overrightarrow{\boldsymbol{h}}_i$ can capture the context information of $\boldsymbol{e}_i$ and its left side. $\overleftarrow{\boldsymbol{h}}_i$ can capture the context information of $\boldsymbol{e}_i$ and its right side. Hence, by

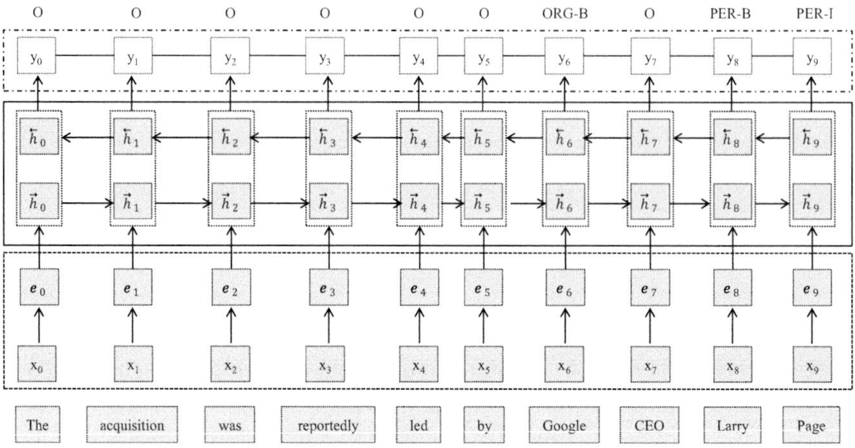

Fig. 10.6 Named entity recognition based on the LSTM-CRF model

concatenating $\overrightarrow{\boldsymbol{h}}_i$ and $\overleftarrow{\boldsymbol{h}}_i$, Bi-LSTM captures the global features centered on $\boldsymbol{e}_i$ through $\boldsymbol{h}_i = [\overrightarrow{\boldsymbol{h}}_i; \overleftarrow{\boldsymbol{h}}_i]$.

If Bi-LSTM is directly used to identify named entities, the following formula can be employed to calculate the probability of each label $y_i \in$ {ORG-B, ORG-I, LOC-B, LOC-I, PER-B, PER-I, O} for each word x_i.

$$p(y_i) = p(\boldsymbol{e}_{y_i}) = \text{softmax}(\boldsymbol{e}_{y_i}) = \frac{\boldsymbol{h}_i \cdot \boldsymbol{e}_{y_i}}{\sum_k \boldsymbol{h}_k \cdot \boldsymbol{e}_{y_i}} \tag{10.15}$$

where $\boldsymbol{e}_{y_i}$ denotes the distributed representation corresponding to the label y_i. The label of word x_i corresponds to the y_i with the maximum probability. Since Bi-LSTM cannot utilize the relationship between category labels and cannot rule out unreasonable combinations such as ORC-B and PER-I, the CRF model can be used for global optimization on top of the Bi-LSTM model; we call this the Bi-LSTM-CRF model.

The Bi-LSTM-CRF model also directly models conditional probability $p(\boldsymbol{y}|\boldsymbol{x})$.

$$p(\boldsymbol{y}|\boldsymbol{x}) = \frac{\prod_{t=1}^{T} \varphi_t(y_{t-1}, y_t, \boldsymbol{x})}{\sum_{\boldsymbol{y}'} \prod_{t=1}^{T} \varphi_t(y'_{t-1}, y'_t, \boldsymbol{x})} \tag{10.16}$$

where $\varphi_t(y_{t-1}, y_t, \boldsymbol{x}) = \exp(\boldsymbol{W}_{y_{t-1},y_t} \times \boldsymbol{h}_i + b_{y_{t-1},y_t})$, $\boldsymbol{W}_{y_{t-1},y_t}$ and b_{y_{t-1},y_t} are parameter weights and biases, respectively. The above formula is actually a generalization of the CRF model in feature modeling.

In the CRF model, $\varphi_t(y_{t-1}, y_t, \boldsymbol{x}) = \exp(\sum_k \lambda_k f_k(y_{t-1}, y_t, \boldsymbol{x}, t))$ can be transformed into $\varphi_t(y_{t-1}, y_t, \boldsymbol{x}) = \exp(\Lambda_{y_{t-1},y_t} F(y_{t-1}, y_t, \boldsymbol{x}, t))$, where $F(y_{t-1}, y_t, \boldsymbol{x}, t)$ and Λ_{y_{t-1},y_t} are eigenvectors and eigenvalues, respectively.

Therefore, BI-LSTM is equivalent to automatically learning a set of eigenvectors $F(y_{t-1}, y_t, \boldsymbol{x}, t) = \boldsymbol{h}_i$.

The training and decoding of the BI-LSTM-CRF model is similar to that of the CRF model. For example, the Viterbi algorithm can obtain globally optimal label sequences according to the input sequence $\boldsymbol{x} = x_0 x_1 \ldots x_T$.

10.2.3 Semisupervised Named Entity Recognition Method

Given large-scale annotated data, supervised NER methods can achieve acceptable performance. However, the annotation corpus of named entities is very limited in practice. Most of the training sets contain only approximately 100,000 sentences, and areas such as financial fields may not be covered. This leads to serious limitations in the performance of NER, especially in domain adaptation. From another perspective, there are massive unlabeled corpora in various languages and areas in reality. If these unlabeled data can be fully exploited in addition to the limited annotated data, NER will be much better. Following this direction, researchers resort to semisupervised NER methods.

Formally, we let $D_l = \{(\boldsymbol{x}^{(m)}, \boldsymbol{y}^{(m)})\}_{m=1}^{M}$ represent the limited labeled data, and $D_u = \{(\boldsymbol{x}^{(n)})\}_{n=1}^{N}$ denote the unlabeled corpus where $N \gg M$. The semisupervised NER method aims at fully exploring these two kinds of data (D_l, D_u) (l and u denote labeled and unlabeled data, respectively). In the remainder of this subsection, we introduce semisupervised NER from the perspectives of models and features.

From the model perspective, the CRF model can be extended to account for unlabeled data, i.e., semisupervised CRF model (semi-CRF) (Suzuki and Isozaki 2008). We know that in supervised CRF, the objective function is the conditional likelihood of the labeled data:

$$\mathbb{L}(\boldsymbol{\Lambda}|D_l) = \sum_{m=1}^{M} \log(p(\boldsymbol{y}^{(m)}|\boldsymbol{x}^{(m)}, \boldsymbol{\Lambda})) + \log p(\boldsymbol{\Lambda}) \tag{10.17}$$

For unlabeled data $D_u = \{(\boldsymbol{x}^{(n)})\}_{n=1}^{N}$, marginal likelihood $p_u = \sum_{n=1}^{N} \log p(\boldsymbol{x}^{(n)}, \theta)$ can be optimized. Since $\log p(\boldsymbol{x}^{(n)}, \theta) = \sum_{\boldsymbol{y}\in\mathbb{Y}} \log p(\boldsymbol{x}^{(n)}, \boldsymbol{y}, \theta)$, where $\mathbb{Y}$ represents all possible tag sequences, $p(\boldsymbol{x}^{(n)}, \boldsymbol{y}, \theta)$ can also be calculated in a manner similar to $p(\boldsymbol{y}|\boldsymbol{x}) = \prod_{t=1}^{T} \varphi_t(y_{t-1}, y_t, \boldsymbol{x})/Z_x$. Therefore, the following objective function can be designed on the unlabeled dataset:

$$\mathbb{L}(\boldsymbol{\Theta}|D_u) = \sum_{n=1}^{N} \sum_{\boldsymbol{y}\in\mathbb{Y}} \log(P(\boldsymbol{x}^{(m)}, \boldsymbol{y}, \boldsymbol{\Theta})) + \log p(\boldsymbol{\Theta}) \tag{10.18}$$

Since parameter $\boldsymbol{\Theta}$ contains $\boldsymbol{\Lambda}$, $\mathbb{L}(\boldsymbol{\Lambda}|D_l, \boldsymbol{\Theta})$ and $\mathbb{L}(\boldsymbol{\Theta}|D_u, \boldsymbol{\Lambda})$ can be optimized through iterative optimization. For example, $\mathbb{L}(\boldsymbol{\Lambda}|D_l, \Theta)$ can be optimized on the labeled data (parameters ($\boldsymbol{\Theta} \setminus \boldsymbol{\Lambda}$) that are unique to unlabeled data can be initialized with a uniform distribution), and then $\mathbb{L}(\boldsymbol{\Theta}|D_u, \boldsymbol{\Lambda})$ can be optimized on the unlabeled data with updated $\boldsymbol{\Lambda}$. In the second loop, the new $\boldsymbol{\Theta}$ optimizes $\boldsymbol{\Lambda}$ again, and this process repeats until the parameters converge.

From the feature perspective, unlabeled data can be used in many ways, either employed for mining more features according to the similarity of language units or utilized for extracting diverse context patterns. To exploit more features, the typical method is to use the distributed similarity of language units (such as character, word, etc.) in large-scale unlabeled data to discover effective features (Ratinov and Roth 2009). Specifically, Brown clustering and other algorithms can be adopted to cluster language units in unlabeled data. For example, *say* and *tell* should be grouped into one cluster. Assuming that the cluster is C_u, C_u can be used as a feature for NER. If *say* is the important context of the named entity in the annotated corpus, while *tell* only appears in the unlabeled data, *tell* and its context can be employed to correctly predict named entities because similar contexts are shared between *say* and *tell* in the same cluster C_u.

Algorithm 6: Semisupervised named entity recognition algorithm

Input : D_l: Small-scale annotated data; D_u: Large-scale unlabeled data.
Output: $D_{\text{l-new}}$: New labeled training data.

#Initialization
$D_{\text{l-new}} = D_l$
for $k = 1, \ldots, K$ **do**
  Step 1: Use the CRF model to train the NER model C_k on $D_{\text{l-new}}$
  Step 2: Use C_k to construct new annotation D_{new} from D_u
  Step 3: $D_{\text{l-new}} = D_{\text{l-new}} + D_{\text{new}}$; $D_u = D_u - D_{\text{new}}$
end

Another method for NER is to mine the diversity of context patterns. The basic idea is that we can select representative samples with high confidence and low redundancy from the unlabeled data and treat them as labeled samples to enlarge the supervised training data (Liao and Veeramachaneni 2009). The corresponding algorithm is shown in Algorithm 6. In each iteration, the NER model C_k is trained with the most recent labeled data $D_{\text{l-new}}$, and the unlabeled data D_u is automatically annotated with C_k. Intuitively, samples with high confidence (e.g., label prediction probability $p > 0.9$) can be selected as gold samples to be added into the training data. However, this kind of simple method cannot effectively improve performance because the samples with high confidence basically have the same context pattern as the training samples, which results in the newly added instances failing to enrich the context features for NER. In contrast, we should pay attention to the redundancy of new samples while considering high confidence.

Therefore, step 2 is the most important in the above algorithm. First, C_k is used to automatically label the samples in D_u, and the confidence value of each language unit in the samples is calculated. If a sequence seq_u in a sample is labeled as a named entity $NE \in$ {PER, LOC, ORG} and the confidence value is greater than T (e.g., $T = 0.9$), this shows that seq_u can very likely be a named entity. Then, we search all the samples s_u containing seq_u from D_u. If the confidence value of seq_u in s_u is low (e.g., less than 0.5), it indicates that the features learned by model C_k cannot correctly identify seq_u in sample s_u, although seq_u is predicted to be a named entity in other samples. This indicates that the contextual features of seq_u in sample s_u are quite different and further that s_u contains entity seq_u but records richer and diverse contextual patterns about the entity seq_u. Therefore, sample s_u is more informative about entity seq_u and can be added to $D_{\text{l-new}}$ as a labeled instance.

Furthermore, for the three types of named entities, namely, person, location, and organization names, if a sample $s_u^{'}$ in D_u contains a high-confidence entity $seq_u^{'}$ and the context of $seq_u^{'}$ is highly indicative, for example, appellation words such as *professor*, *Mr.*, and *chairman* indicating the person's name, and *company* and *center* indicating the organization's name, then $s_u^{''}$ is obtained by removing the appellation words in the context of $seq_u^{'}$ in $s_u^{'}$. Then, $s_u^{''}$ is automatically labeled with C_k. If the confidence value of $seq_u^{'}$ is relatively low, it tells us that $s_u^{''}$ without the indicator can provide a diverse context pattern for identifying entity $seq_u^{'}$. As a result, $s_u^{''}$ can be added to $D_{\text{l-new}}$.

In addition to the above methods, co-training algorithms can also be employed. This kind of method adopts a multiview model[6] and designs two groups of independent and sufficient features f_1 and f_2, which are used to construct two classifiers C_1 and C_2, respectively. Then, C_1 automatically labels the unlabeled samples in D_u and regards the samples with high confidence as labeled instances to be added to the training data, which is used to train C_2. Then, C_2 automatically labels the remaining unlabeled samples in D_u and treats high-confidence samples as annotated data. The above process iterates until convergence. However, because it is very difficult to design two independent and sufficient sets of features in the NER task, multiview-based methods are generally not as effective as the previously introduced two methods.

10.2.4 Evaluation of Named Entity Recognition Methods

NER methods are usually evaluated in an objective manner. First, a test set D_T (no overlap with the training data) is selected and manually labeled with entities such as person, location, and organization names according to the annotation specification used for the training data, resulting in a reference D_R. Suppose one method automatically recognizes the named entities in the test data D_T and obtains the system output D_s. Then, the performance of the NER method can be calculated by comparing the system output D_s to the gold reference D_R.

[6]Multiview refers to multiple views of data, such as speech and vision views in videos. The two views are independent of each other and can be regarded as two dimensions of the data.

The calculation process involves three variables, count($correct$), count ($spurious$), and count($missing$), which are explained as follows:

count($correct$): the number of named entities correctly recognized in D_s, that is, the overlap between D_s and D_R.

count($spurious$): the number of named entities recognized by the method in D_s but not considered as named entities in the gold reference D_R.

count($Missing$): the number of named entities that exist in the reference D_R that are not recognized by the system in D_s.

Based on the above three variables, the precision, recall, and F_1 can be calculated:

$$\text{precision} = \frac{\text{count}(correct)}{\text{count}(correct) + \text{count}(spurious)} \times 100\% \tag{10.19}$$

$$\text{recall} = \frac{\text{count}(correct)}{\text{count}(correct) + \text{count}(missing)} \times 100\% \tag{10.20}$$

$$F_1 = \frac{2 \times \text{precision} \times \text{recall}}{\text{precision} + \text{recall}} \tag{10.21}$$

F_1 is usually used to measure the overall performance of the NER methods.

10.3 Entity Disambiguation

Entity ambiguity refers to the problem in which a name mention may correspond to multiple entities in the real world. Please look at the following two sentences about the mention *Michael Jordan*.

① Michael Jordan is a leading researcher in machine learning and artificial intelligence.

② Michael Jordan wins the NBA MVP.

In the first sentence, *Michael Jordan* refers to a professor, while in the second sentence, *Michael Jordan* refers to the basketball player. Obviously, the same name mention maps into different real-world entities. The process of predicting the mapping from a mention into a real-world entity is called entity disambiguation or entity linking.

In documents, webpages, research papers, and other text sets, all entities exist in the form of mentions, which include named entities, pronouns, noun phrases, and other formats. In this chapter, we focus on named entities for disambiguation. However, even if only named entities are considered, the same named entity will appear in different forms, such as full names, abbreviations, and nicknames. Thus, the entity disambiguation task faces several major challenges. For example, *United*

States of America appears in the form of the abbreviation *USA* in many cases, and *IFA* can be either *International Franchise Association* or *International Factoring Association*.

Taking the name mention *Michael Jordan*, which is commonly used in the literature as an example, we introduce the typical methods for entity disambiguation.

The entity disambiguation task can be formalized with a quadruple: $ED = \{M, E, K, f\}$, where $E = \{e_1, e_2, \ldots, e_T\}$ denotes the set of all entities in the real world.

$M = \{m_1, m_2, \ldots, m_N\}$ represents the mentions that need disambiguation in the documents.

K denotes the knowledge source, or background knowledge, that can be utilized for entity disambiguation. It can be the social network related to one person's name or knowledge bases such as Wikipedia and WordNet.

$f : M \times K \rightarrow E$ is the entity disambiguation function, which maps a mention to the entity in the real world. For example, *Michael Jordan* in the first sentence above will be mapped to the entity *Michael Jordan (Professor)*. *Michael Jordan* in the second sentence above will be mapped to the entity *Michael Jordan (Basketball Player)*.

According to whether the real-world entity set E is given or not, entity disambiguation methods can be classified into clustering-based and linking-based methods.

10.3.1 Clustering-Based Entity Disambiguation Method

When the entity set E is unknown, the entity disambiguation function f becomes a clustering algorithm that clusters the name mention set $M = \{m_1, m_2, \ldots, m_N\}$ in the documents into different groups. In each group, all the mentions refer to the same entity in the real world.

Currently, most entity disambiguation methods are data-driven. The available knowledge mainly includes context information C and background knowledge K. Context information C refers to the context in which the mention is located, such as all the words in a window centered on the mention. Background knowledge K refers to social networks, Wikipedia, ontology, and so on. According to the background knowledge used, the clustering-based methods can be further roughly divided into vector space-based, social network-based, and Wikipedia-based clustering methods.

(1) Clustering methods based on the vector space model

The vector space-based clustering method does not use any background knowledge but only uses the context information around the mention (Bagga and Baldwin 1998; Mann and Yarowsky 2003; Fleischman and Hovy 2004; Pedersen et al. 2005). This kind of method is based on a distributed assumption: mentions that share the same entity should have similar context distributions, while mentions referring to different entities must have quite different contexts.

The process behind this kind of method can be generally divided into three steps: (a) obtaining a real-valued vector representation for each mention in $M = \{m_1, m_2, \ldots, m_N\}$ by using a vector space model; (b) calculating the distance between mentions in M; (c) clustering based on the distance between mentions and determining which mentions can be mapped into the same entity.

Taking the bag-of-words (BOW)-based context representation model as an example, let the context of m_i be $c_i = \{c_{i_1}, c_{i_2}, \ldots, c_{i_m}\}$, where c_{i_k} is a word in the context window with m_i as the center or a word in the document where m_i is located. Generally, c_{i_k} does not include stop words. For instance, in the two example sentences about *Michael Jordan* given earlier, the mention *Michael Jordan* in the first sentence can be represented by {*researcher, machine learning, artistic intelligence*}, and the mention *Michael Jordan* in the second sentence can be denoted by {*NBA, MVP*}. Then, TFIDF can be used to calculate the real-valued context vector $\boldsymbol{x}_i = \{x_{i_1}, x_{i_2}, \ldots, x_{i_{|V|}}\}$ of m_i. The vocabulary V contains all the nonstop words of the documents and shapes the feature space for each mention m_i. x_{i_k} is the TFIDF weight of the k-th word.

$$x_{i_k} = \text{tf\_idf}(c_{i_k}) \tag{10.22}$$

In addition to the TFIDF method, we can also use a variety of distributed representation methods introduced in the text representation chapter. For example, the weighted average method of word vectors or the distributed representation method based on convolutional neural networks can be used to learn the context representation $\boldsymbol{x}_i$ of m_i.

Given the context vector representation of m_i, cosine similarity is used to calculate the distance between two mentions:

$$\text{sim}(m_i, m_j) = \text{sim}(\boldsymbol{x}_i, \boldsymbol{x}_j) = \text{cosine}(\boldsymbol{x}_i, \boldsymbol{x}_j) \tag{10.23}$$

Hierarchical agglomerative clustering (HAC) is a commonly used method for clustering based on cosine distance and has been employed many times in evaluations of entity disambiguation tasks. HAC adopts a bottom-up merging clustering strategy. First, each mention is taken as a cluster, and then the two clusters with the highest similarity are merged until the maximum similarity score is less than some threshold or only one cluster remains. The distance between two clusters u and v can be calculated as follows:

$$\text{sim}(u, v) = \frac{\sum_{m_i \in u, m_j \in v} \text{sim}(m_i, m_j)}{||u|| \times ||v||} \tag{10.24}$$

Finally, mentions in the same cluster correspond to the same entity in the real world.

(2) Clustering method based on social networks

The clustering method based on social networks is mainly used to disambiguate name mentions. The method assumes that the entity corresponding to a mention is determined by its associated entity network.[7] This method takes the social network between entities as background knowledge (Bekkerman and Mccallum 2005; Malin et al. 2005; Minkov et al. 2006). For example, the mentions in the social network of *Michael Jordan (Basketball Player)* include { Scottie Pippen, Dennis Rodman, Magic Johnson, Shaquille O'Neal, Kobe Bryant, ... }; The mentions in the social network of *Michael Jordan (Professor)*" include { Yoshu Bengio, David Blei, Andrew Ng, Geoffrey Hinton, Yann LeCun, ... }.

The core idea of the social network method is based on the observation that the webpages about the persons who know each other well or have similar backgrounds are very likely to be linked with each other, while persons with the same name mention but different backgrounds rarely have linkages between them. The basic idea of name disambiguation is that for name mention t_h whose entity is h, the background knowledge K is the mention set $T_H = \{t_{h_1}, \ldots, t_{h_N}\}$ (T_H contains t_h), each of which has a social relationship with h. $t_{h_1}, \ldots, t_{h_N}$ are used as query terms to retrieve from the search engine, and each query retains the top L returned webpages, resulting in the set D which contains $h_N \times L$ webpages. The webpage returned by retrieving t may link to the entity of the person name h or the entity of the same name h'. Therefore, the goal of person name disambiguation is to learn the function f. With the help of background knowledge K, we can predict whether a webpage $d \in D$ containing name mention t_h links to the specific entity of the person name h.

The clustering method based on a social network aims to construct a connection graph $G_{LS} = (V, E)$ for the webpage set D, and each node in V corresponds to a webpage in D. If there is an edge between d_i and d_j, then the two webpages d_i and d_j have a link relationship. Given G_{LS}, it is easy to find the connected subgraph with the most nodes, which is called central clustering C_0. The remaining clusters (connected subgraphs) are $C_1, \ldots, C_B (B < h_N \times L)$. Then, whether a webpage d in D links to h will be determined by the following function f:

$$f(d, h) = \begin{cases} 1, \text{ if } & d \in C_i : \|C_i - C_0\| < \delta, \quad i = 0, \ldots, B \\ 0, \text{ Otherwise} \end{cases} \tag{10.25}$$

There are three problems to be solved: (a) how to judge whether two webpages have a link relationship; (b) how to measure the distance between two clusters; and (c) how to determine the distance threshold δ. Bekkerman and Mccallum (2005) proposed the following solutions to these three problems:

[7]This idea is very similar to the PageRank algorithm in which the importance of a webpage is determined by the pages linking to it.

For the first question, a hyperlink set LS(d) for each webpage d is constructed. If d_i and d_j satisfy $\mathrm{LS}(d_i) \cap \mathrm{LS}(d_j) \neq \emptyset$, then there is a link relationship between them. Otherwise, d_i and d_j have no link relationship. LS(d) consists of three parts:

$$\mathrm{LS}(d) = \mathrm{url}(d) \cup (\mathrm{links}(d) \cap \mathrm{TR}(D)) \tag{10.26}$$

In which, url(d) retains the first-level directory of the URL corresponding to d. For example, if the URL of d is http://www.ia.cas.cn/yjsiy/zs/sszs/, url(d) returns http://www.ia.cas.cn/yjsjy. If the website of d is http://www.ia.cas.cn, url(d) will return http://www.ia.cas.cn. $\mathrm{TR}(D) = \{\mathrm{url}(d_i)\} \setminus \mathrm{POP}$, where POP denotes a collection of popular websites, such as www.google.com. Then, TR(D) becomes the website after all results returned by url(d_i) excluding the popular websites. links(d) represents a collection of all webpage addresses in d.

For the second question, the distance between two clusters is measured by the cosine distance between two real-valued vectors. Each element in the vector is represented by a specific tf_idf(w):

$$\mathrm{tf\_idf}(w) = \frac{\mathrm{tf}(w)}{\log \mathrm{google\_df}(w)} \tag{10.27}$$

where google_df(w) indicates the number of webpages returned by the Google search engine according to query w, which can be estimated through the Google API.

Concerning the final question, δ is not predefined, and it is usually dynamically determined. For example, one-third of webpages in D should satisfy δ.

(3) Clustering method based on wikipedia

Wikipedia is currently the largest semistructured knowledge base in the world, containing large-scale concepts and rich semantic knowledge about concepts. The vast majority of these concepts include persons, organizations, locations, occupations, publications, etc. Each article in Wikipedia describes a concept, and the title of the article corresponds to a mention of the concept, such as *artificial intelligence*. Moreover, the article contains rich link information between concepts, which can directly reflect the correlation between them. For example, the webpage of *artificial intelligence* contains several hyperlinks, which link to the concepts of *computer science*, *machine learning*, *natural language processing*, and so on. Therefore, Wikipedia can be used as powerful background knowledge for the entity disambiguation task.

Taking the following three sentences as examples, we introduce the method of using Wikipedia to disambiguate the name mention *Michael Jordan*.

MJ_1: *Michael Jordan is a leading researcher in machine learning and artificial intelligence.*

MJ_2: *Michael Jordan has published over 300 research articles on topics in computer sciences, statistics, and cognitive science.*

MJ_3: *Michael Jordan wins the NBA MVP.*

The entity disambiguation process using Wikipedia is also divided into three steps. First, we represent a mention with a vector of relevant concepts in Wikipedia. Then, we calculate the similarity between any two mentions. Finally, HAC is employed to cluster the mentions. Since the HAC algorithm is introduced in the previous section, only the first two steps will be detailed here.

The idea behind the Wikipedia-based method is that if two mentions point to the same entity, the Wikipedia concepts should be highly relevant to the context of the mentions. Otherwise, the concepts in their context will be quite different. Therefore, mention m can be represented by the Wikipedia concept vector in its context:

$$m = (c_1, w(c_1, m)), (c_2, w(c_2, m)), \ldots, (c_n, w(c_n, m)) \tag{10.28}$$

where $w(c_i, m)$ indicates the relevance score between Wikipedia concept c_i and mention m in the context, which can be calculated by the following formula:

$$w(c_k, m) = \frac{1}{|m|} \sum_{c_k \in m, c_k \neq c} \text{sr}(c, c_k) \tag{10.29}$$

where $\text{sr}(c, c_k)$ denotes the correlation score between the two Wikipedia concepts and is calculated using the following formula:

$$\text{sr}(c_i, c_j) = \frac{\log(\max(|A|, |B|)) - \log(|A \cap B|)}{\log(|W|) - \log(\min(|A|, |B|))} \tag{10.30}$$

where A and B represent the collection of all concepts linked to c_i and c_j in Wikipedia, respectively, and $|W|$ is the total number of concepts in Wikipedia. According to the calculation results obtained by the above formula, MJ1, MJ2, and MJ3 can be expressed in the following forms:

MJ_1: *Researcher (0.42) Machine Learning (0.54) Artificial Intelligence (0.51)*
MJ_2: *Research (0.47) Statistics (0.52) Computer Science (0.52) Cognitive Science (0.51)*
MJ_3: *NBA (0.57) MVP (0.57)*

Next, we need to calculate the similarity between any two mentions m_i and m_j. First, the concepts in m_i and m_j are aligned. For example, for any concept c in m_i, the most similar concept is searched in m_j by:

$$\text{align}(c, m_j) = \operatorname*{argmax}_{c_k \in m_j} \text{sr}(c, c_k) \tag{10.31}$$

Then, the semantic relevance score in the $m_i \rightarrow m_j$ direction is calculated:

$$\mathrm{SR}(m_i \rightarrow m_j) = \frac{\sum_{c \in m_i} w(c, m_i) \times w(\mathrm{align}(c, m_j), m_j) \times \mathrm{sr}(c, \mathrm{align}(c, m_j))}{\sum_{c \in m_i} w(c, m_i) \times w(\mathrm{align}(c, m_j), m_j)} \tag{10.32}$$

$\mathrm{SR}(m_j \rightarrow m_i)$ can be calculated in a similar way.

Finally, the similarity between mentions m_i and m_j can be obtained by using the following formula:

$$\mathrm{Sim}(m_i, m_j) = \frac{1}{2}(\mathrm{SR}(m_i \rightarrow m_j) + \mathrm{SR}(m_j \rightarrow m_i)) \tag{10.33}$$

Based on the similarity between any two mentions, the HAC algorithm is employed to perform entity disambiguation. Details of this method can be found in the literature (Han and Zhao 2009a).

10.3.2 Linking-Based Entity Disambiguation

Linking-based entity disambiguation is also called entity linking, and its goal is to learn a mapping function $f : M \times K \rightarrow E$ to accurately link each name mention $m \in M = \{m_1, m_2, \ldots, m_N\}$ in the document to its referent entity in the entity set $E = \{e_1, e_2, \ldots, e_T\}$. Wikipedia is commonly used as background knowledge K.

Suppose a document consists of the following sentence:

EL_1: *Michael Jordan* is a leading *researcher* in *machine learning* and *artificial intelligence*, and he also plays *basketball* in his free time.

The mentions include {Michael Jordan, Researcher, Machine Learning, Artificial Intelligence, Basketball}, of which *Michael Jordan* is the most ambiguous. The candidate entities consist of {Michael Jordan (basketball player), Michael Jordan (football player), Michael Jordan (mycologist), Michael Jordan (professor), ...}. Entity linking aims to link the mention *Michael Jordan* in this document to the entity *Michael Jordan (professor)*.

A typical entity-linking method includes two steps: (1) determine the candidate entity set and (2) rank the candidate entities. The first step is to determine the possible candidate set E_m from E for a given mention m. The second step attempts to score all the entities in the candidate set E_m and select the entity that is ranked first as the final answer.

(1) Determine the candidate entity set

The candidate entity set directly affects the candidate space of entity linking. If the correct entity is not included in the candidate space, the entity linking will fail regardless of how accurate the subsequent entity ranking algorithm is. Therefore, it is very important to generate the appropriate set of candidate entities.

The most popular method is to produce the candidate set by resorting to the search engine. Shen et al. (2015) summarized a variety of methods for determining candidate entity sets, and the approach of constructing a (mention, entities) dictionary is widely used. This method employs Wikipedia as the knowledge source to construct a dictionary of (mention, entities) and ultimately generates a dictionary $Dic = \{key, value\}$ in the form of key–value pairs, where the key represents mentions (such as Michael Jordan) and the value denotes the corresponding candidate entity set (such as {Michael Jordan (basketball player), Michael Jordan (football player), Michael Jordan (mycologist), Michael Jordan (professor),..}). Dictionary D is constructed mainly by exploring various features of Wikipedia pages, such as entity pages, redirection pages, disambiguation pages, bold phrases in the first paragraph, and hyperlinks in pages.

In Wikipedia pages with entity descriptions, the titles are usually the most common mentions corresponding to entities. For example, the Wikipedia page with the title *Microsoft* describes the entity of *Microsoft Corporation*. Therefore, (title, entity) can be added to the dictionary as <key, value>.

The redirection page links different mentions of the same entity, which generally represent synonyms, abbreviations, etc. For example, *Edson Arantes do Nascimento* is redirected to *Pele*. Therefore, *Edson Arantes do Nascimento* and *Pele* can be added to the dictionary as <key, value>, respectively.

The disambiguation page contains different entities corresponding to the same mention. For example, the disambiguation page of *Michael Jordan* contains multiple links to different entities. Therefore, the title of the disambiguation page can be used as key, and all entities in the page can be added to the dictionary as values.

The first paragraph of a Wikipedia page is often a summary of the whole article and usually contains some phrases in bold font. These phrases are often aliases, full names, or abbreviations of corresponding entities. For example, the first paragraph of the page of *Michael Jordan* contains the phrases *Michael Jeffrey Jordan* and *MJ* in bold font. The former is the full name, and the latter is the abbreviation. Therefore, each bold phrase can be used as a key, and the entity described in the page can be added to the dictionary as a value.

Each Wikipedia page contains several hyperlinks. For example, the *Michael Jordan* page contains a hyperlink *ACC* pointing to the page *Atlantic Coast Conference*. These hyperlinks generally provide information such as aliases or abbreviations for entities. Therefore, the hyperlink can be used as a key, and the linked entity can be added to the dictionary as a value.

Through the above operations, a comprehensive mapping dictionary Dic from name mention to candidate entities can be constructed. According to Dic, each mention m in the document can obtain the corresponding candidate entity set E_m by exact or partial matching of character strings. According to the statistics, each mention corresponds to an average of more than 10 candidate entities.

Alternatively, Han and Zhao (2009b) proposed a method that submits the mentions and their contextual words to search engines such as Google, taking the returned entities described in Wikipedia pages as candidate entity sets.

(2) Rank candidate entity concepts

Given the candidate entity set E_m for the mention m, the next step is to rank the entities in E_m to find the correct entity that has a linking relationship with m. The classical methods can be divided into two categories, namely, independent entity ranking and joint entity ranking. The independent entity ranking method assumes that different mentions in a document are independent of each other. When ranking candidate entities of a mention, only the context of the mention and the semantic information of the candidate entity are taken into consideration. In contrast, the joint entity ranking method assumes that the mentions in the document are related to each other and belong to the same topic, so they should influence each other during the entity-linking process.

We introduce each of them in the following:

(a) Independent Entity Ranking

The core problem of this method is calculating the semantic correlation between the mention and the candidate entities. The contexts and semantic knowledge base are the main sources for ranking.

The context-based ranking method assumes that there is a link between a mention and an entity if they share a similar context. The key of this method is to measure the context similarity between mentions and candidate entities. The vector space model is the most widely used context representation method. First, context vectors are constructed for the mentions and candidate entities. For example, the words in a window (e.g., $K = 50$) that centers the mention in the text are considered to be the context. Similarly, all the words in the Wikipedia page where the entity is located can also be used as context. Then, the BOW model is employed to represent the context of mentions and entities, and cosine similarity based on TFIDF is used to calculate the distance between mentions and entities $\mathrm{Sim}_{\mathrm{TFIDF}}$ (Chen et al. 2010).

Han and Zhao (2009b) leveraged the Wikipedia concepts appearing in the context to construct contextual vectors and then calculated the semantic distance $\mathrm{Sim}_{\mathrm{wiki}}$ between the mention and the entity. This is similar to the method that calculates $\mathrm{Sim}(m_i, m_j)$ in clustering-based entity disambiguation.

With the development of deep learning methods in recent years, entity-linking algorithms based on neural networks have become popular. Their core idea is to use a distributed text representation model to calculate semantic similarity between the mention and candidate entities $\mathrm{Sim}_{\mathrm{distri}}$ (He et al. 2013; Sun et al. 2015). Figure 10.7 demonstrates the basic framework of the entity-linking algorithm based on neural networks (Sun et al. 2015). The goal of this algorithm is to calculate the similarity between the mention and the candidate entity. First, the words and positions in the context are represented by real-value distributed vectors, and then convolutional neural networks are employed to learn the distributed vector representation $\boldsymbol{v}_c$ of the context. Meanwhile, the vector representations of the mention, entity, and entity

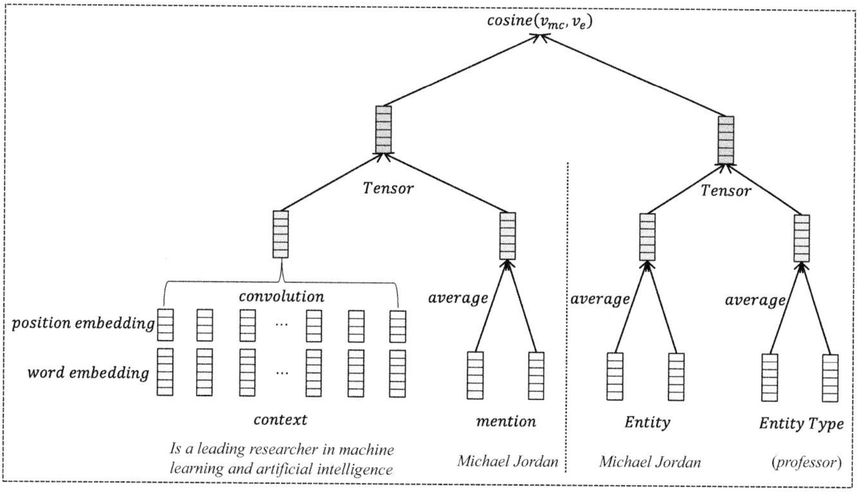

Fig. 10.7 Neural network-based entity linking

category[8] are obtained by averaging the word vectors, respectively denoted as $\boldsymbol{v}_m$, $\boldsymbol{v}_{ew}$, and $\boldsymbol{v}_{el}$. Then, the tensor model is leveraged to combine $\boldsymbol{v}_c$ and $\boldsymbol{v}_m$, $\boldsymbol{v}_{ew}$, and $\boldsymbol{v}_{el}$, resulting in $\boldsymbol{v}_{mc}$ and $\boldsymbol{v}_e$, which denote the contextualized mention and the candidate entity, respectively. Finally, cosine distance $\text{Sim}_{\text{distri}} = \text{cosine}(\boldsymbol{v}_{mc}, \boldsymbol{v}_e)$ is utilized to measure the similarity between the entity and the candidate entity.

For distributed text representation, convolutional neural networks and word vector averaging methods have been introduced in detail in Chaps. 3 and 4. We introduce how the tensor model works below.

Given the vectors $\boldsymbol{v}_c \in \mathbb{R}^d$ and $\boldsymbol{v}_m \in \mathbb{R}^d$, the formula for calculating $\boldsymbol{v}_{mc}$ using the tensor model is as follows:

$$\boldsymbol{v}_{mc} = [\boldsymbol{v}_c; \boldsymbol{v}_m]^T [\boldsymbol{M}_i]^{1:L} [\boldsymbol{v}_c; \boldsymbol{v}_m] \tag{10.34}$$

where $[\boldsymbol{v}_c; \boldsymbol{v}_m]$ represents the concatenation of the context vector and the mention vector, $\boldsymbol{M}_i \in \mathbb{R}^{d \times d}$ is a tensor that performs an operation on $[\boldsymbol{v}_c; \boldsymbol{v}_m]$ to obtain an element, and L tensors will lead to a L-dimensional vector output, namely, $\boldsymbol{v}_{mc} \in \mathbb{R}^L$. The vector representation $\boldsymbol{v}_e$ of the entity can be obtained by the same method. Note that the word embeddings, position vectors, and tensor matrix $\boldsymbol{M}_i$ are all neural network parameters and need to be optimized in the training process.

The training process optimizes an objective function on the supervised data. In ranking-based entity linking, max-margin loss (MML) is generally employed as the objective function. It requires that the correct mention–entity pair (m, e) has a higher

[8]The entity category can be retrieved from the knowledge base and is generally expressed by a phrase. For example, Donald Trump's entity category is *president of the United States*.

similarity score than the incorrect mention–entity pair (m, e') and the similarity gap should be greater than a certain threshold ϵ:

$$\text{loss} = \sum_{(m,e)\in T} \max\{0, \text{score}(m, e') + \epsilon - \text{score}(m, e)\} \tag{10.35}$$

where T denotes the annotated dataset consisting of all the correct mention–entity pairs. (m, e) is a positive sample, and (m, e') is a negative sample. $e' \neq e$ can be randomly selected from the entity set E. The score function $\text{score}(m, e)$ can be calculated through $\text{Sim}_{\text{distri}}$, which denotes the similarity of distributed representations between the mention and entity. In addition, $\text{score}(m, e)$ can be a weighted sum of various similarities, such as $\text{Sim}_{\text{TFIDF}}$, Sim_{Wiki}, $\text{Sim}_{\text{distri}}$. In the weighed sum algorithm, the popularity $\text{Pop}(e_i)$ of the candidate entity can also be used:

$$\text{Pop}(e_i) = \frac{\text{count}_{\text{m}}(e_i)}{\sum_{e_j\in E_m} \text{count}_{\text{m}}(e_j)} \tag{10.36}$$

where $\text{count}_{\text{m}}(e_i)$ indicates the number of times that m links to e_i in all Wikipedia pages.

(b) Joint Entity Ranking

To make full use of the topic consistency in the whole document, the joint entity ranking method infers the linking relations of all the mentions in the document. Taking the graph-based ranking algorithm as an example (Han et al. 2011), we next introduce the joint entity ranking method.

The method consists of two steps: (1) construct the referent graph RG for all the mentions and their candidate referent entities in the document; (2) perform global inference of entity linking on the referent graph RG. Taking the sentences EL_1 given at the beginning of this Sect. 10.3.2 as an example, we first introduce the construction method of the referent graph and then detail the global inference algorithm of entity linking.

The referent graph RG between mentions and candidate entities in a document is a weighted undirected graph $G = (V, E)$, in which V contains all mentions and their candidate referent entities in the document, and E includes two types of edges. One is the *mention–entity* edge, which depicts the correlation between the mention and a candidate entity. The other is the *entity–entity* edge, which depicts the semantic correlation between two candidate entities. Figure 10.8 is a semantic referent graph corresponding to the example EL_1. The core problem is to calculate the weights of *mention–entity* and *entity–entity* edges during RG construction. The *mention–entity* edge weight in the figure can be calculated based on the context similarity using the BOW model (i.e., $\text{Sim}_{\text{TFIDF}}(m, e)$, as introduced earlier). The *entity–entity* edge weight can be computed with $\text{sr}(\text{e}_\text{i}, \text{e}_\text{j})$, which calculates the relevance between entities based on Wikipedia, as introduced earlier.

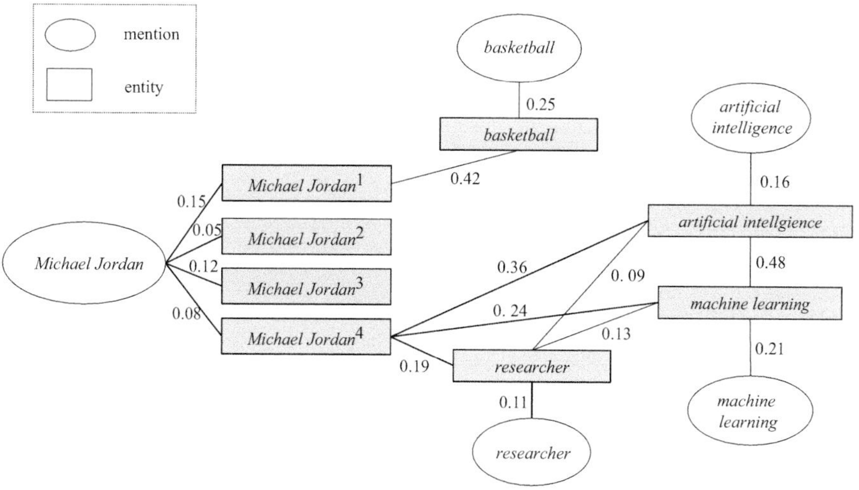

Fig. 10.8 Relevance graph between mentions and entities

In Fig. 10.8, the candidate entities *Michael Jordan*$^{(1)}$, *Michael Jordan*$^{(2)}$, *Michael Jordan*$^{(3)}$, and *Michael Jordan*$^{(4)}$ respectively denote *Michael Jordan (basketball player)*, *Michael Jordan (football player)*, *Michael Jordan (mycologist)*, and *Michael Jordan (professor)*.

After the construction of the mention–entity referent graph RG, the next step is to jointly infer entity linking. The joint inference process can be divided into three steps: (1) assign a confidence score for each candidate entity; (2) perform a random walk through the "mention–entity" and "entity–entity" edges in the graph to pass confidence; and (3) globally infer the entity linking according to the entity confidence.

In the initialization stage of step (1), the confidence score of the candidate entity is approximated by the importance score of the corresponding mention. The importance score of each mention m in the document is calculated by the normalized TFIDF value:

$$\text{Importance}(m) = \frac{\text{tf\_idf}(m)}{\sum_{m' \in D} \text{tf\_idf}(m')} \tag{10.37}$$

m' denotes any mention in the document D.

The key of step (2) is to calculate the final confidence score $r_D(e)$ of each candidate entity. The calculation of $r_D(e)$ involves three variables: $\boldsymbol{s}$, $\boldsymbol{r}$, and $\boldsymbol{T}$. $\boldsymbol{s}$ is the initial confidence vector, $\boldsymbol{s}_i = \text{Importance}(m_i)$; $\boldsymbol{r}$ denotes the score vector of the final confidence of the candidate entities and $\boldsymbol{r}_i$ represents the confidence of the i-th entity, namely, $r_D(e_i)$. $\boldsymbol{T}$ is the transition matrix, and $\boldsymbol{T}_{ij}$ indicates the transition weight of the confidence from node j to node i. $\boldsymbol{T}_{ij}$ can be either the mention-to-entity transition weight $p(m \rightarrow e)$[9] or the entity-to-entity transition weight $P(e_i \rightarrow e_j)$, which can be calculated by the following formulas:

$$p(m \rightarrow e) = \frac{\mathrm{Sim}_{\mathrm{TFIDF}}(m, e)}{\sum_{e' \in E_m} \mathrm{Sim}_{\mathrm{TFIDF}}(m, e')} \tag{10.38}$$

$$p(e_i \rightarrow e_j) = \frac{sr(e_i, e_j)}{\sum_{e_k \in N_m} sr(e_i, e_k)} \tag{10.39}$$

where N_m represents the adjacent entities of e_i in RG. Based on $\boldsymbol{s}$ and $\boldsymbol{T}$, $\boldsymbol{r}$ is calculated through the following iterative process:

$$\boldsymbol{r}^0 = \boldsymbol{s} \tag{10.40}$$

$$\boldsymbol{r}^{t+1} = (1 - \lambda) \times \boldsymbol{T} \times \boldsymbol{r}^t + \lambda \times \boldsymbol{s} \tag{10.41}$$

The results can be obtained by solving the above formula:

$$\boldsymbol{r} = \lambda \times (\boldsymbol{I} - (1 - \lambda)\boldsymbol{T})^{-1}\boldsymbol{s} \tag{10.42}$$

where $\boldsymbol{I}$ is the identity matrix. Finally, the entity linking of each mention in the document can be optimized by solving the following formula:

$$e^* = \underset{e}{\mathrm{argmax}}\, \mathrm{Sim}_{\mathrm{TFIDF}}(m, e) \times r_D(e) \tag{10.43}$$

10.3.3 Evaluation of Entity Disambiguation

Researchers have designed different automatic evaluation methods for clustering-based and linking-based entity disambiguation.

For the clustering-based entity disambiguation task, the evaluation method is to evaluate the effect of mention clustering[10] for the same name in the document set. Assume that the gold clustering result on the set of n mentions is $L = \{L_1, L_2, \ldots, L_M\}$, and the clustering result given by the system is $C = \{C_1, C_2, \ldots, C_N\}$. The automatic evaluation method is mainly evaluated based on the two aspects of clustering purity and inverse purity. For cluster C_i in the system, cluster L_j sharing the largest intersection with C_i can be found in L. Accordingly,

[9]The transition is unidirectional and there is no transition from the entity to the mention.

[10]The mention is usually a person's name that is ambiguous because it corresponds to different entities in different contexts. For example, *Michael Jordan* corresponds to entities in different documents. The clustering-based entity disambiguation method performs clustering on all appearances of *Michael Jordan* in the document set.

$\frac{|C_i \cap L_j|}{|C_i|}$ is called the precision of C_i, and the weighted sum of the precision of all system clusters is called the purity. The calculation of inverse purity is similar except that inverse purity focuses on the recall of clustering. The calculation formulas of the two values are as follows:

$$\text{Purity} = \sum_i \frac{|C_i|}{n} \max_j \text{Precision}(C_i, L_j) \tag{10.44}$$

$$\text{Precision}(C_i, L_j) = \frac{|C_i \cap L_j|}{|C_i|} \times 100\% \tag{10.45}$$

$$\text{Inverse Purity} = \sum_i \frac{|L_i|}{n} \max_j \text{Precision}(L_i, C_j) \tag{10.46}$$

$$\text{Precision}(L_i, C_j) = \frac{|L_i \cap C_j|}{|L_i|} \times 100\% \tag{10.47}$$

Generally, the performance of clustering is measured by the harmonic average $F_{\alpha=0.5}$ of purity and inverse purity:

$$F_\alpha = \frac{1}{\alpha \frac{1}{\text{Purity}} + (1-\alpha) \frac{1}{\text{Inverse Purity}}} \tag{10.48}$$

For entity-linking tasks, the evaluation method is similar to that of the classification task. The performance of entity linking is directly measured by precision and recall. For document D, assume that the manually annotated mention list is $M = \{m_i, m_j, \ldots, m_M\}$ and the gold entity-linking result is $E = \{e_i, e_j, \ldots, e_M\}$. Suppose the system identified mention list is $M' = \{m'_{i'}, m'_{j'}, \ldots, m'_{N'}\}$, and the automatic entity-linking result generated by the system is $E' = \{e'_{i'}, e'_{j'}, \ldots, e'_{N'}\}$, where i and i' respectively denote the positions of the mentions in the document. Then, the intersection of the automatic and gold results can be obtained as follows:

$$M^* = \left\{m_k | \forall k, m_k = m'_k\right\} \tag{10.49}$$

$$E^* = \left\{e_k | \forall k, m_k \in M^*, e'_k = e_k\right\} \tag{10.50}$$

where M^* indicates the set of correctly recognized mentions by the system, and E^* is the set of correctly linked entities for mentions in M^*. The precision, recall, and F_1 value are calculated by the following formulas:

$$\text{Precision} = \frac{|E^*|}{|E'|} \times 100\% \tag{10.51}$$

$$\text{Recall} = \frac{|E^*|}{|E|} \times 100\% \tag{10.52}$$

$$F_1 = \frac{2 \times \text{Precision} \times \text{Recall}}{\text{Precision} + \text{Recall}} \tag{10.53}$$

10.4 Relation Extraction

In natural language texts, only entities cannot provide abundant semantic information. For example, in the following two sentences, it is difficult to reveal the core information contained in the text by just identifying the person names of *John* and *Mary* and the location name of *London*.

Example: [John] is from [London]. [John] and [Mary] officially registered for marriage in 2007.

Generally, our real world can be seen as a network composed of nodes and edges. Nodes represent various entities, and edges denote the relationship between entities. Therefore, in addition to identifying and disambiguating the entities, another important task is identifying the semantic relationship between them. In the above example, *John* is a citizen of *London*, and the relation can be expressed as citizen_of(John, London). *John* and *Mary* are husband and wife, and the relation can be expressed as spouse(John, Mary).

Relation extraction is a task that aims to identify the entities in texts and determine the relationships between these entities. This technology plays a key role in many downstream tasks, such as knowledge graph construction, social network analysis, and automatic question answering.

Formally, the entity relation can be expressed as an $n + 1$ tuple $t = (e_1, e_2, \ldots, e_n, r)$, where $e_1, e_2, \ldots, e_n$ denotes n entities in natural language texts, and r denotes the relationship among these n entities, which is called the n-ary relation. At present, the binary relations (the relationship between two entities) are the research focus, and in most cases, the two entities are in the same sentence. Therefore, we focus on binary relations in this section and detail the relation extraction methods for entities in a single sentence, namely, recognizing the triple $t = (e_1, e_2, r)$ in a sentence. In the above example, each sentence contains a relation triple: [John, London, citizen_of] and [John, Mary, spouse].

Assuming that the entities in the sentence have been identified and disambiguated using the aforementioned NER and entity disambiguation techniques, entity relation extraction becomes a task for predicting relation types. In the open domain scenarios, there are thousands of relation types. To simplify the problem, we take the guidelines of popular IE evaluations, such as MUC, ACE, and Semeval, as an example to introduce the methods of implementing relation extraction. All the evaluation campaigns provide manually labeled training data of entity relations (denoted as $D_{train} = \{s_i, (e_{i_1}, e_{i_2}, r_i)\}_{i=1}^N$), the set of relation types (denoted as

Table 10.2 Distribution of relationship categories in the ACE 2003 training set

Relation type	Subtype	Frequency
AT(2781)	Based-in	347
	Located	2126
	Residence	308
NEAR(201) PART(1298)	Relative-location[a]	201
	Part-of	947
	Subsidiary	355
	Other	6
ROLE(4756)	Affiliate-partner	204
	Citizen-of	328
	Client	144
	Founder	26
	General-staff	1331
	Management	1242
	Member	1091
	Owner	232
	Other	158
SOCIAL(827)	Associate[a]	91
	Grandparent	12
	Other-personal	85
	Other-professional[a]	339
	Other-relative[a]	78
	Parent	127
	Sibling[a]	18
	Spouse[a]	77

[a]In the table means this relationship is symmetrical; e.g., in the relation type *spouse*, A is the spouse of B, and B is also the spouse of A

$R = \{r_k\}_{k=1}^{K}$) and the test dataset (denoted as $D_{test} = \{s_j, (e_{j_1}, e_{j_2})\}_{j=1}^{M}$). The training data include N sentences (s_i for the i-th sentence) and K relation types. For example, in ACE 2003 and 2004, the annotated data included 16,771 instances from 1000 English documents, 5–7 main relation types, and 23−24 subtypes. Table 10.2 gives the statistics for all kinds of relation types in the ACE 2003 training data. The relation extraction system needs to learn a model from the training data and predict the appropriate relation type from the type set R for each entity pair (e_{j_1}, e_{j_2}) in the test set.

As seen from the statistics in Table 10.2, the distribution of relation types is very unbalanced. For example, the *Founder* subtype in the *ROLE* relation and the *Grandparent* subtype in the *SOCIAL* relation occurred only 26 and 12 times, respectively. In contrast, the *Locate* subtype in the *AT* relation appeared more than 2100 times. In addition, the ACE relation extraction task also defines some subtype categories that are very difficult to identify, such as *Based-in*, *Located*, and *Residence*. In the example sentence *China's Huawei Company has business all over the world*, *Huawei* and *China* have the *Based-in* relation. In the sentence *John went to Beijing for business trips*, *John* and *Beijing* have a *Located* relationship. In the sentence *John moved to Beijing*, *John* and *Beijing* belong to the *Residence* relation type. It can be seen that the nuances between these relation types are sometimes difficult to distinguish even by human experts.

Since the set of relation types is given, the relation prediction task is usually converted into a supervised relation classification problem. The basic idea is to extract informative features from the contexts of two entities and the whole sentence. Then, machine learning methods can be employed to train the classification models on the annotated training corpus $f(s, (e_1, e_2)) \in R$. Finally, the classifier predicts the relation type between entities. Classification methods can be generally divided into methods based on discrete feature engineering and methods using distributed representation learning. We will detail these two kinds of methods in the following sections.

10.4.1 Relation Classification Using Discrete Features

The key to predicting the relation type between entities is to fully mine and utilize the contexts of these entities. For example, *marriage* in the context is one of the most important indicators for the *Spouse* relation type. In the example *John and Mary officially registered for marriage in 2007*, if the key contextual feature *registered for marriage* can be effectively used, we can easily predict the relation between *John* and *Mary*.

There are several kinds of methods for relation classification using discrete features. The main difference lies in feature selection and classification models. In terms of feature selection, different levels of granularity such as lexicon, syntax, and semantics can be used. In terms of classification models, maximum entropy, perceptron, and support vector machines can be employed. Since classification models are detailed in the previous sections, we introduce two typical methods for feature selection: explicit discrete feature engineering and implicit kernel function features.

(1) Relation classification using explicit discrete features

Explicit discrete features are symbol strings representing lexical, syntactic, and semantic structures. Taking the entity pair (*John*, *Mary*) in the sentence *John and Mary officially registered for marriage in 2007* as an example, we introduce which discrete features can be explored. Zhou et al. (2005) conducted a very detailed study on the selection of discrete features. According to their research, the following discrete features are proven effective.

(a) Word feature

There are four main types of such features: ① words contained in the entity pair (e_1, e_2); ② words between the entities e_1 and e_2; ③ words in front of the entity e_1;

and ④ words after the entity e_2. Corresponding to the above example, the specific features are listed below:

WE1: The word in entity e_1. In the above example, it is *John*;
HE1: The head word in entity e_1. If e_1 is a phrase, **HE1** corresponds to the head word of the phrase. If e_1 is just a word, **HE1** is e_1 itself. In this case, it is *John*;
WE2: The word in entity e_2. In this case, it is *Mary*;
HE2: The head word in entity e_2. In this case, it is *Mary*;
HE12: The concatenation of **HE1** and **HE2**. In this case, it is *John-Mary*;
WBNULL: A Boolean variable that is True if there is no word between e_1 and e_2 and False otherwise. It is False in the above example;
WBFL: If there is only one word between e_1 and e_2, **WBFL** indicates that word. In this case, it is *and*;
WBF: If there are multiple words between e_1 and e_2, **WBF** indicates the first word;
WBL: If there are multiple words between e_1 and e_2, **WBL** represents the last word;
WBO: If there are multiple words between e_1 and e_2, **WBO** represents words other than **WBF** and **WBL**;
BM1F: The first word before e_1. In this case, there is no word before *John*, so **BM1F** is NULL;
BM1L: The second word before e_1. In this case, **BM1L** is NULL;
AM1F: The first word after e_2. In this case, **AM1F** is *officially*;
AM1L: The second word after e_2. In this example, **AM1L** is *registered*.

(b) Entity type feature

Entity types offer a strong indication of the relation between entities. If e_1 is a person name and e_2 is an organization name, then we can determine that e_1 and e_2 belong to one of the relations in {Client, Founder, General-staff, Management, Member, Owner}. Obviously, the entity type is an important feature. The entity types mainly include PERSON, ORGANIZATION, LOCATION, FACILITY, and GPE (geopolitical entity, such as country names). We can use the following entity type feature:

ET12: Entity type combination of e_1 and e_2. In the above example, **ET12** is *PERSON-PERSON*.

(c) Mention feature

The mention feature refers to the mention type of the entity in the text, which is a specific NAME, NOMIAL, or PRONOUN.

ML12: Mention type combination of e_1 and e_2. In the above example, **ML12** is *NAME-NAME*.

(d) Overlap features

The overlap feature refers to the statistics for word overlapping between two entities e_1 and e_2. Specific features include the following:

#EB: The number of entities between e_1 and e_2, and **#EB** $= 0$ in the above example;

#WB: The number of words between two entities e_1 and e_2, and **#WB** $= 1$ in the above example;

E1 > E2: A Boolean variable where if e_1 contains e_2, the value is *True*; otherwise, it is *False*. In the above example, this Boolean variable is *False*. Similar features include **E2 > E1**, **ET12+E1>E2**, **ET12+E1 < E2**, **HE12+E1>E2**, and **HE12+E1<E2**.

(e) Base phrase chunking features

Before using this kind of feature, we first need to obtain the phrase parse tree of the sentence. Base phrase chunking features mainly include three categories: ① The phrase head words between (e_1; e_2), including the head words of the first phrase, the last phrase, and the middle phrase; ② The phrase head words in front of the entity e_1, including the head words of the first two phrases; ③ The head words after the entity e_2, including the head words of the latter two phrases. In addition, the path from one entity to the other in the parse tree can also be considered. The specific features are as follows:

CPHBNULL: A Boolean variable where if there is no phrase between e_1 and e_2, the value is *True*; otherwise, the value is *False*. In the above example, this variable is *False*;

CPHBFL: If there is only one phrase between e_1 and e_2, then **CPHBFL** represents the head word of the phrase; otherwise, it is empty. In the above example, **CPHBFL** is *and*;

CPHBF: If there are multiple phrases between e_1 and e_2, then **CPHBF** denotes the head word of the first phrase; otherwise, it is empty. In the above example, **CPHBF** is NULL;

CPHBL: If there are multiple phrases between e_1 and e_2, then **CPHBL** represents the head word of the last phrase; otherwise, it is empty. In the above example, **CPHBL** is NULL;

CPHBO: If there are multiple phrases between e_1 and e_2, then **CPHBO** represents the head word of the phrase except for the first and last phrases; otherwise, it is empty. In the above example, **CPHBO** is NULL.

CPHBE1F: The head word of the first phrase before e_1. In the above example, **CPHBE1F** is NULL;

CPHBE1L: The head word of the second phrase before e_1. In the above example, **CPHBE1L** is NULL;

CPHAE1F: The head word of the first phrase after e_2. As shown in Fig. 10.9, there are four base phrases after e_2, namely, ADVP, VBD, and two

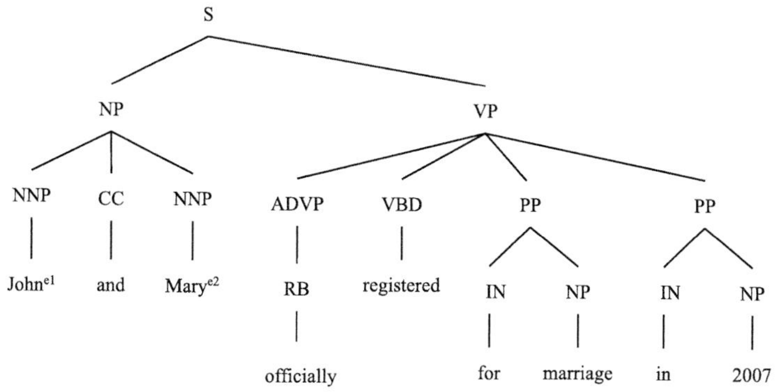

Fig. 10.9 The phrase structure parsing corresponding to the second clause in the example at the beginning of Sect. 10.4

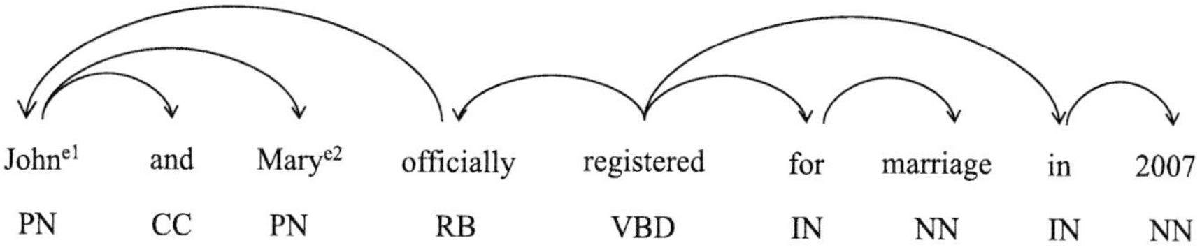

Fig. 10.10 The dependency parsing corresponding to the second clause in the example at the beginning of Sect. 10.4

PPs. The head word of ADVP is *officially*, and therefore, the value of **CPHAE1F** is *officially*;

CPHAE1L: The head word for the second phrase after e_2. In the above example, **CPHAE1L** is *registered*;

CPP: A path connecting two entities e_1 and e_2 in a phrase structure tree. As seen from Fig. 10.9, **CPP** is NNP-NP-NNP;

CPPH: If there are at most two phrases between e_1 and e_2, then **CPPH** denotes the phrase path between e_1 and e_2 associated with head words; otherwise, it is empty. In the above example, **CPPH** is NNP(*John*)-NP-NNP(*Mary*).

(f) Dependency tree feature

Before using this feature, we need to obtain the dependency parse tree for the sentence as shown in Fig. 10.10.

The dependency features are as follows:

ET1DW1: The combination of entity type and dependent word of e_1. In the above example, **ET1DW1** is *PERSON-registered*;

H1DW1: The combination of head and dependent words of e_1. In the above example, **H1DW1** is *John-registered*;

ET2DW2: The combination of entity type and dependent word of e_2. In the above example, **ET2DW2** is *PERSON-John*;

H2DW2: The combination of head and dependent words of e_2. In the above example, **H2DW2** is *Mary-John*;

ET12SameNP: the combination of **ET12** and a Boolean value indicating whether e_1 and e_2 are included in the same noun phrase. In the aforementioned example, **ET12SameNP** is *PERSON-PERSON-True*;

ET12SamePP: the combination of **ET12** and a Boolean value indicating whether e_1 and e_2 are included in the same preposition phrase. In the aforementioned example, **ET12SamePP** is *PERSON-PERSON-False*;

ET12SameVP: the combination of **ET12** and a Boolean value indicating whether e_1 and e_2 are included in the same verb phrase. In the aforementioned example, **ET12SameNP** is *PERSON-PERSON-False*;

(g) Phrase parse tree feature

Phrase parse tree features include the following two:

PTP: Phrase label path from e_1 to e_2 (removing duplicate labels). As shown in Fig. 10.9, **PTP** is NNP-NP-NNP;

PTPH: The phrase label path from e_1 to e_2 combined with the head word of the top-level phrase. In Fig. 10.9, **PTPH** is NNP(*John*)-NP(*John*)-NNP(*Mary*).

(h) Semantic features

In addition to the lexical and various syntactic features, many semantic resources can also be used to enhance the feature representation. A list of trigger words indicating the relation between a country name and a person name is a commonly used resource. The country names are easy to collect, and the trigger words can be obtained in two ways: they can be collected from semantic dictionaries such as WordNet and HowNet or obtained from the training data. The specific features are used as follows:

- Country name list features:

  ET1Country: If e_2 is a country name, then **ET1Country** represents the entity type of e_1;

  CountryET2: If e_1 is the country name, then **CountryET2** denotes the entity type of e_2.

- Trigger word features for relations between people

  ET1SC2: If e_2 triggers the personal social relation, **ET1SC2** represents the combination of the entity type of e_1 and the semantic class of e_2;

  SC1ET2: If e_1 triggers the personal social relation, **SC1ET2** denotes the combination of the entity type of e_2 and the semantic class of e_1.

For an entity pair (e_1, e_2) and the sentences they are in, various discrete features such as lexicon, syntax, and semantics can be extracted in the above manner, and

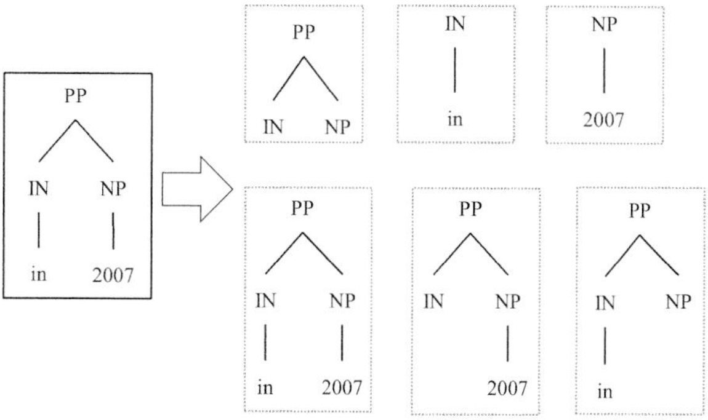

Fig. 10.11 Phrase structure tree and its subtree collection

then the semantic relation between e_1 and e_2 can be predicted using a classifier such as a support vector machine.

(2) Relation classification using kernel functions

Explicit discrete features usually capture the local contexts, and it is difficult to model the similarity between syntactic structures. In many cases, if the syntactic structure of the test sentence s_{test} is very similar to the syntactic structure of sentence s_{train} in the training data, then the entity pairs in s_{test} are very likely to have the same relation type as the entity pairs in s_{train}. Therefore, it becomes a challenge to extract structural features and effectively calculate the similarity between two syntactic structures.

Intuitively, we can extract all the subtrees in a syntactic tree and determine the shared subtrees between the syntactic trees of the two sentences. As shown in Fig. 10.11, six subtrees are extracted from a phrase structure tree containing two leaf nodes. All subtrees can be exhaustively extracted from the tree corresponding to each sentence in the training data. Assume that n different subtrees appear in all sentences and are recorded in the order of occurrence as $\text{subt}_1, \text{subt}_2, \ldots, \text{subt}_n$. Then, the syntactic tree corresponding to any sentence can be represented as an n-dimensional vector, where the i-th element represents the number of times subt_i appears in the tree. If $h_i(T)$ is used to indicate the number of occurrences of subt_i in the syntactic tree T, T can be expressed as $h(T) = (h_1(T), h_2(T), \ldots, h_n(T))$.

Based on the above analysis, the structural similarity between any two sentences can be obtained by calculating the inner product $h(T_1) \cdot h(T_2)$. The method is simple and easy, but the number n of subtrees is very large. The number of nodes increases exponentially with the size of the tree, and it is not easy to exhaustively extract all subtrees. Therefore, efficiently calculating the structural similarity becomes a challenge. Collins and Duffy (2002) proposed a method to calculate $h(T_1) \cdot h(T_2)$ based on tree kernel and convolutional tree kernel and applied the structural similarity into syntax parsing. Later, researchers introduced this method to the relation classification task and further proposed kernel methods based on phrase structure trees (Zelenko et al. 2003), dependency trees (Culotta and Sorensen 2004), convolutional kernels (Zhang et al. 2008), and so on.

We take the phrase structure tree as an example to introduce the calculation method of the inner product $h(T_1) \cdot h(T_2)$ based on tree kernels. Assume that N_1 and N_2 are the sets of nodes in trees T_1 and T_2, respectively. If the subtree rooted at n matches the i-th element in the subtree set, then $I_i(n) = 1$; otherwise, $I_i(n) = 0$. Since $h_i(T_1) = \sum_{n_1 \in N_1} I_i(n_1)$, $h_i(T_2) = \sum_{n_2 \in N_2} I_i(n_2)$, therefore, $h(T_1) \cdot h(T_2)$ can be calculated by the following kernel function $K(T_1, T_2)$:

$$\begin{aligned} K(T_1, T_2) = h(T_1) \cdot h(T_2) &= \sum_i h_i(T_1) h_i(T_2) \\ &= \sum_{n_1 \in N_1} \sum_{n_2 \in N_2} \sum_i I_i(n_1) I_i(n_2) \\ &= \sum_{n_1 \in N_1} \sum_{n_2 \in N_2} C(n_1, n_2) \end{aligned} \tag{10.54}$$

where $C(n_1, n_2) = \sum_i I_i(n_1) I_i(n_2)$ can be solved by the following recursive method:

(1) If the CFG rule[11] with n_1 as the root node in T_1 is different from the CFG rule rooted at n_2 in T_2, then $C(n_1, n_2) = 0$;
(2) If the CFG rule rooted at n_1 in T_1 is the same as the CFG rule with n_2 as the root node in T_2, and both n_1 and n_2 are part-of-speech nodes, then $C(n_1, n_2) = 1$;
(3) If the CFG rule with n_1 as the root node in T_1 is the same as the CFG rule rooted at n_2 in T_2, but n_1 and n_1 are not part-of-speech nodes, then,

$$C(n_1, n_2) = \sum_{j=1}^{nc(n_1)} (1 + C(\mathrm{ch}(n_1, j), \mathrm{ch}(n_2, j))) \tag{10.55}$$

where $nc(n_1)$ denotes the number of child nodes of n_1, and $\mathrm{ch}(n_i, j)$ represents the j-th child node of $n_i (i = 1, 2)$. Since the CFG rule rooted at n_1 in T_1 is the same as the CFG rule with n_2 as the root node in T_2, $nc(n_1) = nc(n_2)$. Collins and Duffy (2002) proved that the above recursive calculation method is equivalent to the naive method that directly calculates $h(T1) \cdot h(T2)$ by exhausting all subtrees, and the computation complexity of the kernel function $K(T_1, T_2)$ is only $O(|N_1| \cdot |N_2|)$.

The above recursive algorithm can be applied to any tree structure, regardless of whether the tree structure is a whole syntactic tree or a subtree. Based on this property, researchers further proposed a method to tackle relation classification using convolutional kernel functions. The basic idea is to select a number of subtrees according to a specified strategy. For example, for the relation classification task,

[11] CFG represents context-free grammar, e.g., VP → PP VP.

the subtrees around the entity pair are selected from the tree structure, and the kernel function calculation can be performed for each subtree according to the above recursive algorithm. Finally, the results of all the kernel functions are summed together to obtain the structural similarity between the two sentences. Unlike the tree kernel method, since the kernel function is calculated between the tree segments of the two syntactic trees, the number of nodes between the tree segments may vary greatly. For example, the tree segment in T_1 contains 10 nodes, while T_2 contains only 3. As a result, the convolutional tree kernel function needs to consider the difference in the node number. Usually, a hyperparameter λ ($0 < \lambda \leqslant 1$) is used to balance the node numbers. The calculation formulas in steps (2) and (3) of the above recursive algorithm are modified to:

$$C(n_1, n_2) = \lambda \tag{10.56}$$

$$C(n_1, n_2) = \lambda \sum_{j=1}^{nc(n_1)} (1 + C(\text{ch}(n_1, j), \text{ch}(n_2, j))) \tag{10.57}$$

Accordingly, the convolutional kernel function is defined as:

$$h(T_1) \cdot h(T_2) = \sum_k \lambda^{\text{size}_k} h_k(T_1) \cdot h_k(T_2) \tag{10.58}$$

where size_k is the number of CFG rules for the k-th subtree fragment.

After determining the kernel function $h(T_1) \cdot h(T_2)$, the relation classification can be modeled using a support vector machine or other classification models. Zhang et al. (2008) proved through experiments that kernel methods can lead to better performance.

10.4.2 Relation Classification Using Distributed Features

Although the methods based on discrete features improve relation classification, they suffer from two issues. First, these methods rely on the quality of part-of-speech tagging and syntactic analysis. Second, discrete features confront sparseness problems and cannot capture the potential semantic similarities between features. To overcome these problems, many researchers have recently proposed relation classification methods based on distributed feature representation and achieved much better classification results. In this section, we introduce the method using convolutional neural networks for relation classification that was proposed in Zeng et al. (2014).

The main idea is that instead of using discrete features, distributed feature representations are employed to overcome the data sparseness problem and bridge the semantic gap. We first learn local distributed representation to capture the

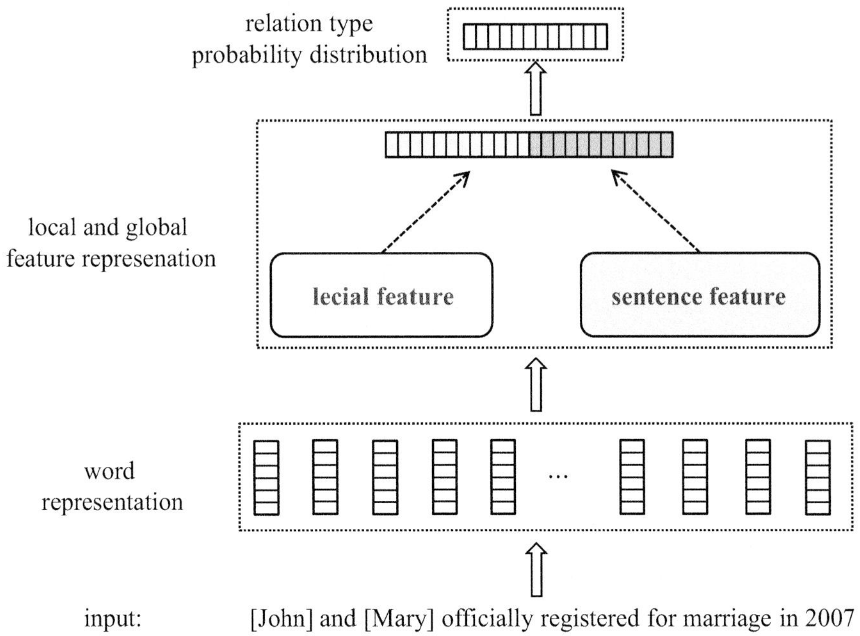

Fig. 10.12 Relation classification method based on distributed feature representation

surrounding contexts of the entity pair. Then, we employ convolutional neural networks to model the global information of the sentence where the entity pair is located. The overall framework of the method is shown in Fig. 10.12.

The input of the model is a sentence $s = (w_1, w_2, \ldots, w_n)$ in which two entities e_1 and e_2 are given. The model first maps each word w_i into a low-dimensional real-valued vector (word embedding) $\boldsymbol{x}_i \in \mathbb{R}^d$, resulting in a list of vectors $\boldsymbol{X} = \{\boldsymbol{x}_1, \boldsymbol{x}_2, \ldots, \boldsymbol{x}_n\}$. Then, word-level and sentence-level representations are learned: ① lexicalized distributed feature representation $\boldsymbol{X}_{\text{lex}} \in \mathbb{R}^{d_1}$; ② distributed feature representation of sentences $\boldsymbol{X}_{\text{sen}} \in \mathbb{R}^{d_2}$. The word-level and sentence-level representations are concatenated to obtain a global feature representation: $\boldsymbol{X}_{\text{final}} = [\boldsymbol{X}_{\text{lex}}; \boldsymbol{X}_{\text{sen}}] \in \mathbb{R}^{d_1+d_2}$. Finally, the linear transformation and the softmax function are used to calculate the probability distribution of the relation types, wherein the one corresponding to the maximum probability is chosen as the relation type between two entities (e_1, e_2):

$$\boldsymbol{O} = \boldsymbol{W}_o \cdot \boldsymbol{X}_{\text{final}} \tag{10.59}$$

$$p(l_i|s, e_1, e_2) = \text{softmax}(O_i) = \frac{e^{O_i}}{\sum_{k=1}^{n_l} e^{O_k}} \tag{10.60}$$

where $W_o \in \mathbb{R}^{n_l \times (d_1+d_2)}$ is the weight matrix, n_l represents the number of relation types, and l_i represents the i-th type. As introduced in Chaps. 3 and 4, word embeddings can be learned through pretraining and fine-tuning. The initial word embeddings can be obtained by pretraining on large-scale unlabeled data using methods such as skip-gram and continuous bag-of-words (CBOW). Fine-tuning optimizes word embeddings on the limited relation classification training set.

The following sections describe the methods for learning word-level distributed representations X_{lex} and sentence-level distributed representations X_{sen}.

(1) Word-level distributed representations

The word-level features are the key clues for predicting the relation type between entity pairs. Three kinds of word-level features can be considered: ① the entities themselves (e_1, e_2); ② the contexts of the two entities; ③ the hypernym of each entity in the semantic knowledge base (such as WordNet in English, HowNet in Chinese, and so on). Since these three kinds of features are all specific words, we can concatenate all these word embeddings to obtain the word-level distribution representation X_{lex}.

(2) Sentence-level Distributed representations

Since the word-level features only consider the local contextual information of the entities, it is usually impossible to capture key clues that indicate the relation type. For example, the key word *marriage* in Fig. 10.12 is far from the two entities *John* and *Mary*, and it is hard to capture using local contexts. Therefore, learning the sentence-level distributed feature representations is a promising solution. Figure 10.13 shows a sentence-level representation learning framework based on convolutional neural networks. Because Chap. 3 has already detailed the CNN-based sentence representation method, the following parts will introduce the core module that is adapted for the relation classification task.

In the relation classification task, the dependence between words (especially between the entity and other words) is a very important feature, and the traditional neural network method cannot capture this dependency information. Therefore, it is often necessary to adapt convolutional neural networks to explicitly model word dependency. To this end, the input word feature WF is represented by the context in a fixed window. For example, WF of the i-th word w_i corresponding to window size 3 is $[x_{i-1}; x_i; x_{i+1}]$, that is, the concatenation of the corresponding word embeddings of the three words. In addition, the position feature (PF) of the word in the sentence is taken as the input, and PF is the vector representation of the relative distance between the word and the two entities (e_1, e_2). For example, the relative distances between *marriage* and the two entities *John* and *Mary* in Fig. 10.12 are 6 and 4, respectively. In the model, the relative distances can also be mapped into continuous real-valued vectors. Assuming that the distances between the words w_i and (e_1, e_2) are d_{i_1} and d_{i_2}, then **PF** is the concatenated representation of the vectors corresponding to the relative distances d_{i_1} and d_{i_2}: PF=$[x_{d_{i_1}}; x_{d_{i_1}}]$. The

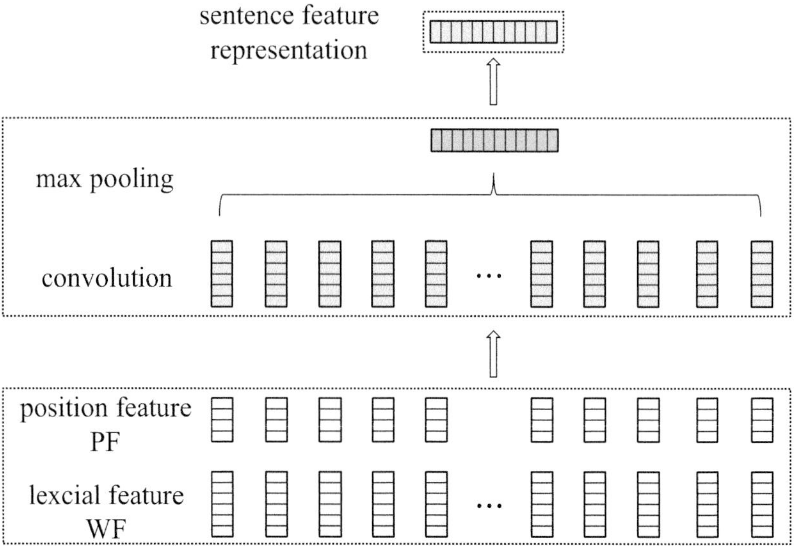

Fig. 10.13 Sentence-level feature representation based on convolutional neural networks

concatenation of the word feature $\boldsymbol{WF}$ and the position feature $\boldsymbol{PF}$ can be used as the final input to the convolutional neural network.

Zeng et al. (2014) found that without using any syntactic and semantic features, the best relation classification performance can also be achieved with word-level and sentence-level distributed representations, where the position feature $\boldsymbol{PF}$ was proven to play a very important role.

10.4.3 Relation Classification Based on Distant Supervision

The previously introduced methods using discrete and distributed features are both supervised models that suffer from two issues in practical use. First these supervised models require a manually annotated corpus of entity relations, which is time-consuming and labor-intensive to create, and the training data size is usually limited. In addition, the annotated corpus is domain specific, and the relation classification performance will decrease considerably for out-of-domain data. Accordingly, researchers resort to the distant supervision method, which can automatically obtain high-confidence samples from massive unlabeled data and treat these samples as annotated data. Then, supervised methods can be employed to learn a relation classification model (Mintz et al. 2009).

Existing open semantic knowledge bases (such as Freebase and HowNet) are important resources used by distant supervised methods. A large number of entity relation examples (e_1, e_2, r) are provided in the semantic knowledge base, such as (John, Mary, spouse). Mintz et al. (2009) extracted 1.8 million entity relation examples that include 940,000 English entities and 102 relations from Freebase. Distant supervision uses these instances as seeds to automatically label unlabeled data.

The basic idea is as follows. For a given entity relation example (e_1, e_2, r) in the semantic knowledge base, if a sentence s from the massive unlabeled data contains the entity pair (e_1, e_2), then the relation between e_1 and e_2 in that sentence s is very likely to be r. Therefore, some distinct features for this relation r can be extracted from s. Taking the entity relation instance (John, Mary, spouse) as an example, the entity pair (John, Mary) appears in the unlabeled sentences below:

John and his wife Mary attended the ceremony. Mary successfully gave birth to a girl, and John became a father on that date.

Here, the distant supervised model will assume that there is a *spouse* relation type between the two entities in the two sentences. From these sentences, complementary features such as various lexicalized contexts and syntactic information can be extracted to enrich the features used in the original model.

To minimize the impact of noise, the distant supervised model uses a feature merging technique. For an entity relation instance (e_1, e_2, r), if there are n sentences in the unlabeled data including the same entity pair (e_1, e_2), then the features extracted from these n sentences are combined and used in combination as an additional feature. For example, the feature extracted from the sentence *John and Mary are both famous basketball players* has nothing to do with the *spouse* relation. If the features extracted from this sentence are directly used alone as features, it will become noise. Feature combination will somewhat lower the influence of the irrelevant features.

Mintz et al. (2009) proved in experiments that distant supervision could perform with 67.6% accuracy in relation classification.

10.4.4 Evaluation of Relation Classification

The evaluation of relation classification methods generally focuses on the precision, recall, and F_1 scores. Given a test set, assuming that the manually labeled entity relation result is R and the automatically recognized result is O, then the precision, recall, and F_1 scores are calculated as follows:

$$\text{precision} = \frac{|O \cap R|}{|O|} \times 100\% \tag{10.61}$$

$$\text{recall} = \frac{|O \cap R|}{|R|} \times 100\% \tag{10.62}$$

$$F_1 = \frac{2 \times \text{precision} \times \text{recall}}{\text{precision} + \text{recall}} \tag{10.63}$$

where $|O|$ and $|R|$ denote the number of entity relations in the system output and the reference, respectively. $|O \cap R|$ represents the number of entity relations where the system output matches the reference.

10.5 Event Extraction

An event is a specific occurrence involving participants. An event has several properties, such as event type, participants, time, location, reasons, consequences, and so on. Compared to entity recognition and relation classification, event extraction is a more complicated task, and different types of events correspond to different structures. For example, the *company acquisition* event includes *acquirer*, *acquiree*, *price*, etc. *end-position* events include *position*, *employer*, *person*, *time*, etc. The variability of events makes event extraction a major challenge in the open domain. In this section, we introduce event extraction tasks in specific domains.

10.5.1 Event Description Template

In the evaluations of event extraction organized by MUC, ACE, and TAC, the event definition and the event types are slightly different. We use the annotation standard of ACE 2005, which defines a total of 8 major event types and 33 subtypes. Contestants are required to train models using the given annotated corpus, use the trained model to identify specific event types from unseen test data, and finally fill the slots of predefined event templates.

Each event type corresponds to a template. Table 10.3 lists the event types labeled by ACE 2005.

An event is usually described within a sentence, in which there is always a keyword (e.g., *born* and *leave* in the examples below) indicating the event types. Such words are called *triggers*.

[Andy Mike] was born in [New York] [in 1969].

[On March 22], [Baidu Chief Scientist Andrew] announced on Twitter that he will leave [Baidu].

The main task of event extraction is to determine the event type and extract the corresponding elements of this event type. Trigger words are the core elements that determine the event type, and thus they are the key to event extraction. The event elements belong to two categories: participants and attributes, which are usually referred to as event arguments.

The event participants are the entities that are involved in this event and usually named entities such as persons and organizations. The event attributes consist of

Table 10.3 Annotated event type in ACE 2005

Event type (major-type)	Event type (subtype)
Life	Be-Born, Marry, Divorce, Injure, Die
Movement	Transport
Transaction	Transfer-ownership, Transfer-money
Business	Start-Org, Merge-Org, Declare-Bankruptcy, End-Org
Conflict	Attack, Demonstrate
Contact	Meet, Phone-Write
Personnel	Start-Position, End-Position, Nominate, Elect
Justice	Arrest-Jail, Release-Parole, Trial-Hearing, Charge-Indict, Sue, Convict, Sentence, Fine, Execute, Extradite, Acquit, Appeal, Pardon

Table 10.4 Template for the *Be-Born* event

Trigger	Born
Person-Arg	Andy Mike
Time-Arg	1969
Place-Arg	New York

Table 10.5 Template for *End-Position* event

Trigger	Leave
Person-Arg	Andrew
Entity-Arg	Baidu
Position-Arg	Chief Scientist
Time-Arg	March 22

two categories: general attributes and event-specific attributes. Since the location, time, and duration appear in almost all events, such attributes are called general event attributes. The event-specific attributes are determined by the specific event type, such as the *crime* attribute in the *Convict* event and the *position* attribute in the *Start-Position* event. Considering all participants and attributes, there are a total of 35 different arguments in the annotation system of ACE 2005.

The triggers and arguments for events of each type can be organized and represented by a template, which can be a universal template or a specific template related to the event type. The universal template contains 36 slots; one of these needs to be filled with trigger words, and the remaining slots correspond to 35 different arguments. Since the event arguments for various types of events are quite different, and each event type only activates a few of the 36 slots in the universal template, specific templates are usually adopted for each specific event type.

Tables 10.4 and 10.5 are specific templates for the two events *Be-Born* and *End-Position*, respectively. The specific template corresponding to the remaining event types can be referred to LDC (2005).

After the templates describing events are determined, the event extraction task is converted into a slot filling task. That is, it finds the trigger word and the event arguments and then fills them into the slots corresponding to a specific template. In addition to event types and event arguments, the overall attributes of an event are often useful in IE. The overall attributes mainly include the following four categories: polarity (*positive* or *negative*), modality (*known* or *unknown*), genericity (*specific* or *universal*), and tense (*past*, *present*, *future*, or *unspecified*).

10.5.2 Event Extraction Method

(1) Pipeline event extraction method

Ahn (2006) proposed a pipeline method for event extraction, which divided event extraction tasks into four subtasks: (a) trigger detection, that is, to detect the trigger and determine the event type; (b) argument identification, detecting which entity mentions and values are arguments of the specific event; (c) attribute assignment, determining the specific attribute values of the modality, polarity, genericity, and tense of the event; and (d) event coreference, determining whether different event mentions refer to the same event. Ahn treats each subtask as a classification problem, designed corresponding features for each subtask, and then trained the model using the same classifiers, such as the maximum entropy model and the support vector machine.

Compared with the latter two subtasks, the first two are more important. Accordingly, we mainly introduce the methods for the first two subtasks of trigger detection and argument identification. The main contribution of Ahn's pipeline approach lies in its feature design. Therefore, we focus on the features used in Ahn's method.

Events are occasionally triggered by multiple words (or phrases), but researchers find through analysis that more than 95% of the trigger words are single words, so the detection of trigger words can be simply regarded as word classification. Furthermore, trigger words are often verbs, nouns, and pronouns. Therefore, trigger detection is further simplified into the multiclass classification problem of specific parts-of-speech. There are a total of 34 categories, of which 33 are event types and the remaining one is a *None* class, indicating that it is not a trigger for any event. For example, in the aforementioned examples, *Andy Mike*, *1969*, *born*, and *New York* can be used as trigger candidates. The ideal model can classify *Andy Mike*, *1969*, and *New York* into the *None* class and correctly identify *born* as the *Be-Born* event. To achieve high classification performance, Ahn designed the following features:

(a) Lexical features: including the full word, lowercased word, lemmatized word, POS tags, depth of a word in the parse tree;
(b) Semantic features: for the detection of trigger words in English, with the help of WordNet, if the candidate word belongs to the type verbs, nouns, adjectives, or adverbs and there is a corresponding entry in WordNet, the synset of the first sense will be regarded as a feature;
(c) Contextual features: including three words before and after the candidate word and their POS tags;

(d) Dependency features: if the candidate word is dependent on a certain dependency relationship, then the relation label, the dependent word, the part-of-speech, and the entity category will all be regarded as features.

According to the above features, training instances can be extracted from the ACE annotated corpus to optimize the classifier. Through analyzing the ACE data, it can be found that trigger words account for less than 3% of all words; that is, most words are not trigger words. As a result, the 34-class classification faces a serious imbalance problem. To address this issue, a two-step strategy is more appropriate: the first step is to train a binary classifier to filter out non-trigger words, and the second step is to train a multiclass classifier to determine which event type the trigger word belongs to. Experiments have shown that the two-step strategy helps to achieve better performance.

For argument identification, it is usually assumed that candidate entities such as name entities, time, and proper nouns have been recognized (can be implemented using the entity recognition and disambiguation methods introduced in the earlier section), such as the name *Andy Mike*, time *1969*, and location *New York*. The argument identification task can be converted into a multiclassification problem with 36 types. Since there are 35 argument types in the ACE annotation data, there are a total of 36 types, including a *N*one type. The argument identification task can be converted into a multiclassification problem with 36 types. Similar to the trigger detection task, it also faces a serious class imbalance problem: more than 70% of the candidate entities do not belong to any argument. That is, the *None* type occupies more than 70%. In addition, there is another phenomenon that needs to be addressed. Each event type involves far fewer than 36 types of arguments. For example, the *Be-Born* event includes only three arguments. Therefore, the argument identification task can be converted to a multiclassification problem for specific event types. For example, after the trigger word *born* is correctly detected, the event type is confirmed, and the candidate entities *Andy Mike*, *1969*, and *New York* can only be classified into four argument categories (*person*, *time*, *place*, and *None*).

No matter whether it is a 36-class classification model or a multiclassification model for specific event types, feature design is still the key component. Ahn designed the following features:

(a) Event trigger and type features: the trigger word itself, its POS tag, the depth of the trigger word in the parse tree, and the event type;
(b) Entity head features: entity head word, its POS tag, and its depth in the parse tree;
(c) Entity determiner if any;
(d) Entity type and mention type: mention types include name, pronoun, and other nouns; entity types include person, location, organization, time, and so on;
(e) Dependent path between entity head word and trigger word; the dependency path is a sequence of words, POS tags, and dependency labels.

Experiments on the ACE 2005 dataset show that multiclassification models for specific event types can achieve better classification results than the 36-class classification model.

(2) Joint event extraction model

Pipeline-based event extraction inevitably faces error propagation problems: the errors of the previous modules propagate to the subsequent modules and will continue to be amplified. Meanwhile, the subsequent modules cannot help the decision process of the previous modules.

For example, if the trigger detection is incorrect, the subsequent argument identification cannot be correct. At the same time, the results of the argument identification cannot be employed to help in trigger detection. In fact, trigger words and event arguments influence each other in many cases. In the following examples, *fired* is the trigger for the *Attack* event in the first example, while in the second example, *fired* is the trigger for *End-Position*. In the first example, if the model correctly identifies *Peter* as the *victim* argument in the *Die* event, this result can be helpful to determine *fired* in the sentence as the trigger for the *Attack* event. Similarly, if the *Cleveland Cavaliers* in the second example is correctly identified as the NBA club, it will help to predict that *fired* is the trigger for the *End-Position* event.

[*P*eter] died when [gunmen] fired on the [crowd] with AK-47.

[*T*he Cleveland Cavaliers] fired the head coach [Henry].

In addition, there may be multiple events in the same sentence. As shown in the first example, there are two events of *Attack* and *Die*. The pipeline method cannot capture the dependencies between triggers and arguments belonging to different events. Figure 10.14 illustrates the correct results for the trigger words and event arguments for the two events in the first example. The pipeline method independently extracts these two events, and it is very likely that *Peter* cannot be identified as the goal of the *Attack* event. Ideally, we should make full use of the global information to pass the *victim* argument *Peter* of the *Die* event to the *Attack* event.

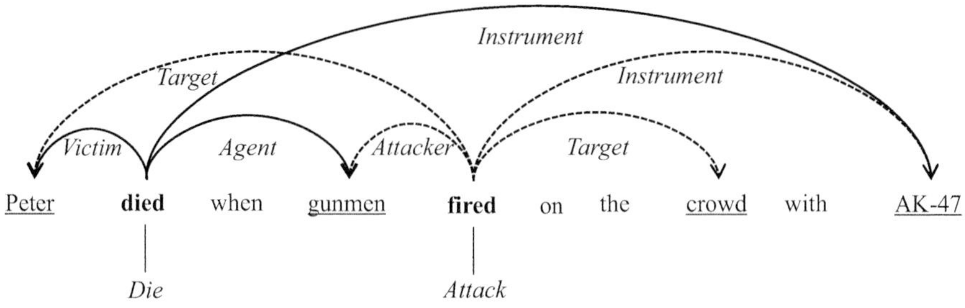

Fig. 10.14 *Attack* and *Death* share the two arguments *victim* and *instrument*

To solve the above problems, Li et al. (2013b) proposed a joint labeling algorithm for both trigger detection and argument identification. The event extraction task is regarded as a structural learning problem, and the structural perceptron model is employed to simultaneously identify both triggers and event arguments. This approach not only captures the dependencies between triggers and arguments across different events but also leverages global information. The details of this algorithm are elaborated below.

First, we formalize the joint labeling task. The event types are represented by $L \cup \{\varnothing\}$, where L contains 33 event types, and $\varnothing$ indicates that the candidate word is not a trigger for any event type. $R \cup \{\varnothing\}$ represents the argument set. R contains 35 event arguments, and $\varnothing$ indicates that the candidate is not an argument for the focal trigger word.

The input of the algorithm is a sentence consisting of n words or phrases $x = (x_1; x_2; \cdots; x_n)$ and a list of candidate arguments $\epsilon = \{e_k\}_{k=1}^m$. For the sentence in the first example above, $n = 10$, $\{e_k\}_{k=1}^m = \{Peter,\ gunmen,\ crowd,\ AK\text{-}47\}$. Therefore, the input can be expressed by $x = \langle (x_1; x_2; \cdots; x_n), \{e_k\}_{k=1}^m \rangle$.

The output y of the algorithm can be represented by the following equation:

$$y = \langle t_1, (a_{11}, \ldots, a_{1m}), \ldots, t_n, (a_{n1}, \ldots, a_{nm}) \rangle \tag{10.64}$$

where $t_i \in L \cup \{\varnothing\}$ is the trigger marker (event type) of the i-th word or phrase x_i, and $a_{ij} \in R \cup \{\varnothing\}$ indicates that the candidate argument e_j is an argument of the event type t_i. Take the simplest event *Bill Gates founded Microsoft* as an example; the input and correct output are:

$$x = \langle (\textit{Bill Gates}, \textit{founded}, \textit{Microsoft}), \{\textit{Bill Gates}, \textit{Microsoft}\} \rangle \tag{10.65}$$

$$y = \langle \varnothing, (\varnothing, \varnothing), \text{Start\_Org}, (\text{Agent}, \text{Org}), \varnothing, (\varnothing, \varnothing) \rangle \tag{10.66}$$

where $n = 3$, $m = 2$, $\{e_k\}_{k=1}^m = \{\textit{Bill Gates, Microsoft}\}$. In the output result y, Start_Org indicates that the second word *founded* is a trigger belonging to the *Start_Org* event. Agent indicates that *Bill Gates* is the founder and Org indicates that *Microsoft* is the company founded in the *Start_Org* event. The goal of the joint labeling algorithm is to accurately generate the result y for any x, which can be solved by the following objective function:

$$y = \underset{y' \in \mathscr{Y}(x)}{\operatorname{argmax}}\, \boldsymbol{W} \cdot \boldsymbol{F}(x, y') \tag{10.67}$$

where $\boldsymbol{F}(x, y')$ represents a feature vector and $\boldsymbol{W}$ is a corresponding feature weight. The feature vector can be defined as the features in the pipeline method and some global features. The parameter $\boldsymbol{W}$ can be optimized based on a perceptron model with an online update algorithm. If z is the gold result on x, and y is the model result, then parameter $\boldsymbol{W}$ can be updated by the following formula:

$$\boldsymbol{W} = \boldsymbol{W} + \boldsymbol{F}(x, z) - \boldsymbol{F}(x, y) \tag{10.68}$$

Algorithm 7: Parameter training algorithm for joint event extraction

```
Input  : Training Dataset D = {x^i, z^i}_{i=1}^N, Maximum iteration T.
Output: Feature weight parameter W.
#Initialization
W = 0
for t = 1, ..., T do
    #Online update for each training instance.
    foreach (x, z) ∈ D do
        y = beamSearch(x, z, W)
        #Parameter update
        if y ≠ z then
            W ← W + F(x, z_{1:|y|}) − F(x, y)
        end
    end
end
```

The detailed training procedure is given in Algorithm 7 (Huang et al. 2012).

For each training instance (x, z) in the training dataset D, the model predicts result y for x using a beam search algorithm. If the predicted output y is inconsistent with the gold result z, the parameter $\boldsymbol{W}$ is updated using a perceptron algorithm. This training process can be repeated T times on the training dataset D. The beam search algorithm is the core component, and the details can be found in Algorithm 8.

The beam search algorithm initializes with an empty stack B and then focuses on each position of the input sentence from left to right (lines 2–3 in the algorithm). For the word or phrase x_i to predict, we enumerate the possible results of being a trigger word and retain the best K candidates (lines 5–6). If this is the parameter training process, it is necessary to compare whether the current model output matches the gold result. If not, the process terminates early (lines 8–9). After candidate trigger prediction, we classify the candidate arguments (lines 12–22): for each candidate argument e_k, examine all of the trigger candidates in the stack B (line 14). If x_i is the trigger word, put all possible argument types into the buffer (line 19) and retain the best K candidates (line 22). Lines 23–24 are similar to lines 8–9. The algorithm iterates until the last position of the sentence. Finally, the model outputs the best prediction in stack B (line 28).

(3) Event extraction model based on distributed representation

The joint event extraction method not only considers the relationship between trigger detection and argument classification but also fully exploits the dependency information between multiple events. However, similar to all methods based on discrete features, the joint event extraction model is also unable to capture the semantic similarity between words, and it is difficult to utilize the deep semantic features on the sentence level. Therefore, joint event extraction delivers limited performance.

Algorithm 8: Beam search algorithm for joint event extraction

---

Input : Sentence and candidate event element $x = \langle (x_1, x_2, \dots x_n), \{e_k\}_{k=1}^m \rangle$; real label z in
training procedure; beamSize k, event type set $L \cap \{\varnothing\}$, event argument set $R \cap \{\varnothing\}$.

Output: Optimal prediction sequence of x.

1 #Set storage space for beam, initialized with empty
2 $B \leftarrow [\epsilon]$
3 **for** $i = 1, \dots, n$ **do**
4   #Trigger word prediction
5   buf $\leftarrow \{y'\Diamond l | y' \in B, l \in L \cup \{\varnothing\}\}$
6      $B \leftarrow$ Kbest(buf)
7   #Early update
8   **if** $z_{1:t_i} \notin B$ **then**
9   return $B[0]$
10   **end**
11   #Argument prediction
12   **foreach** $e_k \in \{e_k\}_{k=1}^m$ **do**
13   buf $\leftarrow \varnothing$
14   **foreach** $y' \in B$ **do**
15   buf $\leftarrow$ buf $\cup \{y'\Diamond\varnothing\}$
16   #x_i is trigger word
17   **if** $y'_{t_i} \neq \varnothing$ **then**
18   #Consider all argument types
19   buf $\leftarrow$ buf $\cup \{y'\Diamond r | r \in R\}$
20   **end**
21   **end**
22         $B \leftarrow$ Kbest(buf)
23   **if** $z_{1:a_{ik}} \notin B$ **then**
24   return $B[0]$
25   **end**
26   **end**
27 **end**
28 return $B[0]$

In recent years, an increasing number of methods based on distributed feature representation have been employed for event extraction tasks, and they have proven to be much better than methods using discrete features. The main idea behind these is that distributed real-valued vectors are first used to represent words to overcome sparsity problems and capture semantic similarities between words. Then, deeper and global features are further learned. Finally, the classification algorithm is utilized to perform the trigger detection and argument classification tasks.

This section takes the method proposed in Chen et al. (2015b) as an example to introduce the application of distributed feature representations in event extraction tasks. From the perspective of machine learning, the method based on distributed representations also follows the data-driven methodology for event extraction and continues to decompose this task into trigger detection and argument classification. Moreover, the two cascaded subtasks are both considered as a multiclassification

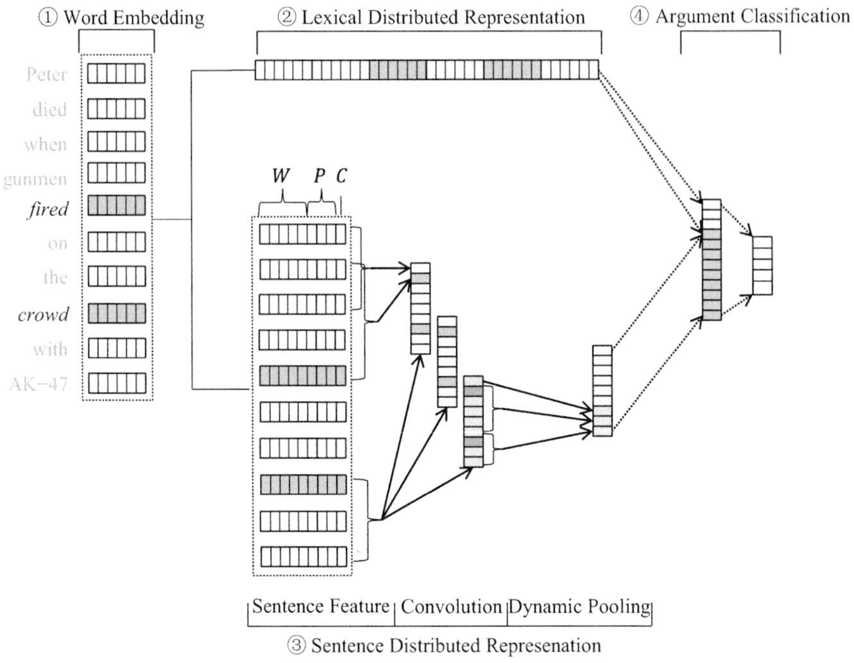

Fig. 10.15 Argument classification model based on distributed feature representation

problem. Thus, the same neural network architecture can be employed for each subtask.

Compared with the trigger detection task, the event argument classification task is more complicated. Therefore, we mainly introduce the methods of distributed feature representation in the argument classification task and then discuss the model adaptation required for the trigger detection task.

We assume that the event is expressed by a sentence. Once the trigger word t is detected, the argument classification task aims to determine whether an argument candidate is a true argument of the trigger t and further decide which type of argument. Suppose we now need to classify the argument type of *crowd* with respect to the trigger *fired* for the example sentence introduced in the previous section. Figure 10.15 shows the flow chart of the process of argument classification for this example.

As shown in Fig. 10.15, the argument classification task is divided into three parts. The first part is the word embedding, which maps each word into a low-dimensional real-valued vector. The second part is the distributed representation of lexical and sentence-level features, including (a) distributed representations for combinations of different levels of lexicalized features and (b) distributed representations of global features at the sentence level. The third part is the argument classification module that determines the type of argument candidate based on lexical and sentence-level distributed representations.

Word embedding is the foundation of the whole process. Due to the limited scale of the training data for event extraction, it is difficult to obtain high-quality word embeddings using limited annotated data. Therefore, word embeddings can be pretrained on large-scale unlabeled monolingual data, e.g., Wikipedia pages. In the previous chapter of this book, we introduced several efficient methods to learn word embeddings, such as CBOW and skip-gram. These pretrained word embeddings can be further fine-tuned together with optimizing the model, as shown in Fig. 10.15.

The second part is the core of the argument classification model. After each word in the sentence is mapped into a low-dimensional real-valued vector, we will learn two types of feature representations, namely, lexicalized and sentence-level features. For the lexicalized features, we model the context centering upon the trigger word t and the candidate event argument e. Specifically, if window size K_l is used, the word embeddings of t, e, and the K_l words before and after them are concatenated to form a long vector representation R_l, serving as the lexicalized distributed features.

Since the local context cannot capture global information, it is easy to ignore important clues. For example, in Fig. 10.15, if the context window $K_l = 1$, the indicative information *AK-47* cannot be contained in the context of the argument candidate *crowd*. Therefore, it is necessary to learn the global sentence-level features. Convolutional neural networks are typically employed for sentence representation learning and have been successfully applied to tasks such as text classification, sentiment analysis, and machine translation. However, classical convolutional neural networks are not suitable for direct application to the argument classification task because there may be multiple trigger words and candidate arguments in a sentence. The convolutional neural network does not model position information, so it is difficult to obtain a global sentence representation that is sensitive to trigger words and candidate arguments. To address this issue, Chen et al. (2015b) proposed a dynamic convolutional neural network model, as shown in Fig. 10.15. The dynamic convolutional neural network is divided into three modules: ① surface feature input; ② convolutional operation; ③ dynamic pooling.

Surface features include word embeddings, relative position information, and event types. For the i-th word w_i in the sentence, we first obtain the corresponding word embedding $\boldsymbol{x}_i \in \mathbb{R}^{d_w}$. Then, we calculate the distance dis_{it} between w_i and the trigger word t and the distance dis_{ie} between w_i and the candidate event argument e. For example, in Fig. 10.15, the relative distance between the word *gunmen* and the trigger word *fired* is 1, and the distance between *gunmen* and the candidate argument *crowd* is 4. The distances are then mapped into continuous vectors $\boldsymbol{x}_{it}$ and $\boldsymbol{x}_{ie} \in \mathbb{R}^{d_d}$. Next, the event type corresponding to the trigger word t is mapped into a vector $\boldsymbol{x}_c \in \mathbb{R}^{d_c}$. Finally, $\boldsymbol{x}_i$, $\boldsymbol{x}_{it}$, $\boldsymbol{x}_{ie}$, and $\boldsymbol{x}_c$ are concatenated as the input $\boldsymbol{L}_i \in \mathbb{R}^d$ corresponding to the i-th word w_i, where $d = d_w + 2 \times d_d + d_c$. Given a sentence s containing n words $s = w_1 \cdots w_i \cdots w_n$, the surface feature input is an $n \times d$ matrix $\boldsymbol{L}_{1:n}$.

The convolutional operation aims to summarize the global semantic information from the surface features $\boldsymbol{L}_{1:n}$ with filters. A filter $f_k \in R^{h \times d}$ scans from the first word to the last word of the sentence at the pace of a window including h words, and each window $\boldsymbol{L}_{i:h+i-1}$ leads to an output value.

$$v_{ki} = f(\boldsymbol{W}_k \cdot \boldsymbol{L}_{i:h+i-1} + b_k) \tag{10.69}$$

where $\boldsymbol{W}_k$ and b_k are weight and bias, respectively. f is a nonlinear activation function. By traversing each window, f_k will obtain a $(n-h+1)$ dimensional vector $v_k = [v_{k_1}, \ldots, v_{k_i}, \ldots, v_{k_{n-h+1}}]$. If K filters are used, a matrix of $K \times (m-h+1)$ dimensions will be obtained. Since n is the length of a sentence, the dimensions of the resultant vectors will vary for sentences of different lengths. Thus, a pooling operation is required.

Max pooling and average pooling are the most commonly used pooling operations. Max pooling selects the largest element from the vector v_k as a typical feature. Since conventional max pooling is insensitive to positions, it is difficult to reflect the contributions of the trigger word and the candidate arguments. As a result, the pooling method needs to be adapted. The dynamic max pooling method is proposed for the event extraction task, as it is a pooling method that is sensitive to the positions of trigger words and candidate arguments. As shown in Fig. 10.15, the vector $\boldsymbol{v}_k$ is dynamically divided into three parts according to the trigger word and the candidate argument: $\boldsymbol{v}_{k,1:e}, \boldsymbol{v}_{k,e+1:t}, \boldsymbol{v}_{k,t+1:n-h+1}$, where e and t represent the position of the candidate argument and the trigger word, respectively. If the trigger word is in front of the candidate event argument, then $\boldsymbol{v}_{k,1:t}, \boldsymbol{v}_{k,t+1:e}, \boldsymbol{v}_{k,e+1:n-h+1}$. Finally, we can select the maximum output from the three vectors $\boldsymbol{v}_{k,1:e}, \boldsymbol{v}_{k,e+1:t}$ and $\boldsymbol{v}_{k,t+1:n-h+1}$ and obtain a three-dimensional vector. The dynamic pooling output of the K filters is concatenated, and finally, a vector with a fixed dimension $\boldsymbol{R}_s \in \boldsymbol{R}^{K\times 3}$ is obtained. We can see that each dimension in the vector is sensitive to the trigger word, and the candidate argument and the vector can better represent the sentence semantics.

The third part of the argument classification model uses a feed-forward neural network, which takes as input the representation $[\boldsymbol{R}_l; \boldsymbol{R}_s]$ obtained by concatenating lexical features $\boldsymbol{R}_l$ and sentence-level features $\boldsymbol{R}_s$. The softmax function is employed to calculate the probability distribution of the candidate argument e given the trigger word t.

Returning to the trigger detection task, we can use the same framework as shown in Fig. 10.15, but the model is simpler. Unlike event argument classification, in which the input includes a sentence, a trigger word, and a candidate argument, the input of the trigger detection task is a sentence and a candidate trigger word. Therefore, the learning of lexicalized and sentence-level features in the second part of Fig. 10.15 needs to be adjusted accordingly. For lexicalized feature learning, it is only necessary to take the words centering on the candidate trigger as the context. For sentence-level feature learning, dynamic pooling only takes the candidate trigger word as the segmentation point and chooses the maximum values from the left and right convolution vectors, respectively. The classification model of the third part in Fig. 10.15 remains unchanged.

Experiments show that the event extraction model based on distributed feature representation can achieve much better performance.

10.5.3 Evaluation of Event Extraction

The general method for evaluating the event extraction model is as follows. Given a test set $\text{Test}_{\text{event}}$, human experts correctly label the events in $\text{Test}_{\text{event}}$, and the results serve as reference $\text{Ref}_{\text{event}}$. The event extraction model M automatically extracts events on the test set $\text{Test}_{\text{event}}$ to obtain the results $\text{Model}_{\text{event}}$. By comparing the prediction result $\text{Model}_{\text{event}}$ with the reference $\text{Ref}_{\text{event}}$, the precision, recall, and F_1 score can be calculated.

Almost all the methods divide the event extraction task into two steps, trigger detection and event argument classification, and further divide trigger detection into two subtasks, trigger word location and event type classification; and event argument classification is also divided into argument identification and argument classification. Therefore, an objective evaluation generally tests these four subtasks separately.

If the model correctly locates the trigger word in the event description, the trigger word is correctly recognized. Given the trigger word, if the event type is also correctly predicted, the result of the event type classification is correct. If a candidate argument is correctly identified as the associated attribute of the trigger word, the event argument identification is correct. If the correctly identified event argument is further predicted to be the correct type, the final argument classification result is correct. The precision, recall, and F_1 score can be easily calculated based on these results.

10.6 Further Reading

In summary, IE includes multiple interdependent tasks, such as entity recognition, relation classification, and event extraction. Currently, this community is still dominated by cascaded methods that mainly focus on specific domains.

From the perspective of methods, the deep learning method has become the dominant model for each subtask of IE. There is a trend toward exploring more effective models. For example, Miwa and Bansal (2016) and Peng et al. (2017) used more expressive neural networks to model relation extraction, such as TreeLSTM and GraphLSTM. Narasimhan et al. (2016) and Wu et al. (2017) optimized IE models with advanced methods such as reinforcement learning and adversarial learning. To reduce error propagation, the joint modeling of two or more tasks is also a focus of many researchers. For example, the joint modeling of entity recognition and relation classification (Li and Ji 2014; Zheng et al. 2017) formalizes these two tasks into a unified sequence labeling problem. To utilize more contexts, global optimization and inference for IE tasks has also become a potential research direction (Zhang et al. 2017).

From the view of data, the training data for each task in IE is very limited, and it is difficult to support sophisticated machine learning models. The question of how to automatically generate large-scale high-quality annotated training data for IE has become a hot research topic, and a popular approach proposed in recent years is the distant supervision method based on a knowledge base (Mintz et al. 2009; Riedel et al. 2010; Hoffmann et al. 2011; Surdeanu et al. 2012; Zeng et al. 2015; Lin et al. 2016; Chen et al. 2017b; Luo et al. 2017). However, the distant supervision method faces the problem of noisy and erroneous data. For example, not all sentences containing *Yaoming* and *Yeyi* indicate a *spouse* relationship. Therefore, it is challenging to reduce the influence of noise as much as possible. Recently, some researchers have proposed a multi-instance learning method based on the *at least one positive example* hypothesis (Zeng et al. 2015) and another model based on the selective attention mechanism (Lin et al. 2016). In addition to the use of distance supervision methods in relation classification, this method has also been introduced into event extraction tasks, producing a large number of annotated event data (Chen et al. 2017b). In addition, an efficient crowdsourcing approach has also become a reasonable strategy for expanding training data (Abad et al. 2017).

From the perspective of application, academic research still mainly concentrates on IE technology in specific domains. However, in real applications, especially in the era of big data, the IE technology of the open type and open domain is more demanding. Therefore, research on open domain IE has received increasing attention. Open domain entity extraction tasks focus on entity expansion technology in open text (data in Internet) (Pennacchiotti and Pantel 2009; Jain and Pennacchiotti 2010). Open domain relation extraction focuses on mining entity relations without predefined relation types (Banko et al. 2007; Mausam et al. 2012; Angeli et al. 2015; Stanovsky and Dagan 2016). Event aggregation and new event prediction focus on the aggregation of multiple events and new event detection without predefined event types (Do et al. 2012; Huang and Huang 2013).

Exercises

10.1 Named entity recognition (NER) is usually performed as a sequence labeling task based on the character level. Please comment on the advantages and disadvantages of character-based NER compared to word-based NER.

10.2 Please identify and analyze the reason why the semisupervised method could improve performance in NER.

10.3 In the entity disambiguation task, the entity set is usually known but incomplete in real applications. How can entity disambiguation be performed in this scenario? Could clustering-based methods and linking-based approaches be combined in a unified framework?

10.4 Please discuss the reasons why position embeddings are helpful in relation classification methods using distributed features.

10.5 Please comment on the challenge of distance supervision for relation classification and compare the related work addressing the noise resulting from distant supervision.

10.6 Event extraction is usually based on the assumption that the event occurs in a sentence. Please point out the challenges of the event extraction task if the events occur across the sentences.

Chapter 11
Automatic Text Summarization

11.1 Main Tasks in Text Summarization

Automatic text summarization aims to compress the input texts and generate a condensed summary. Generally, the short summary must satisfy the requirements of sufficient information by covering the main content of the original text and providing low redundancy and high readability. It has become an important means of data mining, information filtering, extraction, and recommendation.

In 1958, H.P. Luhn first proposed the idea of automatic summarization (Luhn 1958), which opened the prelude to the study of automatic text summarization. With the upsurge of the information age, there is an urgent need for a technology that can help users obtain useful information efficiently and quickly and enable them to understand the gist of news events in a short time to reduce reading time. This demand has promoted the rapid development of automatic summarization technology, moving it gradually toward maturity. Since 2000, the evaluation of text automatic summarization technology (DUC[1], TAC[2] etc.) organized by the National Institute of Standards and Technology (NIST) has further accelerated research into this technology, attracting the attention of more researchers and entrepreneurs.

Automatic text summarization technology can be divided into different types from different perspectives. Considering the function of the summary, it can be divided into indicative, informative, and critical summarizations. The indicative summarization provides only key topics of the input document (or collections of documents). It aims to help users decide whether they need to read the original text, such as through title generation. Informative summarization provides the main information of the input document (or collections of documents) so that users do not need to read the original text. Critical summarization provides not only the main

[1] https://duc.nist.gov./.

[2] https://tac.nist.gov/about/index.html.

© Tsinghua University Press 2021

C. Zong et al., *Text Data Mining*, https://doi.org/10.1007/978-981-16-0100-2_11

information of the input document (or document sets) but also key comments on the original text.

Based on the number of input documents, automatic summarization can be divided into single-document and multidocument summarization. Considering the relation between input and output languages, automatic summarization can be divided into monolingual, crosslingual, and multilingual summarization. The input and output of monolingual summarization are in the same language, while the input of crosslingual summarization is in one language (e.g., English), and the output is in another (e.g., Chinese). Multilingual summarization takes documents in multiple languages as input (e.g., English, Chinese, and French) and produces the summary in one of the input languages (e.g., English).

Considering the different applications, automatic summarization technology can be divided into generic summarization and query-based summarization. The former summarizes the main points of the original author, while the latter provides content closely related to the user's query. Based on the methods generating the summary, automatic summarization technology can also be divided into extraction-based summarization, compression-based summarization, and abstraction-based summarization. The extraction-based method produces summaries by extracting important sentences from the original text, and the compression-based approach summarizes the original text by extracting and compressing important sentences, while abstraction-based summarization generates final outputs by rewriting or reorganizing the content of the original text.

Figure 11.1 gives the basic framework of automatic text summarization technology. As shown in the figure, the final summaries can be titles, short summaries, or long summaries based on the length of the required output.

Since multidocument summarization involves a much broader range of technology, it has always attracted the most concern and offered the most challenging research direction in the field of automatic summarization. As noted in the previous definition, multidocument summarization aims to compress the information expressed by multiple documents into a summary according to a compression ratio. From the application point of view, when searching information on the Internet, the search engine usually returns hundreds of pages on the same topic. If these pages

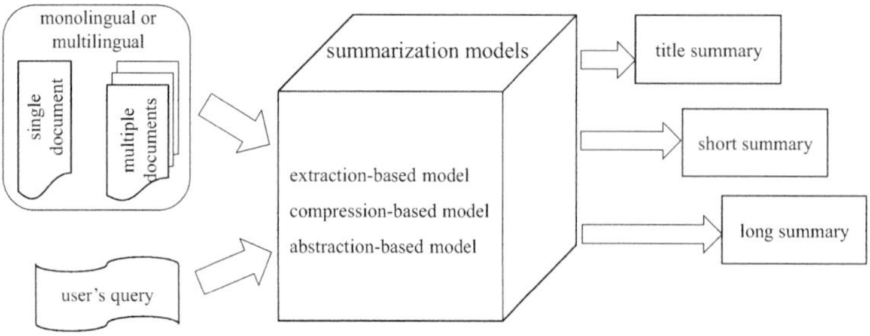

Fig. 11.1 The basic framework of automatic text summarization

are transformed into a concise summary that can reflect the main information, it will greatly improve the efficiency with which users can retrieve information. In addition, several news agencies often report events on the same topic, and the same agency will usually report a series of events on the same topic. If we can extract a concise summary of these highly relevant documents, it will effectively reduce the cost of information storage and allow users to save time while still knowing the events. These are the typical applications of multidocument summarization technology.

In contrast, single-document summarization can be regarded as a special case of multidocument summarization. In recent years, due to the availability of a large number of annotated data, single-document summarization has gradually attracted increasing attention.

11.2 Extraction-Based Summarization

Extraction-based automatic summarization directly extracts sentences from the original text to generate a summary. Although this seems to deviate from the essence of summarization, it is simple and effective in practical applications, and summary sentences can also maintain fluency and readability. Taking multidocument automatic summarization as an example, this chapter will introduce the basic idea and method for implementing of extraction-based automatic summarization.

The task of multidocument automatic summarization can be formalized as follows: given a document set $D = \{D_i\}$, $i = 1, \cdots, N$. Each document $D_i = \left\{S_{i_1}, \cdots, S_{i_j}, \cdots, S_{i_M}\right\}$ consists of a sequence of i_M sentences. Extraction-based summarization chooses K sentences from the document to form a summary, where K is the number of sentences manually set or is obtained through the compression ratio. In many automatic summarization systems, the number of sentences K can be different as long as the final summary is controlled within a limited number of words (e.g., 100 words for English summaries).

From the above formal description, we can see that three key steps are needed to complete automatic summarization: ① finding the most important and informative candidate sentences, ② minimizing the redundancy of candidate sentences, and ③ generating a summary according to a compression ratio or summary length requirements based on sentence order constraints. In the first step, we need to design an evaluation method to estimate the importance of sentences, and in the last two steps, we need to construct a constraint-based summary generation algorithm.

11.2.1 Sentence Importance Estimation

Since Luhn proposed automatic summarization technology in 1958, there have been many algorithms developed to estimate the importance of sentences. Based on whether the algorithm relies on manually labeled data, they can be divided into unsupervised and supervised algorithms.

(1) Unsupervised algorithms

Unsupervised algorithms can be divided into three categories: (a) word frequency-based algorithms, (b) document structure-based algorithms, and (c) graph-based algorithms.

(a) Word frequency-based algorithm

Words are the most commonly used features in sentence importance estimation algorithms. The basic assumption is that the more frequently a word appears in a document, the more important it is. If a sentence contains more high-frequency words, the more important it is. Based on this assumption, the following formula can be used to calculate the importance score of the sentence S_{i_j}:

$$\text{Score}\left(S_{i_j}\right) = \frac{\sum_{w_k \in S_{i_j}} \text{Score}\left(w_k\right)}{\left|\left\{w_k | w_k \in S_{i_j}\right\}\right|} \tag{11.1}$$

$$\text{Score}\left(w_k\right) = \text{tf}_{w_k} = \frac{\text{count}\left(w_k\right)}{\sum_w \text{count}\left(w\right)} \tag{11.2}$$

$\text{Score}\left(S_{i_j}\right)$ denotes the importance score of the j-th sentence in the i-th document, and $\text{count}\left(w_k\right)$ denotes the number of times it appears in the whole document D_i. $\sum_w \text{count}\left(w\right)$ is the occurrence of all words in the entire document. Since some words, such as *in*, *this*, and *for*, make no contribution to evaluating the importance of sentences, in practice, they are usually removed as stop words (see Chap. 2). $\text{Score}\left(w_k\right)$ is usually called the term frequency (TF).

This method is simple and easy, but it has a serious defect: some words are not important for expressing the meaning of a sentence, but they frequently appear in different documents and sentences, and their scores are high in the TF algorithm. To overcome this shortcoming, inverse document frequency (IDF) is widely used:

$$\text{idf}_{w_k} = \log \frac{|D|}{\left|\left\{j | w_k \in D_j\right\}\right|} \tag{11.3}$$

idf_{w_k} is a measure of word universality. If idf_{w_k} is larger, the denominator in formula (11.3) is smaller, indicating that fewer documents contain the word, then the word is more important for those specific documents.

It can be seen that both tf_{w_k} and idf_{w_k} can only represent one aspect of the importance of the word w_k. To more comprehensively describe the importance of the word w_k to document S_{i_j}, the TFIDF calculation method is employed by using the following formula to calculate the final score $\text{Score}\left(w_k\right)$:

$$\text{Score}\left(w_k\right) = \text{tf\_idf}_{w_k} = \text{tf}_{w_k} \times \text{idf}_{w_k} \tag{11.4}$$

The calculation of the sentence importance score given in formulas (11.1) to (11.4) above is simple, but it cannot model the coverage of the final summary. To overcome this shortcoming, topic analysis methods such as the latent semantic analysis (LSA) method (Landauer 2006) and the latent Dirichlet allocation (LDA) model (Blei et al. 2003) have been proposed. For a detailed description of such methods, please see Chap. 7 of this book.

In addition, some clue words (e.g., *in a word, in short*) and named entities are often used as informative features.

(b) Document structure-based algorithm

In addition to content features, document structure clues can often indicate the importance of sentences. Among them, the position of a sentence in the document and the length of the sentence are two common document structure features considered (Edmundson 1969). Some studies have shown that the first sentence in each paragraph can best reflect and express the information of the whole paragraph, especially in English critical articles. Thus, sentence position in the document is very important. In many studies, the importance of sentence position is calculated by the following formula:

$$\text{Score}\left(S_{i_j}\right) = \frac{n - j + 1}{n}, \tag{11.5}$$

where j denotes the position of the sentence S_{i_j} in the document and n represents the number of sentences in the document.

(c) Graph-based algorithm

The importance of a sentence is reflected not only by the internal words of the sentence but also by the relationship between it and other sentences in the documents. The more other sentences there are that support the importance of this sentence, the more important this sentence is. This idea comes from the PageRank algorithm: if a page is linked from thousands of pages or several important pages, this page is important (Page and Brin 1998).

The PageRank algorithm is a ranking model based on a directed graph. For a directed graph $G(V, E)$, V is a set of nodes in which each node represents a web page, and E is a set of directed edges where each edge $e = \left(V_i, V_j\right)$ indicates that it can jump from web page V_i to web page V_j. For a node V_i, $\text{In}(V_i)$ denotes the set of pages linked to V_i, and $|\text{In}(V_i)|$ is the in-degree of V_i. $\text{Out}(V_i)$ denotes the web pages that V_i can link to, and $|\text{Out}(V_i)|$ is the out-degree of V_i. The weight of each page indicates the importance of that page. The weight can be calculated by the following formula:

$$S(V_i) = \frac{1-d}{N} + d \times \sum_{V_j \in \text{In}(V_i)} \frac{1}{\left|\text{Out}\left(V_j\right)\right|} S\left(V_j\right) \tag{11.6}$$

where $d \in [0, 1]$ is the damping factor that assigns V_i a prior probability of jumping to any other node V_j from V_i. d is usually set to 0.85 in page ranking. The graph-based ranking algorithm assigns a random weight to each node during initialization and then iteratively calculates the formula (11.6) until the weight difference of each node between two successive iterations $S^{k+1}(V_i) - S^k(V_i)$ is smaller than a predefined threshold.

Graph-based sentence importance estimation algorithms are an extension of PageRank, for example, the LexRank algorithm (Erkan and Radev 2004) and the TextRank algorithm (Mihalcea and Tarau 2004). The difference is that in the LexRank algorithm, the directed graph $G(V, E)$ becomes the undirected graph, and each edge of the graph $e = \left(V_i, V_j\right)$ carries a weight W_{ij}. In graph $G(V, E)$, V denotes the set of sentences, and E denotes the set of undirected edges. If there is $e = \left(V_i, V_j\right) \in E$, then the two sentences V_i and V_j have relevance or similarity. The degree of correlation or similarity is represented by W_{ij}, which can be computed in multiple ways. We introduce a common cosine similarity method based on TFIDF to calculate the similarity of two sentences:

$$W_{ij} = \frac{\sum\limits_{w \in V_i, V_j} (\text{tf\_idf}_w)^2}{\sqrt{\sum\limits_{x \in V_i} (\text{tf\_idf}_x)^2} \times \sqrt{\sum\limits_{y \in V_j} \left(\text{tf\_idf}_y\right)^2}}$$

where $w \in V_i, V_j$ denotes a word that simultaneously occurs in V_i and V_j. Given a weighted undirected graph $G(V, E)$, the importance score of each node (sentence) is calculated by the following formula:

$$S(V_i) = \frac{1-d}{N} + d \times \sum_{V_j \in \text{adj}(V_i)} \frac{W_{ij}}{\sum\limits_{V_k \in \text{adj}(V_j)} W_{kj}} S\left(V_j\right) \tag{11.7}$$

where $V_j \in \text{adj}(V_i)$ denotes the set of nodes adjacent to V_i, i.e., the set of nodes with edges linked to V_i. The initial value of the node importance score and the condition of convergence are similar to those of PageRank. The TextRank algorithm (Mihalcea and Tarau 2004) and LexRank algorithm have the same basic idea, but the main difference is the way to calculate the similarity between the two sentences. The TextRank algorithm uses the word overlap between two sentences as the similarity metric

$$W_{ij} = \frac{\left|\left\{w_k | w_k \in V_i \ \& \ w_k \in V_j\right\}\right|}{\log |V_i| + \log \left|V_j\right|}$$

where $\left|\left\{w_k | w_k \in V_i \ \& \ w_k \in V_j\right\}\right|$ denotes the number of words co-occurring in two sentences; $|V_i|$ and $\left|V_j\right|$ are the number of words in sentences V_i and V_j. We provide a specific example to illustrate the LexRank algorithm. There are three documents on the same topic, and each document includes multiple sentences, as shown in Table 11.1. d_1s_1 indicates the first sentence in the first document, d_2s_2 is the second

Table 11.1 Sentence set of multiple documents

Number	ID	Sentences
1	d_1s_1	On Tuesday, January 10, FIFA announced that the World Cup had expanded to 48 teams for the first time since 1998. In the 87-year history of the World Cup, the rules and schedules had changed several times. The number of participating teams had increased from 16 to 48
2	d_1s_2	Gianni Infantino has been in charge of FIFA for nearly a year. At the beginning of his tenure, he advocated for reform. The World Cup expansion is his biggest reform in nearly a year
3	d_1s_3	He put forward the idea of expanding the participating teams, which shares his previous motivation for enlarging the number of teams in the European Cup. He did not want the teams who participated in the World Cup finals to always be the same faces. He hoped that more marginal teams could enter the finals and experience the joy of the football festival
4	d_2s_1	Yesterday, the FIFA Council formally voted on the expansion plan. No surprise, the plan of 48 teams divided into 16 groups was approved. The official twitter of FIFA immediately announced the news
5	d_2s_2	Since Infantino was elected president of FIFA in February 2016, the expansion of the World Cup has become imperative. The number of teams and the competition rules are the only uncertainty
6	d_2s_3	At the beginning, Infantino proposed to expand the Word Cup finals to 40 teams. On this premise, he proposed two competition systems, one is divided into eight groups, each of which has five teams, and the other is divided into ten groups, each of which includes four teams
7	d_2s_4	Two months later, Infantino put forward a new plan that includes 48 teams divided into 16 groups, each of which includes three teams. The top two of the group will qualify, and then they go through the knockout rounds to decide the champion
8	d_2s_5	The World Cup teams will be expanded from 32 to 48, which means that nearly a quarter of FIFA members will be able to participate in the future World Cup. Some football-weak countries that could not compete in the World Cup in the past will now have the oppotunity
9	d_2s_6	"There is no better way than getting their national team involved in the World Cup to promote football in the first place." Infantino said before
10	d_3s_1	On January 10, Beijing time, FIFA announced that from the start of the 2026 World Cup, the number of teams will be increased from 32 to 48.
11	d_3s_2	Ultimately, FIFA official announced that after 2026, the teams are divided in to 16 groups in the group stage. Each group includes three teams and will have a single round-robin competition. The top two teams will enter the next round and then go through to the knockout rounds. All the matches will be completed in 32 days
12	d_3s_3	Although parties had different opinions before, it was in the interest of more FIFA members to expand the teams, which was also consistent with Infantino's promise and statement when he was elected president of FIFA last year. Therefore, this expansion reflects the general trend.

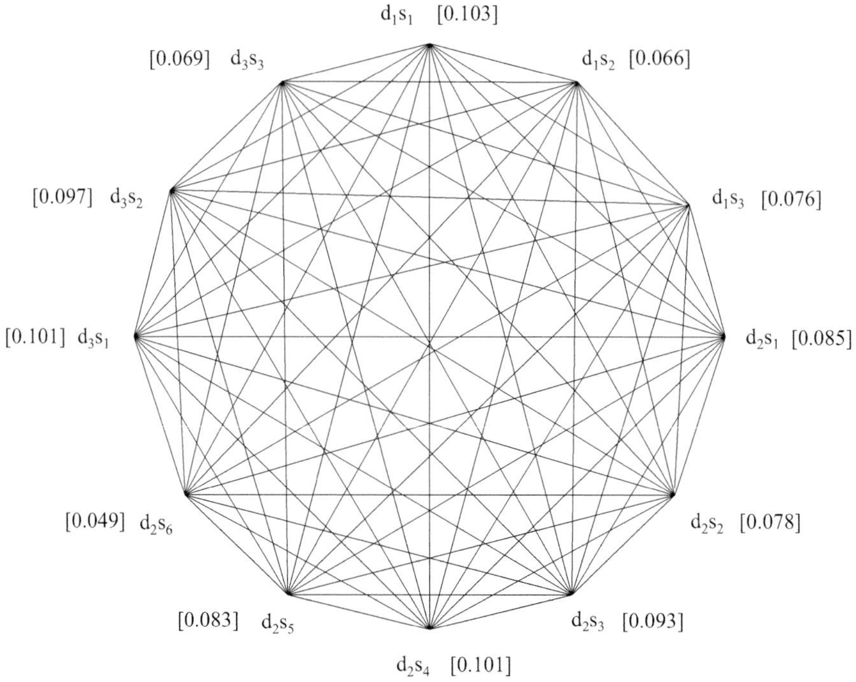

Fig. 11.2 An undirected graph built from multidocument sentences

sentence in the second document, and so on. Figure 11.2 shows an undirected graph based on these 12 sentences from three documents. TFIDF similarities between sentences are calculated and provided in Table 11.2. The graph-based LexRank algorithm iteratively computes equation (11.7) until convergence. This leads to the final importance score of each sentence, as the values shown in the brackets of Fig. 11.2. Among them, the sentence with the highest score is d_1s_1, which is in line with our human judgment. Obviously, the graph-based algorithm is very effective.

(2) Supervised algorithms

There are a large number of summaries generated by specialists in many application scenarios, especially in single-document summarization. Almost every academic paper has a summary given by the author. Obviously, given the labeled training data, sentence importance estimation can not only employ valuable features (e.g., location, word frequency, and graph-based ranking score) but also explore machine learning models (e.g., SVM, log-linear model, and neural networks).

For supervised algorithms, the best reference summaries should consist of sentences that appeared in the original document, and each sentence in the document is assigned a score between 0 and 1, indicating the degree to which that sentence that should be included in the summary, namely, the importance score of the sentence. Generally, a document contains dozens or hundreds of sentences, and sometimes

Table 11.2 The cosine similarity between different sentences' TFIDF values

1	0.129	0.141	0.121	0.187	0.106	0.137	0.173	0.076	0.471	0.266	0.15
0.129	1	0.239	0.04	0.114	0.039	0.032	0.086	0.085	0.137	0.109	0.12
0.141	0.239	1	0.044	0.101	0.094	0.027	0.167	0.052	0.14	0.144	0.199
0.121	0.04	0.044	1	0.152	0.262	0.365	0.197	0.071	0.072	0.138	0.096
0.187	0.114	0.101	0.152	1	0.156	0.196	0.109	0.088	0.091	0.029	0.176
0.106	0.039	0.094	0.262	0.156	1	0.498	0.114	0.082	0.119	0.214	0.051
0.137	0.032	0.027	0.365	0.196	0.498	1	0.135	0.084	0.142	0.282	0.02
0.173	0.086	0.167	0.197	0.109	0.114	0.135	1	0.152	0.206	0.091	0.069
0.076	0.085	0.052	0.071	0.088	0.082	0.084	0.152	1	0.04	0.021	0.043
0.471	0.137	0.14	0.072	0.091	0.119	0.142	0.206	0.04	1	0.369	0.129
0.266	0.109	0.144	0.138	0.029	0.214	0.282	0.091	0.021	0.369	1	0.162
0.15	0.12	0.199	0.096	0.176	0.051	0.02	0.069	0.043	0.129	0.162	1

more, so it is clearly impractical for experts to assign each sentence a specific importance score. Human-generated summaries are usually not exactly the sentences from the documents, and so automatically transforming manually produced summaries into ideal training data suitable for extraction-based summarization becomes an important problem to be solved for supervised algorithms.

Given the set of (Doc, Sum), one can assign a Boolean value of 0 or 1 or a real-valued number between 0 and 1 to each sentence in $Doc = \{s_0, s_1, \cdots, s_n\}$ according to the similarity between it and the reference summary Sum. If a Boolean value is given, the sentence importance estimation becomes a classification problem. If a real-valued number is given, it is transformed into a regression problem. In fact, regardless of the type of value assigned, a sentence is scored to calculate the similarity between s_i and the reference summary Sum. Many methods can be employed to measure the similarity between two sentences, such as the edit distance and concise distance based on distributed representations. According to the similarity matching algorithm, a sentence s_i will obtain a similarity score for its similarity with each sentence in Sum. The highest score can be regarded as the final score of the correlation between s_i and Sum. If only a Boolean value is needed, we check whether the score of s_i is greater than the predefined threshold. If it is larger than the threshold, 1 will be assigned, otherwise, 0. After preprocessing, the final training data can be obtained: document $Doc = \{s_0, s_1, \cdots, s_n\}$ and its corresponding sentence score $SenLabel = \{sl_0, sl_1, \cdots, sl_n\}$.

Taking the Boolean value as an example, the sentence importance estimation problem is transformed into a sequential labeling task. Given a large number of sets of $(Doc, SenLable)$, a classifier F is learned to predict a Boolean value tag for each sentence in the new document $Doc' = \{s'_0, s'_1, \cdots, s'_n\}$, and the probability that the tag is true will be taken as the importance of the sentence. Next, we introduce two algorithms using discrete features and distributed representations.

(a) Sentence importance estimation based on the log-linear model

To estimate sentence importance using machine learning algorithms, the key issue is designing effective features. Different machine learning methods rely on different assumptions. The naive Bayesian method assumes that the features are conditionally independent when the label is given. The hidden Markov model assumes that the first-order Markov property is satisfied between sentences. In contrast, the log-linear model has no independence hypothesis for features. In this section, we present the sentence importance estimation algorithm by taking the log-linear model as an example (Osborne 2002).

The log-linear model is a discriminative machine learning method that directly synthesizes various features to model the posterior probability $P(sl|s)$

$$p(sl|s) = \frac{1}{Z(s)} \exp\left\{\sum_i \lambda_i f_i(s, sl)\right\} \tag{11.8}$$

where $Z(s) = \sum_{sl} \exp\left\{\sum_i \lambda_i f_i(s, sl)\right\}$ is the normalization factor. $f_i(s, sl)$ denotes all kinds of sentence features, and λ_i is the corresponding feature weight. sl is a Boolean-valued sentence tag, and the value *true*(1) or *false*(0) reflects whether the sentence *is* or *is not* a summary sentence. Because the training data are extremely unbalanced (only a few positive examples in a document indicate summary sentences, and the others are all negative examples), many machine learning algorithms, including the log-linear model, tend to predict most of the test sentences will be negative (non-summary sentences). To alleviate this problem, we can add a class priori constraint:

$$sl^* = \text{argmax}_{sl} p(sl) \times p(sl|s) = \text{argmax}_{sl}\left(\log p(sl) + \sum_i \lambda_i f_i(s, sl)\right) \tag{11.9}$$

Generally, the prior probability $p(sl)$ can be obtained by optimizing the above objective function on the training data. Concerning the discrete features, we can exploit various features from surface information to deep knowledge, such as the sentence position in documents, sentence length, TFIDF statistics for words in sentences, ranking score based on the graph model, and text structure information.

The log-linear model is optimized based on the above features, and a posterior probability $p(sl|s)$, which is actually the importance of the sentence, can be obtained for each sentence in the input document during the testing. If it is required to directly answer whether the sentence should be selected into the summary using $p(sl|s)$, the choice can be made according to whether the posterior probability $p(sl|s)$ is larger than a certain threshold (e.g., 0.5).

(b) Sentence importance estimation based on deep neural networks

Although discrete features such as sentence length and word frequency can reflect the importance of a sentence to a certain extent, they cannot model the global semantic information of a sentence. More importantly, discrete features face a serious problem of data sparsity and cannot capture the semantic similarity between words (phrases, sentences). For example, *abstract* and *summary* have similar meanings, but they cannot be reflected by discrete feature representations such as word frequency.

In recent years, to overcome the aforementioned shortcomings, deep learning methods represent linguistic units in different granularities, such as words, phrases, sentences, and documents, to low-dimensional real-valued vectors, expecting that linguistic units with similar semantics will be close to each other in the vector space. This distributed representation paradigm avoids complicated feature engineering work. To predict the sentence importance using deep neural networks, the central problem is deciding which kind of neural network structure should be employed to learn the semantic representation of sentences. Recurrent, recursive, and convolutional neural networks are commonly used in distributed sentence representation learning. Next, we will use convolutional neural networks as an example to introduce sentence importance estimation using distributed feature representations (Nallapati et al. 2017).

Given a sentence $s = w_0 w_1 \cdots w_{n-1}$, each word is mapped into a low-dimensional vector, resulting in a sequence of vectors $\boldsymbol{Xw} = [\boldsymbol{Xw}_0, \boldsymbol{Xw}_1, \cdots, \boldsymbol{X w}_{n-1}]$, as shown at the bottom of Fig. 11.3. The convolutional neural network includes a convolution operator and a pooling operator. The convolution operator extracts local information, while the pooling operator summarizes the global information of a sentence.

The convolution operator consists of L filters $\boldsymbol{W} \in \mathbb{R}^{h\times k}$, and each filter extracts local features along the window of h words $\boldsymbol{Xw}_{i:i+h-1}$

$$\boldsymbol{u}_i = \sigma\left(\boldsymbol{W} \cdot \boldsymbol{Xw}_{i:i+h-1} + \boldsymbol{b}\right) \tag{11.10}$$

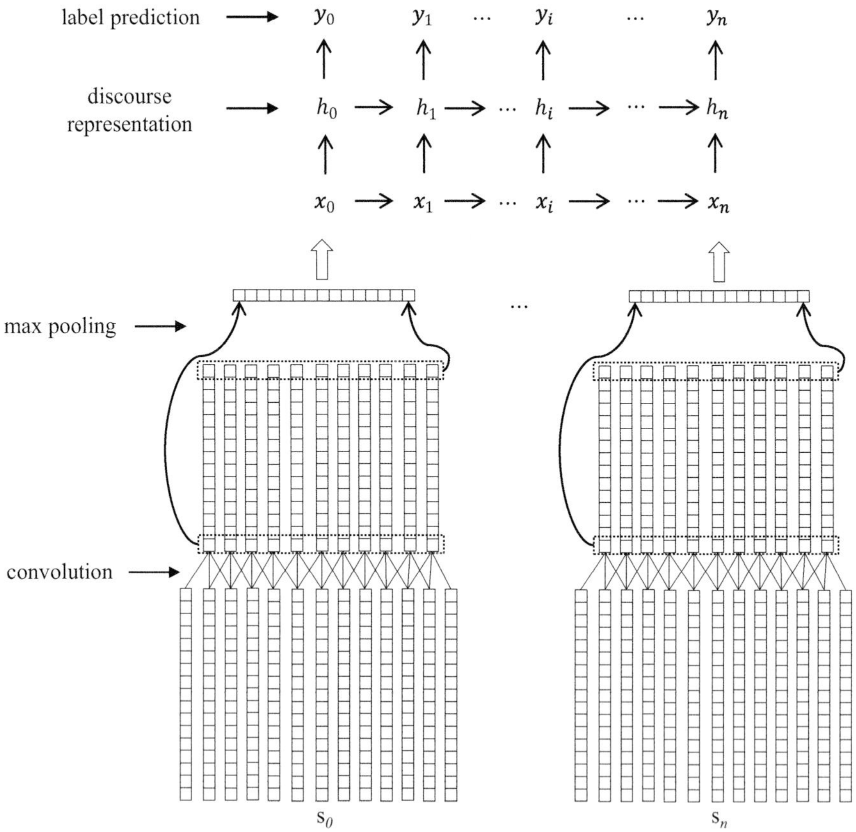

Fig. 11.3 The convolutional neural network for learning sentence semantic representation

where σ is a nonlinear activation function (e.g., ReLU and Sigmoid) and $\boldsymbol{b}$ denotes the bias item. When a filter convolves from $\boldsymbol{X}\boldsymbol{w}_0$ to $\boldsymbol{X}\boldsymbol{w}_{n-1}$, a vector $\boldsymbol{u} = [u_0, u_1, \cdots, u_{n-1}]$ can be obtained. If L different filters are adopted, we will have L local feature vectors of n-dimension.

Because sentences have different lengths, pooling operators are necessary to keep the output of convolutional neural networks within fixed dimensions. Generally, maximum pooling is the most frequently applied. This method chooses the maximum value of a vector as the representative feature: $\hat{u} = \max(\boldsymbol{u})$. Thus, each filter leads to the output of one dimension, and L filters will result in an L-dimensional vector. We can stack multilayer convolution and pooling operators, and ultimately, through a series of linear and nonlinear transformations, we will obtain a fixed dimension output $\boldsymbol{x}_i$, which can be used as the global semantic representation of sentences, as shown in Fig. 11.3.

Given a global sentence representation, various sequence labeling algorithms can be employed. We introduce the model based on a long short-term memory network (LSTM)

$$\boldsymbol{h}_i = \text{LSTM}(\boldsymbol{x}_i, \boldsymbol{h}_{i-1}) \tag{11.11}$$

$$y_i = \text{softmax}(\boldsymbol{h}_i) \tag{11.12}$$

where $\boldsymbol{h}_i = \text{LSTM}(\boldsymbol{x}_i, \boldsymbol{h}_{i-1})$ is calculated as follows:

$$\begin{bmatrix} \boldsymbol{i}_i \\ \boldsymbol{f}_i \\ \boldsymbol{o}_i \\ \hat{\boldsymbol{c}}_i \end{bmatrix} = \begin{bmatrix} \sigma \\ \sigma \\ \sigma \\ \tanh \end{bmatrix} \boldsymbol{W} \begin{bmatrix} \boldsymbol{x}_i \\ \boldsymbol{h}_{i-1} \end{bmatrix}$$

$$\boldsymbol{c}_i = \boldsymbol{i}_i \odot \hat{\boldsymbol{c}}_i + \boldsymbol{f}_i \odot \boldsymbol{c}_{i-1}$$

$$\boldsymbol{h}_i = \boldsymbol{o}_i \odot \tanh(\boldsymbol{c}_i)$$

Formula (11.12) shows that each sentence will eventually obtain a posterior probability of the *True* label, which will be used as the score for the sentence importance.

For a single-document summarization task, choosing sentences with a high importance score can satisfy the requirement of information quantity. Composing the extracted sentences into a summary according to the order in which they appear in the document can ensure the fluency and readability of the output. For the multidocument summarization task, if each document carries a timestamp, composing summaries first according to the sentence order in a document and then according to the chronological order of the multiple documents can also ensure outputs of high readability (Barzilay et al. 2002). However, optimizing the order of sentences in summaries is still an open problem given multiple documents without timestamps.

11.2.2 Constraint-Based Summarization Algorithms

At the beginning of this chapter, we mentioned that wide coverage and low redundancy are the basic requirements for automatic summarization. Generally, summaries are relatively short, limited to no more than K sentences or no more than N words (e.g., 100 Chinese characters for Chinese summarization). Therefore, maximizing coverage is actually equivalent to minimizing redundancy under the constraint of summary length. The widely used algorithm for minimizing redundancy is derived from the idea of maximum marginal relevance (MMR) (Carbonell and Goldstein 1998). The MMR algorithm is mainly proposed for query-related document summarization tasks. It takes the following formula:

$$\mathrm{MMR}(R, A) = \underset{s_i \in R \setminus A}{\operatorname{argmax}} \left\{ \lambda \mathrm{Sim}_1(s_i, Q) - (1-\lambda) \max_{s_j \in A} \mathrm{Sim}_2\left(s_i, s_j\right) \right\} \tag{11.13}$$

where R is the set of all sentences, A denotes the set of selected summary sentences, Q stands for the user query, s_i denotes any sentence in the set of unselected sentences, and s_j means any sentence in the set of selected sentences. $\mathrm{Sim}_2\left(s_i, s_j\right)$ is the similarity between two sentences s_i and s_j. λ is a hyperparameter that weighs correlation and redundancy. The larger lambda is, the more s_i's relevance to the user query is emphasized; otherwise, redundancy is more emphasized. As seen from formula (11.13), the basic idea of the MMR algorithm is to select the sentence from the unselected sentence set $R \setminus A$ that is most relevant to the input query but most irrelevant to the already selected sentences. This operation is performed iteratively until the number of sentences or words reaches the predefined threshold.

In general automatic document summarization tasks, the redundancy calculation method is similar to the MMR algorithm. Basically, the following calculation formulas can be used:

$$\mathrm{MMR}'(R, A) = \underset{s_i \in R \setminus A}{\operatorname{argmax}} \left\{ \lambda \mathrm{Score}(s_i) - (1-\lambda) \max_{s_j \in A} \mathrm{Sim}\left(s_i, s_j\right) \cdot \mathrm{Score}\left(s_j\right) \right\} \tag{11.14}$$

where $\mathrm{Score}(s_i)$ denotes the importance score of sentence s_i. The above formula indicates that sentences that have the highest importance score but are most dissimilar with the already selected sentences will be chosen for each iteration. If the graph-based method is used to calculate the importance score of sentences, the structure information of graphs can be fully utilized to minimize redundancy. We introduce how to generate summaries given the sentence importance score in the framework of the graph-based summarization:

① We first initialize two sets, $A = \varnothing$ and $B = \{s_i | i = 1, \cdots, n\}$, which denote the summary sentence set and the unselected sentence set, respectively. We then initialize the overall score of the importance and redundancy for each sentence $\mathrm{RS}(s_i) = \mathrm{Score}(s_i)$, $i = 1, \cdots, n$ (at the beginning, the redundancy score is unknown, and the overall score only includes the importance score of the sentence).
② Then, we rank set B according to the score from high to low in the results of $\mathrm{RS}(s_i)$.
③ Assuming s_i is the sentence with the highest score or, say, the sentence that ranks first in B, we remove s_i from B and add it to A. Then, we update the overall score of each remaining sentence in B according to the following formula:

$$\mathrm{RS}\left(s_j\right) = \mathrm{RS}\left(s_j\right) - \lambda \mathrm{Sim}\left(s_i, s_j\right) \cdot \mathrm{Score}\left(s_i\right)$$

④ We return to step ② and do the next iteration until set B is empty or set A meets the requirement in terms of sentence number.

The extraction-based automatic summarization methods take sentences as the basic unit. The algorithm is simple and intuitive and can maintain the fluency and readability of the summary. However, these methods confront some difficult problems, such as the irreconcilable contradiction between the coverage and the summary length. The length constraints of summaries are generally expressed by the number of words or sentences contained in the final summary. Length constraints limit the number of generated sentences, while the requirement of summary coverage aims at extracting more information, that is, extracting more sentences. If the sentences selected into summaries are important but contain too much useless information, which leads to longer sentences, it will directly prevent other important sentences from being included in the summary.

11.3 Compression-Based Automatic Summarization

Compression-based automatic summarization can alleviate the aforementioned problem of extraction-based summarization to some extent. The basic idea is to condense sentences by retaining important content and deleting unimportant information so that the final summary can contain more sentences under a fixed length (word count) with improved information coverage. Compressing sentences remains a major challenge.

11.3.1 Sentence Compression Method

Sentence compression can be defined as a word deletion task (Knight and Marcu 2002), namely, deleting the unimportant words to form a compressed version of the sentence. Formally, given a sentence $s = w_0 \cdots w_i \cdots w_{n-1}$, we aim to find a condensed substring $t = w'_0 \cdots w'_j \cdots w'_{m-1}$ of s. The substring t may be continuous or discontinuous with $m < n$. If $w'_j = w_i$, $\forall w'_{j'>j}$, $\exists w_{i'>i}, w'_{j'} = w_{i'}$. That is, the compressed result maintains the same word order as the original sentence.

Simple unsupervised methods or data-driven supervised models can be employed to perform sentence compression. Unsupervised methods generally depend on manually designed rules, and tree-based compression approaches are widely used (Turner and Charniak 2005). For any sentence, the approach first obtains the phrase structure tree corresponding to the sentence and then deletes the unimportant subtrees according to the designed rules, and the remaining tree structure constitutes the condensed sentence. Handcrafted rules usually include removing prepositional phrase subtrees and deleting temporal phrase subtrees and clauses. As shown in Fig. 11.4, after deleting the prepositional phrase and temporal phrase from the left tree, the right tree structure can be obtained. The simplified tree structure leads to the compressed result *The government has not made comments.*

We introduce two types of supervised sentence compression method (Knight and Marcu 2002): the generative model and the discriminant model. Parallel sentence corpora $\{s_k, t_k\}_{k=1}^{K}$ consisting of large-scale original sentences and corresponding consistent results are the basic resource for supervised methods. Some examples of parallel sentences are as follows (S_i denotes the original sentence, and T_i denotes the compressed sentence):

S_1: The two presidents held a 1 h telephone conference yesterday.
T_1: The two presidents held a telephone conference.
S_2: Children are counting stars in the sky at the doorway.
T_2: Children are counting stars.
S_3: The government has not yet commented on this incident.

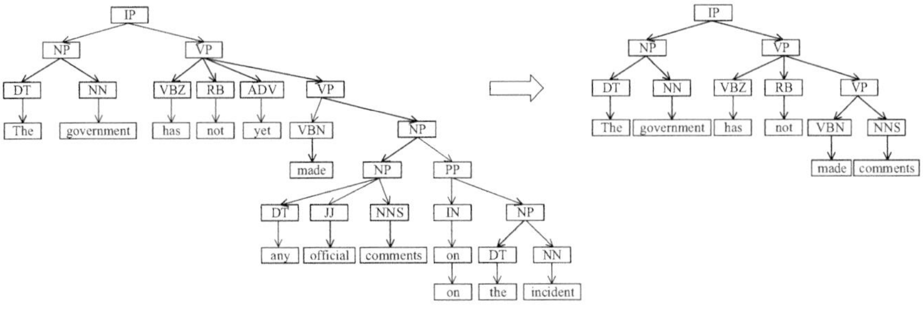

Fig. 11.4 Sentence compression based on the phrase structure tree

T_3: The government has not commented.

Note that the compressed result is only a continuous or discontinuous subsequence of the original sentence. Related studies mainly focus on English data, such as the Ziff-Davis corpus and the parallel corpus[3] manually constructed from British Broadcasting News. The training data are relatively small, including approximately 1000 to 1500 sentence pairs.

Next, we introduce the supervised generative and discriminative methods.

(1) Sentence compression based on the noisy channel model

The noisy channel model assumes that the original sentence is generated by adding auxiliary information to the compressed result. Given the original long sentence s, the goal is to find the best compressed sentence t to maximize the posterior probability $p\,(t|s)$, which can be decomposed by the Bayesian rule:

$$p\,(t|s) = \frac{p\,(t) \cdot p\,(s|t)}{p\,(s)}$$

Since the original sentence is fixed, the denominator $p\,(s)$ on the right side of the above formula can be neglected in the optimization process. Searching for the best compressed result t requires solving the following optimization problem:

$$t^* = \underset{t}{\text{argmax}}\, p\,(t) \cdot p\,(s|t)$$

where $p\,(s|t)$ is called the channel model, which represents the probability that the original sentence is generated from the compressed output, and $p\,(t)$ is called the source model, which measures the fluency of the compressed result.

Typically, the corresponding parsing trees are used to replace the original sentence and the compressed output. As shown in Fig. 11.5, suppose t_1 is the optimal compressed version of the original sentence s under the noisy channel mode. s can be inferred from the observed output t_1 of the noisy channel model. The probability of the source model and the channel model can be calculated as follows:

(a) Source Model

When using parse trees instead of original sentences, the source model will be used to measure the degree to which the compressed parse tree conforms to the grammar. The probabilistic context-free grammars (PCFG) and bigram language model can be employed to calculate the source model

$$\begin{aligned}\hat{p}_{\text{tree}}\,(t_1) = {} & p_{\text{cfg}}\,(\text{TOP} \rightarrow G|\text{TOP}) \cdot p_{\text{cfg}}\,(G \rightarrow HA|G) \cdot p_{\text{cfg}}\,(A \rightarrow CD|A) \cdot \\ & p_{\text{cfg}}\,(H \rightarrow a|H) \cdot p_{\text{cfg}}\,(C \rightarrow b|C) \cdot p_{\text{cfg}}\,(D \rightarrow e|D) \cdot p_{\text{bigram}}\,(a|\text{EOS}) \cdot \\ & p_{\text{bigram}}\,(b|a) \cdot P_{\text{bigram}}\,(e|b) \cdot p_{\text{bigram}}\,(\text{EOS}|e)\end{aligned}$$

[3]http://jamesclarke.net/research/resources.

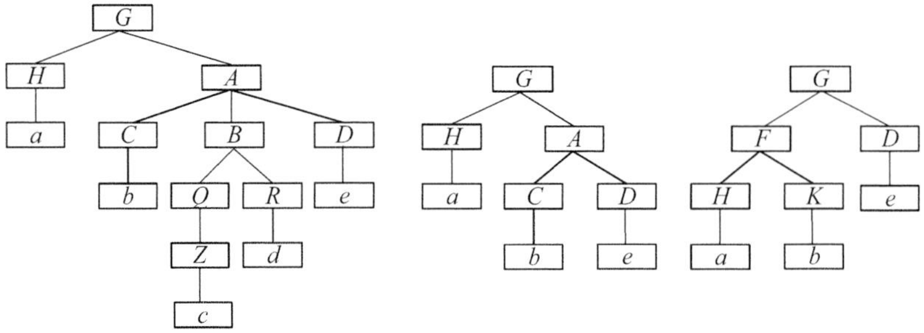

(s) (t_1) (t_2)

Fig. 11.5 The syntactic tree of the original sentence and the output of the noisy channel model

where $p_{\text{cfg}}(\text{TOP} \to G|\text{TOP})$ denotes the probability of generating G from the root node TOP. $p_{\text{cfg}}(G \to HA|G)$ denotes the probability of generating two nodes HA from node G with PCFG. $p_{\text{bigram}}(b|a)$ represents the bigram language model probability between leaf nodes a and b. Other variables in the formula have similar meanings.

(b) Channel Model

There are two parts in the channel model: one is the probability that the compressed tree expands to the original tree, and the other is the context-free grammar probability of the new subtree added in the original tree compared to that of the compressed tree:

$$\begin{aligned} p_{\text{expand\_tree}}(s|t_1) = & \; p_{\exp}(G \to HA|G \to HA) \cdot p_{\exp}(A \to CBD|A \to CD) \cdot \\ & \; p_{\text{cfg}}(B \to QR|B) \cdot p_{\text{cfg}}(Q \to Z|Q) \cdot p_{\text{cfg}}(Z \to c|Z) \cdot \\ & \; p_{\text{cfg}}(R \to d|R) \end{aligned}$$

$p_{\exp}(G \to HA|G \to HA)$ and $p_{\exp}(A \to CBD|A \to CD)$ denote the probability that binary tree $G \to HA$ remains unchanged and binary tree $A \to CD$ expands to ternary tree $A \to CBD$, respectively. The newly expanded node B generates a subtree in four steps: ① Generate nodes Q and R through context-free grammar $B \to QR$; ② New node Q generates node Z; ③ Node Z generates the terminal symbol c ④ Node R directly produces the terminal symbol d. $p_{\text{cfg}}(B \to QR|B)$, $p_{\text{cfg}}(Q \to Z|Q)$, $p_{\text{cfg}}(Z \to c|Z)$, and $p_{\text{cfg}}(R \to d|R)$ measure the probability of each step.

(c) Search the best compressed sentence

According to the source model and channel model, the posterior probability of the compressed tree t_1 can be calculated through the following formula:

$$\hat{p}_{\text{compress\_tree}}(t_1|s) = \frac{\hat{p}_{\text{tree}}(t_1) \cdot p_{\text{expand\_tree}}(s|t_1)}{\hat{p}_{\text{tree}}(s)}$$

In practice, the compressed sentence with the highest posterior probability will be chosen. For two candidate compressed trees, t_1 and t_2, if $\hat{p}_{\text{compress\_tree}}(t_1|s) > \hat{p}_{\text{compress\_tree}}(t_2|s)$, t_1 is selected and the compressed sentence can be obtained by concatenating the leaves of t_1.

(d) Model parameter training

From the probability formula of the source model and the channel model, it can be seen that the parameters of the noisy channel model include three parts: ① the rule deduction probability of context-free grammars, e.g., $p_{\text{cfg}}(G \rightarrow HA|G)$; ② the tree structure extension probability, e.g., $p_{\text{exp}}(A \rightarrow CBD|A \rightarrow CD)$); ③ the bigram language model probability, e.g., $p_{\text{bigram}}(b|a)$.

To estimate the above three kinds of probabilities, we need to parse the training data $\{\langle s_k, t_k\rangle\}_{k=1}^{K}$ to obtain the tree structures of the original sentence and the compressed sentence. For the probabilities of the first kind, the maximum likelihood estimation can be used to obtain the rule deduction probability of context-free grammar in the tree structures of the original and compressed sentences. For example, if $G \rightarrow HA$ appears 20 times while G occurs 100 times, then $p_{\text{cfg}}(G \rightarrow HA|G) = 0.2$.

For the probabilities of the second class, it is necessary to perform node alignment between the tree structures of $\langle s, t\rangle$ and then calculate the extended probability of the tree structure by using the maximum likelihood estimation method. For example, if $(A \rightarrow CD, A \rightarrow CBD)$ occurs 10 times together and $A \rightarrow CD$ appears 100 times, then $p_{\text{cfg}}(G \rightarrow HA|G) = 0.1$. For the language model probabilities, we can simply calculate the bigram probability according to the counts of words and bigrams.

(e) Model decoding

For the original sentence s whose tree structure has n child nodes, there are two choices for each node in the process of sentence compression: delete or retain. Therefore, there are $(2^n - 1)$ choices when compressing the original sentences s. All candidate condensed sentences can be stored in a shared forest, and dynamic programming can be utilized to search for the best compressed candidate.

Since all the probabilities need to be accumulated in the noisy channel model, the longer the candidate is, the smaller the posterior probability. Thus, the shorter candidate compressed sentence will be preferred by the noisy channel model. To overcome this problem, the posterior probability can be normalized by length: $\left(\hat{p}_{\text{compress\_tree}}(t_1|s)\right)^{\frac{1}{\text{length}(t_1)}}$

(2) Decision-based sentence compression method

The decision-based sentence compression method handles sentences from the perspective of tree structure rewriting. For the example given in Fig. 11.5, the goal of this method is to rewrite the structure tree of the original sentence s into the structure tree of the compressed sentence t_2. The rewriting process can be achieved

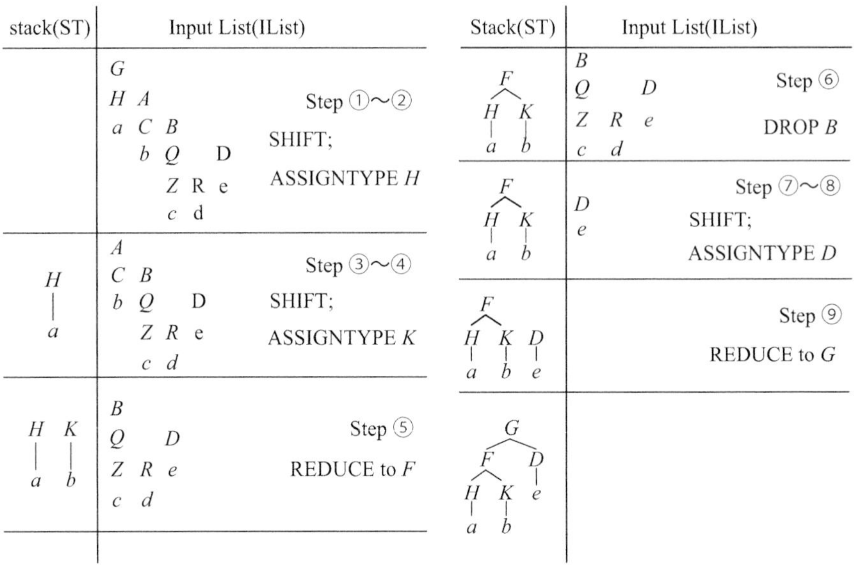

Fig. 11.6 Example of decision-base sentence compression

by a series of *shift*, *reduce*, and *delete* actions (similar to the *shift-reduce* parsing method).

In this algorithm, the stack (ST) and the input list (IList) are two key data structures. The stack is used to store the tree structure fragments corresponding to the compressed sentences so far, and the stack is empty at the beginning of the algorithm. The input list stores the words corresponding to the original sentence and their syntactic structure tags. As shown in Fig. 11.6, each word and all the syntactic tags corresponding to the word are input in sequence. The left part in the second column denotes the tree structure of the original sentence. G is the root and has two children H and A. H has only one child a. A has two children, C and B. B has two children, Q and D. Q has two children, Z and R. Note that each syntactic tag is only assigned to the leftmost word it covers in the subtree. In this example, the root node in the structure tree of the original sentence s is G, and G is only attached to a, the first word of the sentence. Similarly, H is also attached to the word a. As a result, G and H are the syntactic tags related to the word a. When deleting subtrees, if a is to be removed, the whole subtree related to the word a will also be deleted.

The process of rewriting the syntactic structure tree is accomplished by the following four types of actions:

- SHIFT: This action moves the first word in the input list IList into the stack ST;
- REDUCE: Pop K tree fragments up at the top of the ST stack, merge them into a new subtree, and move the new subtree into ST;
- DROP: Delete the complete subtree corresponding to the syntactic tag from the input list IList;

- ASSIGNTYPE: Assign a new syntactic label to the top subtree of the ST stack, which is generally used to rewrite part-of-speech labels of words.

Figure 11.6 shows how to use the four actions mentioned above to rewrite the original sentence s into the compressed sentence t_2 in nine steps. Determining which action should be performed in each step is the key to the decision-based sentence compression model, and this can be regarded as the parameters that the model needs to learn.

According to the training data $\{\langle s_k, t_k \rangle\}_{k=1}^{K}$, the tree structures of each original sentence and the compressed sentence constitute a tree pair, from which the number of times the four kinds of actions occur and their contexts can be counted. Then, using the contexts as input and specific actions as output, an action classifier can be optimized.

11.3.2 Automatic Summarization Based on Sentence Compression

The automatic summarization method based on sentence compression includes two core algorithms: candidate sentence selection and sentence compression. The sentence importance estimation algorithm and sentence compression algorithm have been introduced in previous sections. Here, we discuss how to combine these two algorithms to obtain the final summary.

Usually, there are three ways to combine the two algorithms:

① Select-Compress: We first extract the candidate summary sentences according to the importance score and then simplify them with the sentence compression algorithm. In the end, we can display more information in the summary given the same length constraint.

② Compress-Select: We first simplify all sentences in a document or documents by using a sentence compression algorithm and then select the summary sentences according to the sentence importance scores.

③ Joint-Select-Compress: A unified framework is employed to simultaneously optimize sentence selection and compression and ultimately output the simplified summary sentences.

Generally, the first method is the most efficient, but the original candidate summary sentences may not produce the best result after sentence compression. Furthermore, the compression algorithm only outputs a single result. The latter two methods aim to optimize the quality of the summaries but need to compress all the sentences, which significantly sacrifices the efficiency. Therefore, balancing quality and efficiency becomes the core issue for text summarization methods based on sentence compression.

We now introduce a compression-based summary method that takes into account both the efficiency of the method ① and the quality of the method ③ (Li et al.

2013a). The basic idea behind this approach is to obtain a relatively large set V_s of candidate summary sentences with an extraction-based summarization algorithm and then generate the K-best candidates for each sentence in that set using the sentence compression algorithm. Finally, a unified optimization framework is used to select the best summary sentences among the K-best candidate condensed sentences. Integral linear programming can be used as a unified optimization framework, and the objective function is as follows:

$$\max \quad \sum_i w_i c_i + \sum_j v_j \sum_k s_{jk} \tag{11.15}$$

$$\text{s.t.} \quad \sum_k s_{jk} \leq 1 \ \forall j \tag{A}$$

$$s_{jk} \text{Occ}_{i\_jk} \leq c_i \tag{B}$$

$$\sum_{jk} s_{jk} \text{Occ}_{i\_jk} \geq c_i \tag{C}$$

$$\sum_{jk} L_{jk} s_{jk} \leq L \tag{D}$$

$$c_i \in \{0, 1\} \ \forall i \tag{E}$$

$$s_{jk} \in \{0, 1\} \ \forall j, k \tag{F}$$

w_i denotes the weight of the i-th concept, where the coverage of the concept is used to represent the coverage of the summary. The concept is a tuple of content terms (e.g., telephone, conference). There are several ways to calculate the weight of the concept w_i, such as using the TFIDF method. v_j denotes the weight of compressed sentences and can directly use the weight of the corresponding original sentences, which is calculated through the sentence importance estimation method. c_i and s_{jk} are binary variables that indicate whether a concept or a sentence is selected in the summary. If $s_{jk} = 1$, the k-th candidate compressed result of the j-th sentence is chosen. Formula (11.15) shows the objective function that aims to maximize the coverage of the concepts and the importance score of a sentence. (A)~(F) are constraints. (A) indicates that only one compressed result can be selected for each original sentence. Occ_{i_jk} checks whether the i-th concept appears in sentence s_{jk}. Constraints B and C jointly constrain the concept and the sentence. D constrains the length of the final summary.

The final results that satisfy all the constraints can be generated by solving integral linear programming to account for both efficiency and quality.

Compression-based automatic summarization can remove secondary information or repetitive information in sentences so that a summary with sufficient information and wide coverage can be generated under length constraints. This method is obviously more reasonable than the extraction-based summarization approach. However, it still has some shortcomings: it cannot fuse multiple sentences expressing similar but complementary information. Assume that two sentences have high importance scores and should be included in the summary. Nevertheless, the two sentences are similar, and only one can be selected due to the redundancy constraint. As a result, the complementary and important information in the other sentence will not appear in the final summary if we employ the extraction-based or compression-based summarization methods.

11.4 Abstractive Automatic Summarization

The abstractive summarization method is designed to simulate a human-like summary generation process from document understanding to information compression to summary generation. This section introduces two abstractive summarization methods: one is based on information fusion, and the other is based on the encoder-decoder framework.

11.4.1 Abstractive Summarization Based on Information Fusion

The abstractive summarization method based on information fusion inherits and develops extraction-based and compression-based summarization methods. By considering concepts and facts,[4] it simulates the way a human generates a summary methodology: selecting important concepts and related facts in the process of text reading, reorganizing these concepts and facts, and, finally, generating new summary sentences. Taking Fig. 11.7 as an example, the concept of *Joe's dog* and the facts of *was chasing a cat* and *in the garden* are more important, so they are selected and reorganized to produce a summary sentence, *Joe's dog was chasing a cat in the garden*. This process is actually an information fusion procedure between the two sentences.

According to different definitions of concepts and facts, abstractive summarization methods based on information fusion can also be divided into several classes. The example given in Fig. 11.7 implements abstractive summarization in the following three steps: ① Use the deep semantic analysis method to transform similar sentences into abstract meaning representation (AMR); ② Merge two AMR graphs into one AMR graph; ③ Define concepts and facts using predicate-argument information and generate comprehensive expressions based on core arguments (e.g., *dog*) (Liu et al. 2015a). Because the performance of the automatic analysis from sentence to abstract semantic representations is far from satisfactory, this method is limited to theoretical exploration.

[4]Concepts and facts have different definitions. Generally, concepts correspond to entities (people, objects, institutions, etc.) and facts correspond to actions.

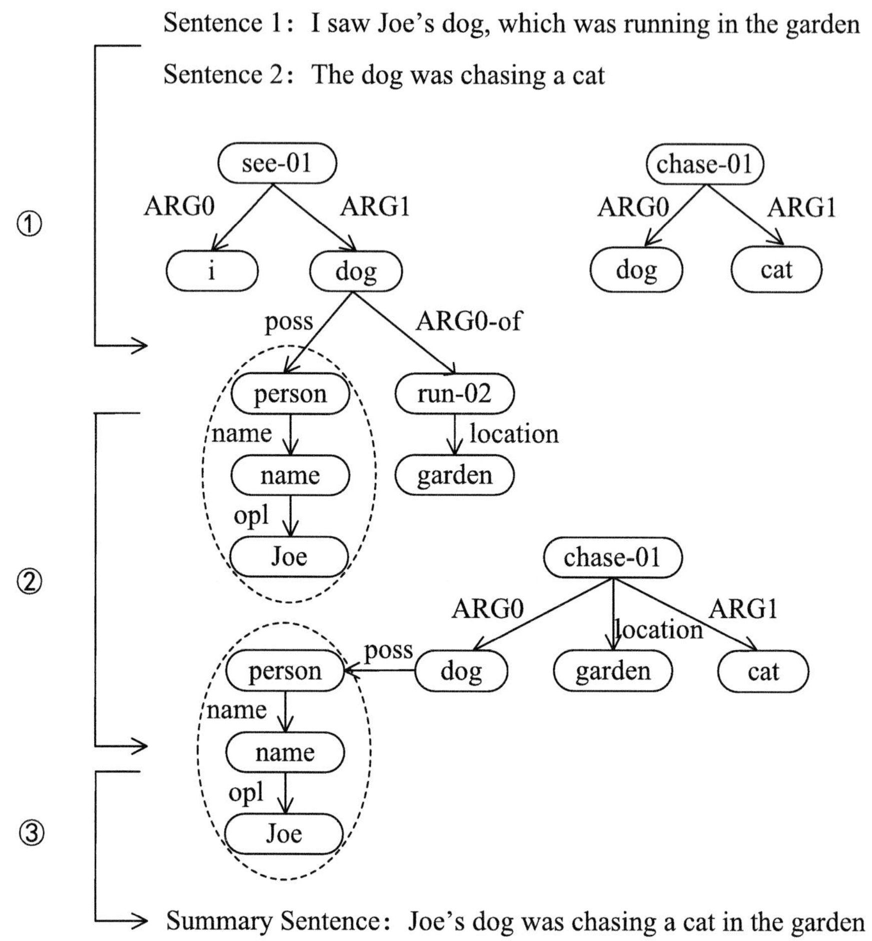

Fig. 11.7 Generative summarization based on abstract meaning representation

We now introduce an information fusion approach based on syntactic parsing technology (Bing et al. 2015). Unlike the above methods based on AMR, this method employs the syntactic structure tree to define and weigh the importance of concepts and facts, measures the compatibility between concepts and facts, and ultimately combines concepts and facts to produce summary sentences.

(1) Definition of concepts and facts

Concepts and facts are determined using parsing trees. Generally, concepts consist of noun phrases (NP), while facts are verb phrases (VP). Since a parsing tree is often very deep and there are many nodes rooted at the NP and VP, it is impossible to use every NP or VP phrase as a candidate for concepts or facts.

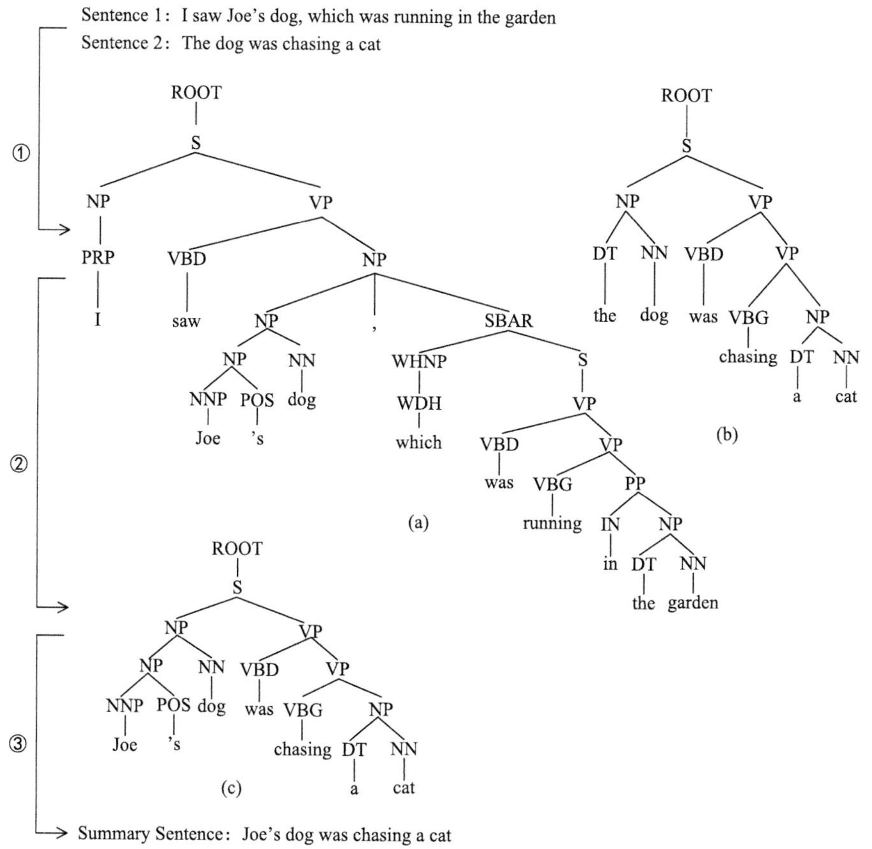
Sentence 1: I saw Joe's dog, which was running in the garden
Sentence 2: The dog was chasing a cat
①
②
③
(a)
(b)
(c)
Summary Sentence: Joe's dog was chasing a cat

Fig. 11.8 Generative summarization based on syntactic parsing tree

Thus, we use the following rules to define the candidate concepts and facts using NP and VP:[5]

(a) If NP/VP is the child node of a complete sentence or clause node (S and SBAR in Fig. 11.8), it is considered a concept/fact candidate, represented by S(NP) and S(VP):
(b) If the parent node of NP/VP is S(NP) (or S(VP)), then it is regarded as a concept/fact candidate and is denoted as S(NP(NP)) and S(VP(VP));
(c) If the parent node of NP/VP is S(NP(NP)) (or S(VP(VP))), then NP/VP is treated as a candidate concept/fact.

[5]In practical application, the scope of candidates for concepts and facts can be expanded or contracted appropriately according to specific needs.

The remaining NP and VP generally cannot fully represent a concept or fact, so they are removed from consideration. Note that some of the nodes' children denote coreference. For example, in Fig. 11.8a, the WHNP node actually denotes the NP phrase on its left side. At this time, this NP phrase is also a concept candidate. The candidate concepts in Fig. 11.8 include *I*, *Joe's dog*, and *the dog*. The candidate facts are *saw Joe's dog which was running in the garden*, *was chasing a cat*, and *chasing a cat*. Given a document or document set, the parsing tree of each sentence can be obtained after syntactic parsing. According to the above definition, all candidate concepts and facts can be extracted.

(2) Importance estimation of concepts and facts

Several sentence importance estimation algorithms have been introduced in extraction-based summarization methods. These algorithms can be used to calculate the importance scores for concepts and facts (NP and VP). For example, position in the documents, TFIDF and graph-based algorithms can be employed. Furthermore, the importance of NP and VP can also be evaluated by named entity-based methods.

Named entities such as person, location, and organization often indicate key information in the text, so the number of named entities contained in a concept or fact will generally reflect its importance. Thus, the importance of NP phrases can be calculated by the following simple formulas:

$$\text{Score}\,(\text{NP}) = \frac{\text{count}\,(\text{NE}_{\text{NP}})}{\text{count}\,(\text{NE}_{\text{doc}})} \tag{11.16}$$

$\text{count}\,(\text{NE}_{\text{NP}})$ denotes the number of named entities contained in the NP phrases, and $\text{count}\,(\text{NE}_{\text{doc}})$ denotes the total number of named entities contained in the documents where NP phrases are located.

Similarly, the importance of VP can be calculated in the same way. Various importance scores using different algorithms (TFIDF, graph-based and named entity-based algorithms) can be integrated by linear combination to obtain a more accurate importance score for the concept and fact.

(3) Definition of the compatibility between concepts and facts

In the information fusion-based methods, summary sentences are obtained by combining NP phrases (concepts) and VP (facts). The compatibility of concepts and facts must be resolved before they are combined, that is, it must be determined which kinds of NP and VP can be combined to form new sentences. Naturally, if the NP_i phrase and the VP_i phrase come from the same sentence node S, the two phrases can be combined. However, with this approach, only the original subsentence can be extracted, and new sentences cannot be generated. Thus, we need to define more relaxed compatibility constraints: if NP_j is compatible with NP_i, NP_j and VP_i can be combined; if VP_j is compatible with VP_i, NP_i and VP_j can also be combined. We now explain how to determine whether NP_i and NP_j (or VP_i and VP_j) are compatible.

Since many NP phrases are composed of entities, the compatibility of two NP can be determined by identifying whether NP_i and NP_j refer to the same entity. First, for a document or document sets, all NP representing entities in documents can be clustered by coreference resolution technology.[6] Any two NP phrases in the same cluster are considered compatible.

VP phrases are more diverse. The compatibility of VP_i and VP_j can be determined by the co-occurrence degree of language units such as words, phrases, and named entities. Specifically, the Jaccard index can be used to calculate whether two VP VP_i and VP_j are compatible

$$J\left(\mathrm{VP}_i, \mathrm{VP}_j\right) = \frac{\left|\mathrm{Set}_{\mathrm{VP}_i} \cap \mathrm{Set}_{\mathrm{VP}_j}\right|}{\left|\mathrm{Set}_{\mathrm{VP}_i} \cup \mathrm{Set}_{\mathrm{VP}_j}\right|} \tag{11.17}$$

where $\mathrm{Set}_{\mathrm{VP}_i}$ and $\mathrm{Set}_{\mathrm{VP}_j}$ denote the sets of words, bigrams, and named entities in VP_i and VP_j, respectively. If the Jaccard index is larger than a predefined threshold, VP_i and VP_j are considered compatible.

(4) Summary generation based on concepts and facts

Under the compatibility constraints of the concepts and facts, abstractive summarization aims to search a set of NP and VP from all the candidate sets of NP (concepts) and VP (facts) to maximize the importance score. The optimization process can be modeled by integer linear programming, and the objective function can be defined as follows:

$$\max \quad \sum_i \alpha_i S_i^N - \sum_{i<j} \alpha_{ij}\left(S_i^N + S_j^N\right) R_{ij}^N + \sum_i \beta_i S_i^V - \sum_{i<j} \beta_{ij}\left(S_i^V + S_j^V\right) R_{ij}^V \tag{11.18}$$

α_i and β_i are Boolean values indicating whether NP_i and VP_i are selected. S_i^N and S_i^V denote the importance scores of NP_i and VP_i, respectively. $\alpha_{ij} \in \{0, 1\}$,$\beta_{ij} \in \{0, 1\}$ indicate whether NP_i and NP_j and VP_i and VP_j coexist in the final summaries. R_{ij}^N and R_{ij}^V denote the similarity score between NP_i and NP_j and VP_i and VP_j. If NP_i and NP_j are coreferential, $R_{ij}^N = 1$. Otherwise, it can be calculated by the above Jaccard index.

Through the above objective function, the first and third terms are used to maximize the importance scores of the selected NP and VP, and the second and fourth terms are utilized to penalize the selection of similar phrases. While optimizing the above objective functions, we need to ensure that the compatibility constraints between concepts and facts are satisfied.

[6] For example, the open-source toolkit released by Stanford University can perform coreference resolution for English entities. https://nlp.stanford.edu/projects/coref.shtml.

To model the compatibility constraints, a Boolean variable $\gamma_{ij}$is introduced. If NP_i and VP_j are compatible, $\gamma_{ij} = 1$. The compatibility constraints of NP and VP are

$$\forall i, j, \alpha_i \geq \gamma_{ij}; \ \forall i, \sum_j \gamma_{ij} \geq \alpha_i \tag{11.19}$$

$$\forall j, \sum_i \gamma_{ij} = \beta_j \tag{11.20}$$

The selection of NP or VP should follow the co-occurrence constraints:

$$\alpha_{ij} - \alpha_i \leq 0 \tag{11.21}$$

$$\alpha_{ij} - \alpha_j \leq 0 \tag{11.22}$$

$$\alpha_i + \alpha_j - \alpha_{ij} \leq 1 \tag{11.23}$$

$$\beta_{ij} - \beta_i \leq 0 \tag{11.24}$$

$$\beta_{ij} - \beta_j \leq 0 \tag{11.25}$$

$$\beta_i + \beta_j - \beta_{ij} \leq 1 \tag{11.26}$$

The first two inequalities mentioned above show that if NP_i and NP_j coexist in the final summary, then $\alpha_{ij} = 1$, indicating that both NP should appear in the summary at the same time. The third inequality indicates the opposite constraint. Similarly, the latter three inequalities are the co-occurrence constraints for VP_i and VP_j.

Of course, the summary length constraints are indispensable

$$\sum_i l\left(\mathrm{NP}_i\right) \times \alpha_i + \sum_j l\left(\mathrm{VP}_j\right) \times \beta_j \leq L \tag{11.27}$$

where L is the predefined length limit for the output summary, such as 100 words. $l\left(\mathrm{NP}_i\right)$ and $l\left(\mathrm{VP}_j\right)$ denote the lengths of NP_i and VP_j, respectively.

To better control the output of the summaries, more constraints can be appropriately added. For example, we can require that NP exclude pronouns (you, I, he, etc.) or that the number of concepts (i.e., NP) should not exceed a certain value. Integer linear programming (ILP) searches for an optimal subset of candidate concepts and facts that not only maximizes the importance score but also satisfies all the above constraints. Based on the selected subset and compatibility variables γ_{ij}, new summary sentences will be generated. The experimental results show that the abstractive summarization method based on information fusion is significantly better than the original extraction-based summarization method.

However, the abstractive summarization method based on information fusion includes several cascading modules, ranging from recognition of the semantic units of sentences (e.g., concepts and facts) to the importance estimation of the semantic units and, finally, to summary sentence generation by fusing the semantic units from various sources. This system is complicated and strongly depends on the quality of the syntactic parsing or semantic analysis. Thus, this method is difficult to widely use.

11.4.2 Abstractive Summarization Based on the Encoder-Decoder Framework

Consider how humans perform text summarization. Loosingly speaking, humans first read the documents and abstractly represent the contents in the brain (like the encoder). Then, humans sum up the gist of the documents and finally generate the summary using their own words (like the decoder).

Inspired by the end-to-end neural machine translation framework, Rush et al. (2015) proposed an abstractive summarization method based on the encoder-decoder framework. It first encodes the text in the semantic vector space to simulate the process of text understanding by humans and then generates the summary word by word through the decoder network to simulate the process of natural language sentence generation.

Unlike equivalent semantic transformation in machine translation, text summarization is a process of semantic compression. That is, the semantic information contained in the summary is only a subset of the semantics in the original text. In theory, we can semantically represent a document or documents in the vector space, condense the content, and, finally, convert the condensed semantic representation into summary output. However, to date, there is no effective way to use a vector in the real-valued semantic space to represent the complete semantic information of the whole text or document set. Mapping methods with a very high compression ratio need to be further studied. Accordingly, at present, the abstractive summarization method based on the encoder-decoder framework is mainly applied to the tasks of microblog summarization, sentence summarization, and title generation.

Next, we introduce an example of sentence summarization generation based on the encoder-decoder framework. Given an original sentence $X = \left(x_1, x_2, \cdots, x_{T_x}\right)$, we plan to generate a simplified version $Y = \left(y_1, y_2, \cdots, y_{T_y}\right)$. $\boldsymbol{x}_j$ and $\boldsymbol{y}_i$ denote the low-dimensional real-valued vector of the j-th and i-th words in the sentence X and Y. It is required that $T_y < T_x$, meaning that the length of the simplified sentence should be shorter than the original sentence. Without loss of generality, a bidirectional recurrent neural network (BiRNN) can be used to encode the original sentence X and obtain the hidden semantic representation $\boldsymbol{C} = \left(\boldsymbol{h}_1, \boldsymbol{h}_2, \cdots, \boldsymbol{h}_{T_x}\right)$. Another recurrent neural network takes the hidden semantic representation $\boldsymbol{C}$ of the original sentence as input and maximizes the conditional probability $p\left(y_i | y_{<i}, \boldsymbol{C}\right)$ to generate the simplified sentences word by word, in which $y_{<i} = y_0, y_1, \cdots, y_{i-1}$. We next detail how to obtain $\boldsymbol{C}$ with the encoder and how to calculate conditional probability $p\left(y_i | y_{<i}, \boldsymbol{C}\right)$ with the decoder.

As mentioned above, the encoder uses forward and backward recurrent neural networks (BiRNN) to obtain the hidden semantic representation $\boldsymbol{C}$ of the original sentence X. The forward neural network encodes the sentence word by word from left to right and generates a hidden semantic representation for each position $\overrightarrow{\boldsymbol{h}} = \left(\overrightarrow{\boldsymbol{h}}_1, \overrightarrow{\boldsymbol{h}}_2, \cdots, \overrightarrow{\boldsymbol{h}}_{T_x}\right)$, where

$$\overrightarrow{\boldsymbol{h}}_j = \mathrm{RNN}\left(\overrightarrow{\boldsymbol{h}}_{j-1}, \boldsymbol{x}_j\right) \tag{11.28}$$

$\overrightarrow{\boldsymbol{h}}_0$ can be generally initialized with $\boldsymbol{0}$ vectors. RNN represents the operator of the recurrent neural network. It is employed to convert two input vectors $\overrightarrow{\boldsymbol{h}}_{j-1}$ and $\boldsymbol{x}_j$ into an output vector $\overrightarrow{\boldsymbol{h}}_j$. GRU and LSTM can be adopted here. Taking GRU as an example, the formula is as follows:

$$\overrightarrow{\boldsymbol{r}}_j = \mathrm{Sigmoid}\left(\boldsymbol{W}^r \boldsymbol{x}_j + \boldsymbol{U}^r \overrightarrow{\boldsymbol{h}}_{j-1}\right) \tag{11.29}$$

$$\overrightarrow{\boldsymbol{z}}_j = \mathrm{Sigmoid}\left(\boldsymbol{W}^z \boldsymbol{x}_j + \boldsymbol{U}^z \overrightarrow{\boldsymbol{h}}_{j-1}\right) \tag{11.30}$$

$$\overrightarrow{\boldsymbol{m}}_j = \tanh\left(\boldsymbol{W}\boldsymbol{x}_j + \boldsymbol{U}\left(\overrightarrow{\boldsymbol{r}}_j \odot \overrightarrow{\boldsymbol{h}}_{j-1}\right)\right) \tag{11.31}$$

$$\overrightarrow{\boldsymbol{h}}_j = \overrightarrow{\boldsymbol{z}}_j \odot \overrightarrow{\boldsymbol{h}}_{j-1} + \left(1 - \overrightarrow{\boldsymbol{z}}_j\right) \odot \overrightarrow{\boldsymbol{m}}_j \tag{11.32}$$

$\overrightarrow{\boldsymbol{r}}_j$ and $\overrightarrow{\boldsymbol{z}}_j$ denote the reset gate and update gate, respectively. $\boldsymbol{W}^r$, $\boldsymbol{U}^r$, $\boldsymbol{W}^z$, $\boldsymbol{U}^z$, $\boldsymbol{W}$, and $\boldsymbol{U}$ are learnable parameter matrices. $\odot$ denotes elementwise multiplication. $\overleftarrow{\boldsymbol{h}}_j$ can be calculated in a similar way. Each element in $\boldsymbol{C}$ ($\boldsymbol{h}_j = \left[\overrightarrow{\boldsymbol{h}}_j; \overleftarrow{\boldsymbol{h}}_j\right]$) denotes the concatenation of two vectors.

The decoder leverages the attention mechanism to dynamically compute the conditional probability $p\left(y_i|y_{<i}, \boldsymbol{C}\right)$

$$p\left(y_i|y_{<i}, \boldsymbol{C}\right) = p\left(y_i|y_{<i}, \boldsymbol{c}_i\right) = g\left(y_{i-1}, z_i, \boldsymbol{c}_i\right) \tag{11.33}$$

where $g\left(\cdot\right)$ denotes the nonlinear activation function. z_i denotes the hidden representation at timestep i, which is jointly determined by the hidden representation z_{i-1}, the output of previous timestep y_{i-1} and $\boldsymbol{c}_i$:

$$z_i = \mathrm{RNN}\left(z_{i-1}, y_{i-1}, \boldsymbol{c}_i\right) \tag{11.34}$$

Note that $\boldsymbol{c}_i$ is not the i-th element of $\boldsymbol{C}$ but computed by an attention module:

$$\boldsymbol{c}_i = \sum_{j=1}^{T_x} \alpha_{ij} \boldsymbol{h}_j \tag{11.35}$$

α_{ij} denotes the correlation between the output of the current timestep y_i and the semantic representation of the j-th input word $\boldsymbol{h}_j$, which is calculated as follows:

$$\alpha_{ij} = \frac{\exp\left(e_{ij}\right)}{\sum_{j'=1}^{T_x} e_{ij'}} \tag{11.36}$$

$$e_{ij} = \boldsymbol{v}_a^{\mathrm{T}} \tanh\left(\boldsymbol{W}_a \boldsymbol{z}_{i-1} + \boldsymbol{U}_a \boldsymbol{h}_j\right) \tag{11.37}$$

where $\boldsymbol{W}_a$, $\boldsymbol{U}_a$, and $\boldsymbol{v}_a$ are the learnable parameter matrices in the attention module. Given N labeled instances (original sentence, simplified sentence) as the training data $D_{\text{Train}} = \{(X_n, Y_n)\}_{n=1}^{N}$, the encoder-decoder framework will optimize all the weight parameters to maximize the conditional log-likelihood:

$$\mathscr{L}(\theta) = \frac{1}{N} \sum_{n=1}^{N} \sum_{i=1}^{T_y} \log p\left(y_i^n | y_{<i}^n, X^n; \theta\right) \tag{11.38}$$

Figure 11.9 illustrates an example of the abstractive sentence summarization method based on the above encoder-decoder framework. Given the original sentence *The life should allow failure*, after adding the end-of-sentence symbol EOS, the hidden semantic representations corresponding to each position can be obtained by using the BiRNN. The recurrent neural network decoder uses the attention mechanism to dynamically calculate the input context (e.g., $\boldsymbol{c}_2$). Then, it predicts the output at each timestep according to the hidden state and the output of the previous timestep in the decoder as well as the dynamic context $\boldsymbol{c}_2$ of the input. For example, the output of the second timestep is the word *tolerate*. This process is iterated until the end-of-sentence symbol EOS is generated and finally the simplified sentence *Life tolerates failure* is obtained.

In addition to the BiRNN, we can employ convolutional neural networks and self-attention neural networks. The abstractive sentence summarization model can be regarded as the core technology of many other summarization tasks. For example, the title generation task can be transformed into a sentence summarization task: the input is an article (usually the first sentence or first several sentences of the article), and the output is a single sentence. However, the arbitrary truncation of the article for the input tends to mean that important information is lost, and thus it is not the ideal solution. To address this issue, we can adopt a coarse-to-fine approach. First, we use an extraction-based summarization method to extract one or several important sentences from the article, and then we use these sentences as input to generate titles with the encoder-decoder framework.

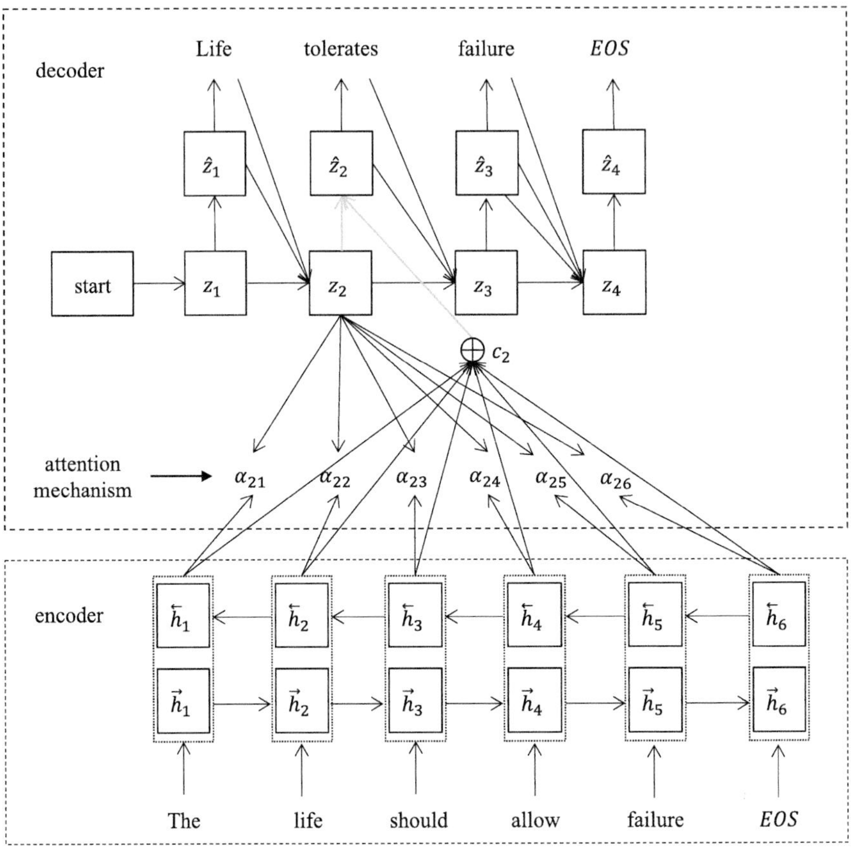

Fig. 11.9 The sentence simplification model based on the encoder-decoder framework

11.5 Query-Based Automatic Summarization

The automatic summarization methods described above focus only on the salience of the content in the document, and they are generally referred to as generic summarization. In many cases, people expect to read a summary that not only contains important information but also is related to a specific topic or query. Therefore, topic-based or query-based automatic summarization methods have gradually become a popular research task.

Query-based automatic summarization methods can be formally defined as follows: given a document or document set D and a query τ expressed in a string or sentence, we attempt to generate a summary that is closely related to the query τ. As seen from the definition, compared with generic summarization, the query-based summarization method emphasizes not only the importance of summary but also the relevance between the result summary and input query. In this section, we introduce three algorithms to calculate the relevance between the sentences and the query.

11.5.1 Relevance Calculation Based on the Language Model

In the early years of research, the query-based automatic summarization method mainly focused on generating summaries of personal profiles (or biographies). It produces summaries according to the queries of *who is X* and *what is X*. Taking the summary generation of a biography as an example, we briefly introduce a method of relevance calculation based on a language model (Biadsy et al. 2008).

For the query *who is* X, this method designs a classifier that identifies whether a sentence in document D is related to the introduction of a biography. First, an unsupervised method is used to extract the biography texts from Wikipedia with information extraction technology, and the extracted biography texts are employed in a language model L_{wiki}. In addition, another general language model L_{news} is trained using news documents. Then, for a sentence s in the test documents, if $L_{\text{wiki}}(s) > L_{\text{news}}(s)$, s is considered to belong to the biography information, while $L_{\text{news}}(s) > L_{\text{wiki}}(s)$ means that it does not belong.

11.5.2 Relevance Calculation Based on Keyword Co-occurrence

Here, we introduce a text summarization method for open queries. This method measures the importance of a sentence by calculating the number of keywords in it, and the keywords are determined by both the query and the original document set.

In a query sentence, not every word deserves attention. The usual approach is to identify all the nouns, verbs, adjectives, and adverbs in the query sentence and regard them as query keywords; these are denoted by the set WS_{query}.

The keywords in the original document set can be obtained by calculating the topic words WS_{topic}. Topic words are those words that are more likely to appear in current documents on a specific topic than in other general texts. For example, *spacecraft* is more likely to appear in aerospace-related news reports and is a topic word, while *today* can appear in any document and is thus not a topic word. Therefore, the proportion of topic words in a sentence basically reflects the importance of the sentence. The set of topic words can be determined by computing the likelihood ratio, mutual information, and TFIDF. For example, we can calculate the TFIDF value of each word in the document set, sort all words, and select the top N words to obtain the topic word set WS_{topic}.

Then, each word in the text is assigned a probability by the following equation:

$$p(w) = \begin{cases} 0.0, \text{ if } w \notin \mathrm{WS_{topic}} \text{ and } w \notin \mathrm{WS_{query}} \\ 0.5, \text{ if } w \in \mathrm{WS_{topic}} \text{ or } w \in \mathrm{WS_{query}} \\ 1.0, \text{ if } \in \mathrm{WS_{topic}} \text{ and } w \in \mathrm{WS_{query}} \end{cases} \tag{11.39}$$

For each sentence in the document, the weighted average score based on the word probability $p(w)$ is used as the final score of the sentence. To some extent, this score reflects not only the importance of the content but also the degree of relevance between the sentence and the query. Finally, the result summary can be generated according to the sentence score in the same way as the generic summarization method.

11.5.3 Graph-Based Relevance Calculation Method

As mentioned earlier, the graph-based PageRank algorithm is widely used in extraction-based generic summarization. This algorithm can be easily adapted to query-based summarization. In the graph-based PageRank algorithm, the importance score of sentences is calculated as follows:

$$S(V_i) = \frac{1-d}{N} + d \times \sum_{V_j \in \mathrm{adj}(V_i)} \frac{W_{ij}}{\sum_{V_k \in \mathrm{adj}(V_j)} W_{jk}} S(V_j) \tag{11.40}$$

To model the relevance between the query and text sentence as well, the cosine similarity between the text sentence and the query can be used as the relevance score $\mathrm{rel}(V_i, \tau)$. It can be further employed as the initial score in the graph-based algorithm: $S(V_i|\tau) = \mathrm{rel}(V_i, \tau)$. Equation (11.7), which focuses only on the sentence importance, can be appropriately adapted to iteratively calculate a comprehensive score that takes into account both the sentence importance and the correlation between the query and this sentence:

$$S(V_i|\tau) = (1-d) \times \frac{S(V_i|\tau)}{\sum_{V_k \in \mathrm{adj}(V_i)} S(V_k|\tau)} + d \times \sum_{V_j \in \mathrm{adj}(V_i)} \frac{W_{ij}}{\sum_{V_k \in \mathrm{adj}(V_j)} W_{jk}} S(V_j|\tau) \tag{11.41}$$

After calculating the score of each sentence with the above formula, the final summary sentences can be generated according to the length and redundancy constraints.

11.6 Crosslingual and Multilingual Automatic Summarization

Quickly and effectively obtaining information in the current complex multilingual environment has become a major concern in both academia and industry. Accordingly, crosslingual and multilingual summarization technology is attracting increasing attention.

11.6.1 Crosslingual Automatic Summarization

Crosslingual automatic summarization uses the documents in source language A as input and outputs a summary in target language B. In this section, we employ English-Chinese crosslingual summarization as an example to introduce some popular methods.

Ideally, if the quality of machine translation is sufficient, crosslingual summarization is not a problem. In this case, English summaries can be generated first and then translated into Chinese using machine translation. Although the performance of machine translation has improved due to deep learning technology, it is still far from satisfactory. Therefore, simultaneously considering both content importance and translation quality is a challenge for crosslingual summarization.

At present, the extraction-based method is dominant in crosslingual summarization, but neither of the two popular methods considers the quality of the machine translation. The first method summarizes before translation, and the second translates before summarization (Wan et al. 2010). As the name implies, the method that summarizes before translation first extracts summary sentences from English documents and then translates the English summary into Chinese. The method that translates before summarization first translates all the sentences in the English documents into Chinese and then extracts the summary sentences from the Chinese translations. Each of the two methods has its advantages and disadvantages. The former method can make full use of English information, but due to the unsatisfactory quality of machine translation, the translated summaries may contain many errors; that is, the original important summary sentence may be incorrectly translated. The latter method can take full advantage of Chinese information, but the summary sentences selected from Chinese translations that contain translation errors may not correspond to the actual important information in the original English texts. That is, the selection of summary sentences according to Chinese features may be negatively influenced by translation errors. Therefore, only using the information of one language is not the best approach. Next, a graph-based crosslingual summarization method is introduced that can simultaneously make full use of information from both languages, which to some extent avoids translation errors (Wan 2011).

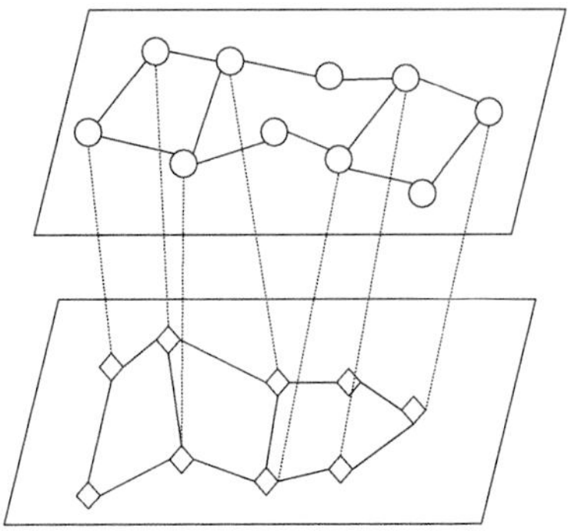

Fig. 11.10 The crosslingual summarization method based on the graph model

Formally, an English dataset D^{en} can be first translated into Chinese D^{cn}. Suppose that $V^{\text{en}} = \left\{s_i^{\text{en}} | 1 \leq i \leq n\right\}$ and $V^{\text{cn}} = \left\{s_i^{\text{cn}} | 1 \leq i \leq n\right\}$ denote sentences in D^{en} and D^{cn}, respectively; n is the number of sentences in the documents, and s_i^{cn} is the Chinese translation of s_i^{en}. As shown in Fig. 11.10, we can construct an undirected graph $G = (V^{\text{en}}, V^{\text{cn}}, E^{\text{en}}, E^{\text{cn}}, E^{\text{encn}})$ with five elements, where E^{en} denotes the relation between any two sentences in the English documents, E^{cn} represents the relation between any two sentences in the translated Chinese documents, and E^{encn} is the relation between any sentence in V^{en} and any sentence in V^{cn}.

Suppose that $W^{\text{en}} = \left(W_{ij}^{\text{en}}\right)_{n \times n}$ is the weight matrix of edges in E^{en}, and W_{ij}^{en} denotes the similarity between the i-th sentence s_i^{en} and the j-th sentence s_j^{en} in the English documents V^{en}:

$$W_{ij}^{\text{en}} = \begin{cases} \text{sim}_{\text{cosine}}\left(s_i^{\text{en}}, s_j^{\text{en}}\right), & i \neq j \\ 0, & \text{others} \end{cases} \tag{11.42}$$

$\text{sim}_{\text{cosine}}\left(s_i^{\text{en}}, s_j^{\text{en}}\right)$ is the cosine similarity between the TFIDF vectors of s_i^{en} and s_j^{en}. The weight matrix $W^{\text{cn}} = \left(W_{ij}^{\text{cn}}\right)_{n \times n}$ of edges in E^{cn} can be calculated in a similar way.

Concerning the weight matrix $W^{\text{encn}} = \left(W_{ij}^{\text{encn}}\right)_{n \times n}$ for E^{encn}, each element W_{ij}^{encn} is related to both English sentence s_i^{en} and Chinese sentence s_j^{cn}. The corresponding TFIDF vectors belong to two languages and are not in the same semantic space, so their similarity cannot be calculated directly with the cosine distance. Since we know that s_i^{cn} is the translated Chinese version of s_i^{en}, and the original English version of s_j^{cn} is s_j^{en}, while s_i^{en} and s_j^{en}, s_i^{cn} and s_j^{cn} are in the same space, $\text{sim}_{\text{cosine}}\left(s_i^{\text{en}}, s_j^{\text{en}}\right)$ and $\text{sim}_{\text{cosine}}\left(s_i^{\text{cn}}, s_j^{\text{cn}}\right)$ can approximate W_{ij}^{encn}:

$$W_{ij}^{\text{encn}} = \sqrt{\text{sim}_{\text{cosine}}\left(s_i^{\text{en}}, s_j^{\text{en}}\right) \times \text{sim}_{\text{cosine}}\left(s_i^{\text{cn}}, s_j^{\text{cn}}\right)} \tag{11.43}$$

Because an undirected graph is adopted here, W^{en}, W^{cn}, and W_{ij}^{encn} are all diagonal matrices, which means $W^{\text{en}} = (W^{\text{en}})^{\text{T}}$, $W^{\text{cn}} = (W^{\text{cn}})^{\text{T}}$, and $W^{\text{encn}} = (W^{\text{encn}})^{\text{T}}$. Normalization of each line in the matrix leads to $\hat{W}^{\text{en}}$ and $\hat{W}^{\text{cn}}$ and $\hat{W}^{\text{encn}}$. If $u\left(s_i^{\text{en}}\right)$ and $v\left(s_i^{\text{cn}}\right)$, respectively, denote the importance score of Chinese sentence s_i^{cn} and English sentence s_j^{en}, then $u\left(s_i^{\text{en}}\right)$ and $v\left(s_i^{\text{cn}}\right)$ can be iteratively updated according to the following formula until convergence:

$$v\left(s_i^{\text{cn}}\right) = \alpha \sum_j W_{ji}^{\text{cn}} v\left(s_j^{\text{cn}}\right) + \beta \sum_j W_{ji}^{\text{encn}} u\left(s_j^{\text{en}}\right) \tag{11.44}$$

$$u\left(s_j^{\text{en}}\right) = \alpha \sum_i W_{ij}^{\text{en}} u\left(s_i^{\text{en}}\right) + \beta \sum_i W_{ij}^{\text{encn}} v\left(s_i^{\text{cn}}\right) \tag{11.45}$$

where $\alpha + \beta = 1$ balances the contribution of two languages. Once $v\left(s_i^{\text{cn}}\right)$ is obtained, summary sentence selection algorithms in the generic summarization method can be used to obtain the final Chinese summary.

To fully consider content importance and translation accuracy, Zhang et al. (2016a) extended the abstractive method based on information fusion in generic summarization to meet the special requirements of crosslingual summarization tasks. The difference is that in generic summarization, NP and NP denote concepts and facts, respectively, while Zhang et al. (2016a) used agent (ARG0) in the predicate-argument structure to represent concepts and used the combination of predicates and arguments (Predicate + ARG1 or Predicate + ARG2) to represent facts. Similar to extraction-based crosslingual summarization, English sentences are first translated into Chinese by means of a machine translation system, and then concepts and facts are extracted.

Figure 11.11 illustrates the crosslingual summarization method based on the information fusion model. It can be seen from the figure that the predicate-argument structure of an English sentence is obtained by semantic role labeling (SRL). Then, the concepts (ARG0) and facts (Predicate+ARG1 or Predicate+ARG2) in English sentences are aligned to Chinese phrases by using the relationship between word translations (i.e., word alignment in machine translation). For example, 美国 总统 布什(*President George Bush*) and 布什 总统(*President Bush*) expressed the same concept. 他 第二 次 访问(*made his second visit*) and 访问 该 地区(*made to the region*), 授权 为 受灾 地区(*authorized federal disaster assistance*), 授权 的 联邦 救灾 援助(*authorized federal disaster assistance*), and 计划 检查 的 状态(*made plans for an inspection tour of the state*) all express facts. Since these two sentences describe the same concept of President Bush, they can be combined into one sentence in the process of abstractive summary generation. Through calculation,

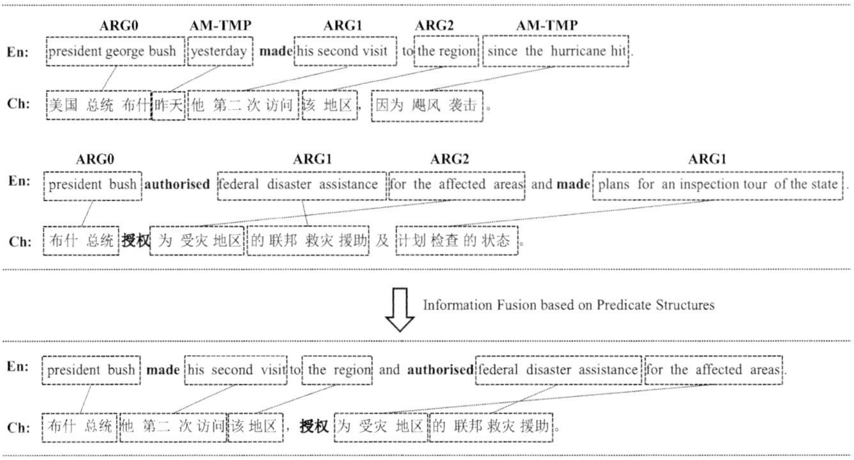

Fig. 11.11 Crosslingual summarization method based on the information fusion model

it is found that the facts of 他 第二 次 访问(*made his second visit*) and 授权 的 联邦 救灾 援助(*authorized federal disaster assistance*) not only have high importance scores but also have good translation quality. By combining compatible concepts and facts, the final summary sentence can be obtained: 布什 总统 他 第二 次 访问 该 地区， 授权 为 受灾 地区 的 联邦 救灾 援助(*President Bush made his second visit to the region and authorized federal disaster assistance for the affected areas*). The two key issues in the whole process are calculating comprehensive scores (importance and translation quality) for the concepts and facts and assessing the compatibility between concepts and facts.

When calculating the comprehensive score of concepts and facts, the importance score S_{im} can be estimated in various ways, as described in previous sections. For example, we can calculate the proportion of named entities or adopt a graph-based bilingual estimation model. In the computation of translation quality, Zhang et al. (2016a) merged the word translation probability p_{lex} and the language model score p_{lm}. Specifically, given an English concept or fact, $\text{ph}_{\text{en}} = e_0 e_1 \cdots e_l$ and its corresponding translation $\text{ph}_{\text{cn}} = c_0 c_1 \cdots c_m$, the word probability can be calculated as follows:

$$p_{\text{lex}}\left(\text{ph}_{\text{cn}}|\text{ph}_{\text{en}}, a\right) = \left\{\prod_{j=0}^{m} \frac{1}{|\{i|\,(i,j) \in a\}|} \sum_{\forall (i,j) \in a} p\left(c_j|e_i\right)\right\}^{\frac{1}{m+1}} \tag{11.46}$$

where a denotes the translation correspondence (word alignment) between the words in ph_{cn} and ph_{en}. For example, $(i, j) \in a$ indicates that c_j and e_i are the word translation pair. The word alignment a and the lexical translation probability

$p(c_j|e_i)$ can be obtained by utilizing the word alignment tool GIZA++[7] in machine translation. The language model probability of the Chinese translation $\mathrm{ph}_{\mathrm{cn}}$ can be calculated through the following n-gram model:

$$p_{\mathrm{lm}}(\mathrm{ph}_{\mathrm{cn}}) = \sum_{j=0}^{m-n+1} p(c_j|c_{j-n+1}\cdots c_{j-1}) \tag{11.47}$$

The translation quality can be calculated by multiplying the word translation probability and the language model probability: $S_{\mathrm{trans}} = p_{\mathrm{lex}}(\mathrm{ph}_{\mathrm{cn}}|\mathrm{ph}_{\mathrm{en}}, a) \times p_{\mathrm{lm}}(\mathrm{ph}_{\mathrm{cn}})$. Finally, the weighted sum of importance score S_{im} and translation quality S_{trans} measures the comprehensive score of a concept or fact: $S_{\mathrm{com}} = \alpha S_{\mathrm{im}} + \beta S_{\mathrm{trans}}$.

The compatibility between concepts and facts includes the compatibility between concepts and the compatibility between concepts and facts. Compatibility assessment between concepts aims to determine whether two concepts described in different sentences belong to the same concept. Compatibility assessment between concepts and facts aims to determine whether a concept is the subject of a fact. The compatibility between concepts can be judged by coreference (anaphora) resolution and similarity calculation. For example, *President George Bush* and *President Bush* are two compatible concepts because they share the same entity *Bush* and title *President.* The concept $\mathrm{concept}_{\mathrm{ch}}$ and the fact $\mathrm{fact}_{\mathrm{ch}}$ satisfy the compatibility constraint if and only if $\mathrm{concept}_{\mathrm{ch}}$ and $\mathrm{fact}_{\mathrm{ch}}$ come from the same sentence or $\mathrm{concept}_{\mathrm{ch}}$ and $\mathrm{fact}'_{\mathrm{ch}}$ come from the same sentence and $\mathrm{fact}'_{\mathrm{ch}}$ is compatible with $\mathrm{fact}_{\mathrm{ch}}$.

Given the comprehensive scores for concepts and facts and their compatibility, algorithms such as the ILP used in generic summarization based on the information fusing method can be employed to obtain a Chinese summary from English documents.

11.6.2 Multilingual Automatic Summarization

Multilingual automatic summarization takes documents on the same topic in different languages as the input and outputs a summary in one of the languages. For example, a Chinese summary can be generated from documents written in a mix of English, Japanese, and Chinese on the same topic. In real life, multilingual summary tasks are very helpful. We know that the world's major media usually report a large number of important events (e.g., the World Cup and US presidential election) in different languages every day. Although the topic is the same, different reports may focus on different aspects. Therefore, compressing multilingual texts

[7] http://www.statmt.org/moses/giza/GIZA++.html.

into a summary in the user's language is very helpful for a user aiming to acquire global information.

In the multilingual summarization task, a typical method is to summarize after translation. First, the texts of all other languages are translated into a specific language through a machine translation system. Then, generic summarization methods are employed to generate the final summary. Many studies have found that extracting summary sentences from machine translation results leads to poor summarization quality. Similar to crosslingual summarization, a key problem arises: how to make full use of the imperfect machine translation results? In the summarization after translation method, the machine-translated Chinese sentences are treated equally to the original Chinese sentences, although machine-translated Chinese sentences are obviously inferior to the originals in terms of information accuracy and text fluency and should be distinguished from the original natural Chinese sentences.

To address these issues, Li et al. (2016a) proposed a multilingual summarization method based on an adaptive graph model, which is an extension of the undirected graph model. It automatically selects some edges among the undirected edges connecting two languages using an adaptive method and converts them into directed edges. Its basic process is as follows:

Given the English and Chinese document sets D^{en} and D^{cn}, $V^{\text{en}} = \left\{s_i^{\text{en}} | 1 \le i \le n\right\}$ and $V^{\text{cn}} = \left\{s_i^{\text{cn}} | 1 \le i \le m\right\}$ denote sentences in D^{en} and D^{cn}. First, an undirected graph $G = (V^{\text{en}}, V^{\text{cn}}, E^{\text{en}}, E^{\text{cn}}, E^{\text{encn}})$ is constructed. Then, before running the graph algorithm, some undirected edges (namely, bidirectional edges) in the set E^{encn} that connects the two languages will be changed into undirected edges if the English sentence only contributes to the importance of its translated Chinese counterpart.

As shown in Fig. 11.12, the Chinese translation S_1^{en2cn} of the English sentence S_1^{en} and the natural Chinese sentence S_1^{cn} are close in meaning, but S_1^{en2cn} is less informative than S_1^{cn}. Under such conditions, the natural Chinese sentence S_1^{cn} is preferred over the translation S_1^{en2cn}. Therefore, the undirected edge between

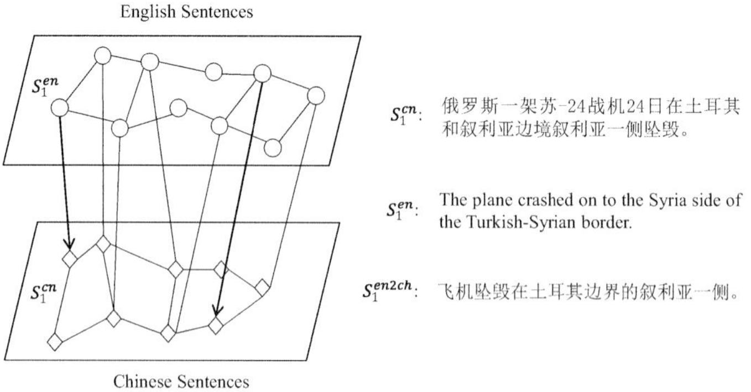
English Sentences
S_1^{en}
S_1^{cn}
Chinese Sentences
S_1^{cn}: 俄罗斯一架苏-24战机24日在土耳其和叙利亚边境叙利亚一侧坠毁。
S_1^{en}: The plane crashed on to the Syria side of the Turkish-Syrian border.
S_1^{en2ch}: 飞机坠毁在土耳其边界的叙利亚一侧。

Fig. 11.12 Multilingual summarization based on the adaptive graph model

English sentence S_1^{en} and Chinese sentence S_1^{cn} needs to be transformed into the undirected edge from S_1^{en} to S_1^{cn}, indicating that S_1^{en} only recommends S_1^{cn} and S_1^{en} is not recommended by S_1^{cn}.

A key issue then arises regarding how to measure whether S_1^{cn} and S_1^{en2cn} are similar in terms of semantics. Various approaches can be employed to address this issue. First, we can calculate the correlation score between S_1^{cn} and S_1^{en2cn} with the cosine similarity. If the result exceeds a prespecified threshold, these two sentences are considered to be similar to each other. Second, we can check if S_1^{cn} entails S_1^{en2cn} with the textual entailment method, where S_1^{cn} and S_1^{en2cn} are similar in semantics if the entailment stands. Third, we can also determine whether the natural Chinese sentence S_1^{cn} is a translation of the English sentence S_1^{en}. The two sentences will be treated similarly if the translation probability is greater than a threshold.

After the edges in E^{encn} are processed, the graph-based algorithm can iteratively calculate the importance scores of the Chinese and English sentences. This method not only avoids choosing English sentences similar to natural Chinese sentences but also retains complementary but important English sentences by including their Chinese translations as part of the final summary.

11.7 Summary Quality Evaluation and Evaluation Workshops

Researchers in the automatic text summarization research community constantly propose new methods and algorithms and determine whether these new methods and algorithms improve the summarization quality compared to the older ones. Therefore, summary quality evaluation has become an important research topic for the rapid iteration of summarization technology.

11.7.1 Summary Quality Evaluation Methods

Compared with the evaluation of text classification and machine translation outputs, quality evaluation for summarization is a more intractable problem because there is no perfect summary in theory. For the same document or document set, different people will generate different summaries. Although summary quality evaluation faces enormous difficulties and challenges, it still attracts the attention of many researchers. Generally, summary quality evaluation methods are divided into manual methods and automatic approaches.

(1) Manual evaluation methods

Manual evaluation is the most intuitive method. Generally, experts are asked to score the summary generated from an automatic summarization system, basing these scores on the consistency, fluency, and information capacity of the summary. In the Document Understanding Conference (DUC) evaluation conducted by NIST in 2005, the manual evaluation focused on the following six aspects: grammaticality, non-redundancy, referential clarity, focus, structure, and coherence. Each score ranged from 1 to 5, in which 1 represents the worst and 5 the best. These six evaluation metrics are now widely accepted in manual evaluations. However, the scores given by different experts in a manual evaluation are quite divergent: a system summary evaluated by one expert as being of good quality may be evaluated by another expert as very poor. Therefore, overcoming the divergency problem between experts has attracted many researchers, and several methods have been proposed to address this issue. Among them, the pyramid method (Nenkova and Passonneau 2004) is one effective way to solve this problem.

In the pyramid method, the summary content unit (SCU) is the key concept, as it represents the important semantic unit at the subsentence level of the summary. Different summaries may share the same SCUs, even though the words used may be quite different. The SCU can be as short as a modifier of a NP or as long as a clause. In the process of analyzing and annotating an SCU, the annotator needs to describe the SCUs shared by different summaries in his/her own words. If the information contained in a sentence only appears in one summary, the sentence can be divided into clauses, and each clause can be treated as an SCU. For a collection of documents (test sets), m experts are first invited to write m reference summaries Sum$(r_0, r_1, \ldots, r_m)$. All reference summaries are then manually analyzed, and a set of SCUs is extracted. If an SCU is mentioned by w reference summaries, the weight w is assigned to this SCU.

The fewest SCUs appear in all m reference summaries, while the number of SCUs mentioned by $(m - 1)$ reference summaries is greater, and the largest number of SCUs occurs in only one reference summary; thus, the SCU values are distributed in a pyramid form, which explains why it is called the pyramid method.

For a summary generated by one system, the first step is to manually analyze the SCUs of the system summary; the second step is to calculate the sum of the scores of all SCUs in the reference summary; and then the third step is to calculate the ratio of the system summary score to an ideal summary score and take the ratio as the evaluation score for the system summary.

We will next use an example to illustrate how to determine the SCUs (Table 11.3).

Table 11.3 An example of summary content units (SCUs)

A_2:	In the 2016 US presidential election, Trump defeated Hillary Clinton and was elected.
B_4:	He won the 45th US presidential election.
C_3:	Trump became the 45th US president.
D_1:	The 2016 US election was full of suspense and Trump finally won the election.

As shown in the table above, we assume that there are four reference summaries. A_2 denotes the second sentence in the first reference summary, B_4 represents the fourth sentence in the second reference summary, and C_3 and D_1 are similar. The human annotator extracts the SCUs containing similar information from these reference summaries. A semantic analysis shows that these sentences contain two summary content units SCU_1 and SCU_2. Since SCU_1 appears in all four reference summaries, the value of SCU_1 is 4, and similarly, the value of SCU_2 is 2.

SCU_1:	Trump was elected the 45th US president
	A_2: Trump defeated Hillary Clinton and was elected
	B_4: He won the 45th US presidential election
	C_3: Trump became the 45th US president
	D_1: Trump finally won the election
SCU_2:	The US held a presidential election in 2016.
	A_2: In the 2016 US presidential election
	D_1: The 2016 US election

The m reference summaries correspond to a pyramid with a height of m. Figure 11.13 shows a pyramid with a height of 4 constructed by four reference summaries. If an SCU appears in all four reference summaries, this SCU is placed at the top tier of the pyramid $W = 4$. An SCU mentioned in only one summary is placed at the bottom tier $W = 1$. The pyramid height may be less than the number of reference summaries; for example, if there is no SCU shared by three reference summaries, the pyramid will not have the layer of $W = 3$.

For an m-tier pyramid, T_i denotes the i-th layer, and all SCUs in T_i have weights of i. That is, these SCUs are mentioned in i reference summaries. $|T_i|$ denotes the number of SCUs in T_i. For an ideal summary that needs to contain X SCUs, it is expected that the SCUs at the top of the pyramid will be included first and then the SCUs at the $m-1$ and $m-2$ levels can be included in that order until X SCUs are included. The formula for calculating the score is as follows:

$$\text{Score}_{\text{ideal}} = \sum_{i=j+1}^{m} i \times |T_i| + j \times \left(X - \sum_{i=j+1}^{m} |T_i| \right) \tag{11.48}$$

where $j = \max_k \left(\sum_{t=k}^{m} |T_t| \geq X \right)$.

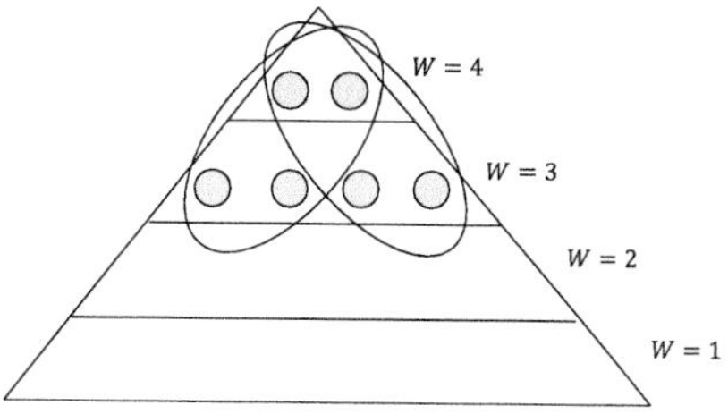

Fig. 11.13 Example of a pyramid constructed by four reference summaries

Figure 11.13 shows a pyramid of four reference summaries. The figure shows that there are six SCUs in the four reference summaries, two of which are mentioned in all reference summaries, with the other four appearing in three reference summaries. For an ideal summary containing four SCUs, there must be two SCUs at the top tier of the pyramid ($W = 4$), and the other two are at tier $W = 3$. The two solid circles in the figure represent two ideal summaries containing four SCUs.

Assume that after comparison between the automatic summary result and the reference summary, it is found that D_i SCUs at tier T_i appear in the system summary, i.e., D_i SCUs in the system summary are mentioned by i reference summaries. Then, the total number of SCUs in Sum_{sys} is $\sum_{i=1}^{m} i \times D_i$, whose score is

$$\text{Score}_{\text{sys}} = \sum_{i=1}^{m} i \times D_i \tag{11.49}$$

Let $X = \sum_{i=1}^{m} D_i$ denote the number of SCUs that should be included in an ideal summary; then, the final evaluation score of the system summary becomes $\text{Score}_{\text{sys}}/\text{Score}_{\text{ideal}}$. Obviously, it is precision oriented. We can also let X be the average number of SCUs in the reference summaries, and then the final evaluation score $\text{Score}_{\text{sys}}/\text{Score}_{\text{ideal}}$ will be recall oriented.

The pyramid method can minimize the influence of the differences in reference summaries. However, manual evaluation always takes considerable time and human resources. For example, it is reported that in a summary evaluation of the DUC, it takes approximately 3000 h to manually evaluate the quality of the summaries. Accordingly, automatic evaluation methods are more attractive.

(2) Automatic evaluation methods

Automatic evaluation methods can be divided into two categories. One is intrinsic evaluation, which directly evaluates the quality of the summary results. The other is extrinsic evaluation, which indirectly evaluates the summary quality by analyzing the performance of the downstream tasks that rely on the results of the summarization.

Intrinsic evaluation methods are intuitive and efficient; thus, they are commonly used. Generally, intrinsic evaluation methods can also be divided into two categories: form metrics and content metrics. Form metrics focus on the grammar, coherence, and organizational structure of the summaries. Content metrics focus on summary content and information, which are believed to be the most important in automatic evaluation. We will here introduce a widely used intrinsic evaluation method based on content metrics, Recall-Oriented Understudy for Gisting Evaluation (ROUGE) (Lin 2004).

ROUGE was proposed by (Lin 2004) and has become almost the de facto automation evaluation metric for summarization methods. The basic idea of ROUGE is borrowed from BLEU (BiLingual Evaluation Understudy) (Papineni et al. 2002), which is an automatic evaluation method for machine translation. In contrast to BLEU, which is precision oriented, ROUGE focuses on recall.

Assuming that a document set corresponds to a reference summary r and that the summary result generated by the system is sum, ROUGE-n (n denotes the number of words contained in the matching unit, namely, n-gram) is calculated as follows:

$$\text{ROUGE-}n(sum, r) = \frac{\sum_{n\text{-gram}\in r} \text{count}_{\text{match}}(n\text{-gram}, sum)}{\sum_{n\text{-gram}\in r} \text{count}(n\text{-gram})} \tag{11.50}$$

where n denotes the length of the phrase n-gram and $\text{count}_{\text{match}}(n\text{-gram}, sum)$ indicates the maximum number of times that n-gram appears in both reference summary r and system summary sum. If n-gram appears a times in the reference summary and b times in the system summary, $\text{count}_{\text{match}}(n\text{-gram}, sum) = \min(a, b)$. It can be seen from the above equation that ROUGE-n is recall oriented.

If there are multiple reference summaries $R = \{r_0, r_1, \ldots, r_m\}$, ROUGE-$n$ can be calculated by comparing the system summary with each reference summary, and the highest matching score is taken as the final result:

$$\text{ROUGE-}n_{\text{multi}}(\text{sum}) = \max_{r\in R} \text{ROUGE-}n(\text{sum}, r) \tag{11.51}$$

There are many other recall-based evaluation metrics within ROUGE, such as ROUGE-L and ROUGE-S. ROUGE-L calculates the matching rate of common substrings. The basic idea is that the longer the common substring is, the more similar the two sentences are. Let $s_0, s_1, \ldots, s_u$ refer to all the sentences in reference summaries R and sum denote the summary (can be viewed as the concatenation of all summary sentences) generated by the system. Then, ROUGE-L can be calculated by the following formula:

$$\text{ROUGE-}L(\text{sum}) = \frac{(1+\beta^2)R_{\text{LCS}}P_{\text{LCS}}}{R_{\text{LCS}} + \beta^2 P_{\text{LCS}}} \tag{11.52}$$

where R_{LCS} and P_{LCS} can be computed by

$$R_{\text{LCS}} = \frac{\sum_{i=1}^{u} \text{LCS}(s_i, \text{sum})}{\sum\limits_{i=1}^{u} |s_i|} \tag{11.53}$$

$$P_{\text{LCS}} = \frac{\sum_{i=1}^{u} \text{LCS}(s_i, \text{sum})}{|\text{sum}|} \tag{11.54}$$

in which LCS(s_i, sum) denotes the longest common substring of s_i and sum. $|s_i|$ denotes the length of the reference summary, while |sum| is the length of the system summary.

ROUGE-S is a special case of ROUGE-n (n=2), which is also called the skip bigram matching rate. For example, in the sentence *Trump became president,*

Trump-president is a skip bigram. ROUGE-S is calculated as follows:

$$\text{ROUGE-}S(\text{sum}) = \frac{(1+\beta^2)R_S P_S}{R_S + \beta^2 P_S} \tag{11.55}$$

where R_S and P_S are the recall and precision of the skip bigram, respectively.

11.7.2 Evaluation Workshops

Evaluation workshops in automatic summarization have greatly promotes the development of summarization technology. Since 2001, an automatic summarization evaluation forum has been held almost every year, including the DUC and Text Analysis Conference (TAC) evaluation organized by the NIST of the United States, the multi-lingual summarization evaluation (MSE) and MultiLing evaluation organized by the Association for Computational Linguistics (ACL), and the automatic summarization evaluation organized by the international conference on Natural Language Processing and Chinese Computing (NLPCC). These evaluation workshops essentially conduct the same process. First, the organizer releases the training data to the participants, which they can use to train their summarization systems. Then, the organizer releases the test data to the participants and asks participants to submit the outputs of their system before a prespecified deadline. Subsequently, the organizer will evaluate the submitted summaries using both manual and automatic evaluation methods. The participants are then invited to introduce their systems and communicate with each other at the evaluation workshop.

(1) The DUC evaluation workshop

NIST started the DUC in 2001, which was launched by Professor Daniel Marcu from the University of Southern California. The main task of the DUC is to evaluate the development trends in text summarization technology. From 2001 to 2007, the DUC attracted over 20 participants each year. In 2001 and 2002, the DUC focused on single- and multidocument news summarization. NIST collected 60 news document sets, each of which discussed a certain topic, and provided several human-generated summaries for each document set as references. Among them, 30 document sets were used as training data, and the other 30 were employed as test data.

In 2003, the DUC introduced new evaluation tasks, such as generating very short summaries for a single document, which is similar to headline generation; multidocument summarization based on events and opinions; and question-oriented summaries, which require the summaries generated by the system to answer specified questions. In 2004, the DUC explored the evaluation of crosslingual text summarization. However, this was more similar to evaluation of the *translation before summarization* method because the organizer only provided machine-translated English documents as input, and the source language information was unknown to the participants. From 2005 to 2007, the DUC mainly evaluated query-based multidocument summarization. The DUC ended in 2008, and the TAC (Text Analysis Conference), also organized by NIST, began to take over the task of text summary evaluation.

(2) The TAC evaluation workshop

Since 2008, the TAC has organized four evaluation tasks, including text summarization, automatic question answering, text entailment, and knowledge base population. The text summarization evaluation was organized five times (2008–2011 and 2014). In 2008 and 2009, the TAC designed the update summarization evaluation; this task assumes that users have read some earlier documents on a topic, and the system aims to produce an updated summary given some new documents on the same topic.

The TAC began to focus on guided summarization in 2010 and 2011. Given multiple documents on the same topic, guided summarization systems generate summaries containing all specified features for the prespecified event categories and aspects. The TAC also explored a language-independent multilingual summarization task in 2011. This task required that the summarization methods proposed by the participants be universal, i.e., not only effective in the summarization task of one language but also achieving good results in other languages. Later, this language-independent summarization evaluation task was continued by the MultiLing workshop and was held every 2 years.

In 2014, TAC organized an evaluation task of automatic summarization for documents from the literature on biomedical science and technology: given a group of papers citing the same literature, the system was required to identify the text blocks describing the references and generate a structured summary for the cited literature.

(3) MSE evaluation workshop

The ACL organized the MSE workshop in 2005 and 2006. The organizers provided a collection of texts on the same topic in Arabic and English, requiring the system to submit an English summary of fewer than 100 words. Most of the participants used machine translation systems to translate Arabic documents into English and then converted these into monolingual text summarization tasks. The evaluation results showed that the summaries generated by this translate-before-summarization method were worse than those produced by monolinguistic summarization methods using only the original English documents. This result may be due to two reasons: on the one hand, the Arabic-to-English machine translation system may not have been good enough at that time or, on the other hand, the multilingual summarization method may not have made effective use of the machine translation results.

(4) The NLPCC evaluation workshop

The DUC, TAC, and other workshops mainly focus on English summarization, while there is almost no evaluation for Chinese texts. The NLPCC conference has been organizing evaluation workshops on Chinese automatic summarization since 2015. In 2015 and 2017, the NLPCC organized an evaluation task of single-document news summarization.[8] In 2015, it was more oriented toward social networks, in which the system is required to generate a summary of news documents in 140 Chinese characters that could be published on Weibo.

In 2016, the NLPCC explored a new text summarization task for sports news.[9] Given a Chinese script file for the live broadcast of a sports event, the system attempted to generate a brief report of the event. We can see from the evaluation tasks that the Chinese evaluation workshops have paid more attention to practical applications of summarization technologies.

11.8 Further Reading

In this chapter, we introduce the typical tasks and methods of automatic summarization. Many other summarization tasks are not involved in this area, for example, comparative summarization (Huang et al. 2011), update summarization (Dang and Owczarzak 2008), timeline summarization (Yan et al. 2011), and multimodal summarization (Wang et al. 2016a; Li et al. 2017b). The purpose of comparative summarization is to generate summaries from multiple perspectives for documents on similar topics, such as comparative summarization for reports on the 2008 and 2016 Olympic Games. Assuming that the user already knows earlier information on a topic, the goal of the update summarization is to generate a brief updated summary based on new documents on the same topic. The timeline summarization generates a series of short summaries for an event or a topic according to the timeline. Detailed descriptions and the most recent progress on these tasks can be found in (Yao et al. 2017).

In addition to the summarization of pure text, text-centered multimodal summarization has recently attracted increasing attention. This approach attempts to answer the following questions: how to make full use of the relevant image information when generating summaries given texts and images (Wang et al. 2016a); how to integrate texts, images, videos, and audio on the same topic; how to model multimodal information to generate a comprehensive but short text summary (Li et al. 2017b); and how to generate summaries presented in multimodal forms (Zhu et al. 2018, 2020).

From a methodological point of view, in recent years, the end-to-end abstractive method has been one of the frontier methods (Rush et al. 2015; Chopra et al. 2016; Gu et al. 2016; Tan et al. 2017; Nema et al. 2017; Zhou et al. 2017). This kind of method involves three key technologies: accurately encoding the original

---

[8] http://tcci.ccf.org.cn/conference/2015/pages/page05_evadata.html. http://tcci.ccf.org.cn/conference/2017/taskdata.php.

[9] http://tcci.ccf.org.cn/conference/2016/pages/page05_evadata.html.

text, precisely finding and attending to the salient information in the original text, and compressing and generating the final summary. Zhou et al. (2017), Gu et al. (2016), Tan et al. (2017), and Nema et al. (2017) discussed these three issues in detail. However, at present, the end-to-end methods mainly focus on sentence summarization and single-document summarization. Applying this paradigm to multidocument, multilingual and multimodal summarizations remains an open problem.

Although ROUGE is currently the de facto evaluation metric for automatic summarization, designing more accurate and reasonable evaluation methods continues to be a hot research topic. Kurisinkel et al. (2016) proposed an evaluation method considering context independence to determine whether the information contained in a summary sentence is complete. Peyrard and Eckle-Kohler (2017) proposed a modified pyramid method to automate manual evaluation. Zhu et al. (2018) designed an automatic evaluation algorithm for multimodal summarization technologies. Although automatic evaluation approaches are relatively behind other areas in research, we believe that they will achieve significant progress with the development of summarization methods.

Exercises

11.1 Please compare extractive and abstractive summarization from the perspectives of resource demand, model complexity, and performance.

11.2 The summary sentences generated by extractive methods are combined without logical order. Please design an algorithm to put the summary sentences in good order.

11.3 Abstractive summarization usually truncates the text to prevent the input to the model from being too long. Please explain why we need to do this and design an algorithm without this constraint to improve the summarization quality.

11.4 Multidocument summarization is usually more difficult than single-document summarization, and extractive methods are most often used. Please analyze the reasons behind this and design an algorithm combining extractive and abstractive methods for multidocument summarization.

11.5 Please list some scenarios where speech and vision modalities are necessary to generate better text summaries.

11.6 Summarization evaluation is very challenging because everyone can provide a different summary result for the same input document. Please comment on whether it is possible to design a good evaluation method without using reference summaries.

References

Abad, A., Nabi, M., & Moschitti, A. (2017). Self-crowdsourcing training for relation extraction. In *Proceedings of ACL*.

Aggarwal, C. C. (2018). *Machine learning for text*. Berlin: Springer.

Ahn, D. (2006). The stages of event extraction. In *Proceedings of TERQAS* (pp. 1–8).

Allan, J. (2012). *Topic detection and tracking: Event-based information organization*. New York, NY: Springer.

Allan, J., Carbonell, J., Doddington, G., Yamron, J., & Yang, Y. (1998a). Topic detection and tracking pilot study final report. In *Proceedings of the DARPA Broadcast News Transcription and Understanding Workshop* (pp. 194–218).

Allan, J., Lavrenko, V., & Jin, H. (2000). First story detection in tdt is hard. In *Proceedings of CIKM* (pp. 374–381).

Allan, J., Papka, R., & Lavrenko, V. (1998b). On-line new event detection and tracking. In *Proceedings of SIGIR* (pp. 37–45).

Andreevskaia, A., & Bergler, S. (2006). Mining wordnet for a fuzzy sentiment: Sentiment tag extraction from wordnet glosses. In *Proceedings of EACL* (pp. 209–216).

Angeli, G., Johnson Premkumar, M. J., & Manning, C. D. (2015). Leveraging linguistic structure for open domain information extraction. In *Proceedings of ACL and IJCNLP*.

Arora, S., Li, Y., Liang, Y., Ma, T., & Risteski, A. (2016). A latent variable model approach to PMI-based word embeddings. In *Transactions on ACL* (pp. 385–400).

Aue, A., & Gamon, M. (2005). Customizing sentiment classifiers to new domains: A case study. In *Proceedings of RANLP*.

Baccianella, S., Esuli, A., & Sebastiani, F. (2010). SentiWordNet 3.0: An enhanced lexical resource for sentiment analysis and opinion mining. In *Proceedings of LREC* (pp. 2200–2204).

Bagga, A., & Baldwin, B. (1998). Entity-based cross-document coreferencing using the vector space model. In *Proceedings of ACL-COLING*.

Banko, M., Cafarella, M. J., Soderland, S., Broadhead, M., & Etzioni, O. (2007). Open information extraction from the web. In *Proceedings of IJCAI*.

Barzilay, R., Elhadad, N., & McKeown, K. R. (2002). Inferring strategies for sentence ordering in multidocument news summarization. *Journal of Artificial Intelligence Research, 17*, 35–55.

Becker, H., Naaman, M., & Gravano, L. (2011). Beyond trending topics: Real-world event identification on twitter. In *Proceedings of ICWSM* (pp. 438–441).

Bekkerman, R., & Mccallum, A. (2005). Disambiguating web appearances of people in a social network. In *Proceedings of WWW* (pp. 463–470).

Bengio, Y., Ducharme, R., Vincent, P., & Janvin, C. (2003). A neural probabilistic language model. *Journal of Machine Learning Research, 3*, 1137–1155.

© Tsinghua University Press 2021

C. Zong et al., *Text Data Mining*, https://doi.org/10.1007/978-981-16-0100-2

Biadsy, F., Hirschberg, J., & Filatova, E. (2008). An unsupervised approach to biography production using wikipedia. In *Proceedings of ACL* (pp. 807–815).

Bickel, S., Brückner, M., & Scheffer, T. (2009). Discriminative learning under covariate shift. *Journal of Machine Learning Research, 10*(9), 2137–2155.

Bing, L., Li, P., Liao, Y., Lam, W., Guo, W., & Passonneau, R. J. (2015). Abstractive multi-document summarization via phrase selection and merging. In *Proceedings of ACL*.

Blair-Goldensohn, S., Hannan, K., McDonald, R., Neylon, T., Reis, G., & Reynar, J. (2008). Building a sentiment summarizer for local service reviews. In *Proceedings of WWW Workshop Track* (pp. 339–348).

Blei, D. M., & Lafferty, J. D. (2006). Dynamic topic models. In *Proceedings of ICML*.

Blei, D. M., Ng, A. Y., & Jordan, M. I. (2003). Latent dirichlet allocation. *Journal of Machine Learning Research, 3*, 993–1022.

Blitzer, J., Dredze, M., & Pereira, F. (2007). Biographies, bollywood, boom-boxes and blenders: Domain adaptation for sentiment classification. In *Proceedings of ACL* (pp. 440–447).

Bojanowski, P., Grave, E., Joulin, A., & Mikolov, T. (2017). Enriching word vectors with subword information. *Transactions of the Association for Computational Linguistics, 5*, 135–146.

Breiman, L. (1996). Bagging predictors. *Machine Learning, 24*(2), 123–140.

Brody, S., & Elhadad, N. (2010). An unsupervised aspect-sentiment model for online reviews. In *Proceedings of NAACL* (pp. 804–812).

Brown, T. B., Mann, B., Ryder, N., Subbiah, M., Kaplan, J., Dhariwal, P., et al. (2020). Language models are few-shot learners.

Carbonell, J., & Goldstein, J. (1998). The use of MMR, diversity-based reranking for reordering documents and producing summaries. In *Proceedings of SIGIR*.

Cataldi, M., Di Caro, L., & Schifanella, C. (2010). Emerging topic detection on twitter based on temporal and social terms evaluation. In *Proceedings of MDM/KDD* (pp. 1–10).

Chang, J., & Blei, D. (2009). Relational topic models for document networks. In *Artificial Intelligence and Statistics*.

Chen, P., Sun, Z., Bing, L., & Yang, W. (2017a). Recurrent attention network on memory for aspect sentiment analysis. In *Proceedings of EMNLP* (pp. 452–461).

Chen, S. F., & Goodman, J. (1999). An empirical study of smoothing techniques for language modeling. *Computer Speech and Language, 13*(4), 359–394.

Chen, X., Xu, L., Liu, Z., Sun, M., & Luan, H. (2015a). Joint learning of character and word embeddings. In *Proceeding of IJCAI*.

Chen, Y., Amiri, H., Li, Z., & Chua, T.-S. (2013). Emerging topic detection for organizations from microblogs. In *Proceedings of SIGIR* (pp. 43–52).

Chen, Y., Liu, S., Zhang, X., Liu, K., & Zhao, J. (2017b). Automatically labeled data generation for large scale event extraction. In *Proceedings of ACL*.

Chen, Y., Xu, L., Liu, K., Zeng, D., & Zhao, J. (2015b). Event extraction via dynamic multi-pooling convolutional neural networks. In *Proceedings of ACL*.

Chen, Y., & Zong, C. (2008). A structure-based model for Chinese organization name translation. *ACM Transactions on Asian Language Information Processing*. https://doi.org/10.1145/1330291.1330292

Chen, Z., Tamang, S., Lee, A., Li, X., Lin, W.-P., Snover, M. G., et al. (2010). Cuny-blender TAC-KBP2010 entity linking and slot filling system description. In *Theory and Applications of Categories*.

Cheng, X., & Zhu, Q. (2010). *Text mining principles*. Beijing: Science Press (in Chinese).

Chernyshevich, M. (2014). IHS R&D belarus: Cross-domain extraction of product features using CRF. In *Proceedings of SemEval* (pp. 309–313).

Cho, K., van Merriënboer, B., Gulcehre, C., Bahdanau, D., Bougares, F., Schwenk, H., et al. (2014). Learning phrase representations using RNN encoder–decoder for statistical machine translation. In *Proceedings of EMNLP*.

Choi, Y., & Cardie, C. (2008). Learning with compositional semantics as structural inference for subsentential sentiment analysis. In *Proceedings of EMNLP* (pp. 793–801).

Chopra, S., Auli, M., & Rush, A. M. (2016). Abstractive sentence summarization with attentive recurrent neural networks. In *Proceedings of ACL*.
Collins, M., & Duffy, N. (2002). Convolution kernels for natural language. In *Proceedings of NIPS*.
Collobert, R., & Weston, J. (2008). A unified architecture for natural language processing: Deep neural networks with multitask learning. In *Proceedings of ICML*.
Collobert, R., Weston, J., Bottou, L., Karlen, M., Kavukcuoglu, K., & Kuksa, P. (2011). Natural language processing (almost) from scratch. *The Journal of Machine Learning Research, 12*, 2493–2537.
Conneau, A., & Lample, G. (2019). Cross-lingual language model pretraining. *Advances in Neural Information Processing Systems, 32*, 7059–7069.
Connell, M., Feng, A., Kumaran, G., Raghavan, H., Shah, C., & Allan, J. (2004). UMass at TDT. In *Proceedings of TDT* (Vol. 19, pp. 109–155).
Cui, H., Mittal, V., & Datar, M. (2006). Comparative experiments on sentiment classification for online product reviews. In *Proceedings of AAAI*.
Culotta, A., & Sorensen, J. (2004). Dependency tree kernels for relation extraction. In *Proceedings of ACL*.
Dai, Z., Yang, Z., Yang, Y., Carbonell, J., Le, Q., & Salakhutdinov, R. (2019). Transformer-XL: Attentive language models beyond a fixed-length context. In *Proceedings of the 57th Annual Meeting of the Association for Computational Linguistics* (pp. 2978–2988).
Dang, H. T., & Owczarzak, K. (2008). Overview of the TAC 2008 update summarization task. In *Proceedings of TAC*.
Das, S., & Chen, M. (2001). Yahoo! for amazon: Extracting market sentiment from stock message boards. In *Proceedings of APFA, Bangkok, Thailand* (Vol. 35, p. 43)
Das, S. R., & Chen, M. Y. (2007). Yahoo! for amazon: Sentiment extraction from small talk on the web. *Management Science, 53*, 1375–1388.
Dave, K., Lawrence, S., & Pennock, D. (2003). Mining the peanut gallery: Opinion extraction and semantic classification of product reviews. In *Proceedings of WWW* (pp. 519–528).
Deerwester, S., Dumais, S. T., Furnas, G. W., Landauer, T. K., & Harshman, R. (1990). Indexing by latent semantic analysis. *Journal of the American Society for Information Science, 41*(6), 391–407.
Devlin, J., Chang, M.-W., Lee, K., & Toutanova, K. (2019). BERT: Pre-training of deep bidirectional transformers for language understanding. In *Proceedings of the 2019 Conference of the North American Chapter of the Association for Computational Linguistics: Human Language Technologies, Volume 1 (Long and Short Papers)* (pp. 4171–4186).
Diao, Q., Jiang, J., Zhu, F., & Lim, E.-P. (2012). Finding bursty topics from microblogs. In *Proceedings of ACL* (pp. 536–544).
Ding, X., & Liu, B. (2007). The utility of linguistic rules in opinion mining. In *Proceedings of SIGIR* (pp. 811–812).
Ding, X., Liu, B., & Yu, P. (2008). A holistic lexicon-based approach to opinion mining. In *Proceedings of the 2008 International Conference on Web Search and Data Mining* (pp. 231–240).
Ding, Y., Yu, J., & Jiang, J. (2017). Recurrent neural networks with auxiliary labels for cross-domain opinion target extraction. In *Proceedings of AAAI* (pp. 3436–3442).
Ding, Z., He, H., Zhang, M., & Xia, R. (2019). From independent prediction to reordered prediction: Integrating relative position and global label information to emotion cause identification. In *Proceedings of AAAI* (Vol. 33, pp. 6343–6350).
Ding, Z., Xia, R., & Yu, J. (2020). ECPE-2D: Emotion-cause pair extraction based on joint two-dimensional representation, interaction and prediction. In *Proceedings of ACL* (pp. 3161–3170). Stroudsburg: Association for Computational Linguistics.
Do, Q., Lu, W., & Roth, D. (2012). Joint inference for event timeline construction. In *Proceedings of IJCNLP and COLING*.
Dong, L., Wei, F., Tan, C., Tang, D., Zhou, M., & Xu, K. (2014). Adaptive recursive neural network for target-dependent twitter sentiment classification. In *Proceedings of ACL* (pp. 49–54).

Dong, L., Yang, N., Wang, W., Wei, F., Liu, X., Wang, Y., et al. (2019). Unified language model pre-training for natural language understanding and generation. In *Proceedings of NeurIPS*.

Dumais, S. T., Furnas, G. W., Landauer, T. K., Deerwester, S., & Harshman, R. (1988). Using latent semantic analysis to improve access to textual information. In *Proceedings of SIGCHI* (pp. 281–285).

Edmundson, H. P. (1969). New methods in automatic extracting. *Journal of the ACM (JACM), 16*(2), 264–285.

Ekman, P., Friesen, W. V., & Ellsworth, P. (1972). *Emotion in the human face: Guide-lines for research and an integration of findings: Guidelines for research and an integration of findings*. Elmsford, NY: Pergamon.

Erkan, G., & Radev, D. R. (2004). Lexrank: Graph-based lexical centrality as salience in text summarization. *Journal of Artificial Intelligence Research, 22*, 457–479.

Esuli, A., & Sebastiani, F. (2007). Pageranking wordnet synsets: An application to opinion mining. In *Proceedings of ACL* (pp. 424–431).

Fang, A., Macdonald, C., Ounis, I., & Habel, P. (2016). Using word embedding to evaluate the coherence of topics from twitter data. In *Proceedings of SIGIR* (pp. 1057–1060).

Feng, W., Zhang, C., Zhang, W., Han, J., Wang, J., Aggarwal, C., et al. (2015). Streamcube: Hierarchical spatio-temporal hashtag clustering for event exploration over the twitter stream. In *Proceedings of ICDE* (pp. 1561–1572).

Firth, J. R. (1957). A synopsis of linguistic theory. In F. R. Palmer (Ed.), *Studies in linguistic analysis*. Oxford: Philological Society.

Fleischman, M., & Hovy, E. (2004). Multi-document person name resolution. In *Proceedings of ACL*.

Forman, G. (2003). An extensive empirical study of feature selection metrics for text classification. *Journal of Machine Learning Research, 3*, 1289–1305.

Freund, Y., & Schapire, R. E. (1996). Experiments with a new boosting algorithm. In *Proceedings of ICML* (Vol. 96, pp. 148–156).

Fung, G. P. C., Yu, J. X., Yu, P. S., & Lu, H. (2005). Parameter free bursty events detection in text streams. In *Proceedings of VLDB* (pp. 181–192).

Gamon, M. (2004). Sentiment classification on customer feedback data: Noisy data, large feature vectors, and the role of linguistic analysis. In *Proceedings of COLING* (pp. 841–847). Stroudsburg: Association for Computational Linguistics.

Gan, Z., Pu, Y., Henao, R., Li, C., He, X., & Carin, L. (2017). Learning generic sentence representations using convolutional neural networks. In *Proceedings of EMNLP* (pp. 2390–2400).

Gers, F. A., Schraudolph, N. N., & Schmidhuber, J. (2002). Learning precise timing with LSTM recurrent networks. *Journal of Machine Learning Research, 3*, 115–143.

Girolami, M., & Kabán, A. (2003). On an equivalence between PLSI and LDA. In *Proceedings of SIGIR* (pp. 433–434).

Graves, A., Jaitly, N., & Mohamed, A.-R. (2013). Hybrid speech recognition with deep bidirectional LSTM. In *2013 IEEE workshop on automatic speech recognition and understanding* (pp. 273–278). New York: IEEE.

Griffiths, T. L., Jordan, M. I., Tenenbaum, J. B., & Blei, D. M. (2004). Hierarchical topic models and the nested Chinese restaurant process. In *Advances in Neural Information Processing Systems* (pp. 17–24).

Griffiths, T. L., & Steyvers, M. (2004). Finding scientific topics. *Proceedings of the National Academy of Sciences, 101*(Suppl. 1), 5228–5235.

Gu, J., Lu, Z., Li, H., & Li, V. O. (2016). Incorporating copying mechanism in sequence-to-sequence learning. In *Proceedings of ACL*.

Gui, L., Wu, D., Xu, R., Lu, Q., & Zhou, Y. (2016). Event-driven emotion cause extraction with corpus construction. In *Proceedings of EMNLP* (pp. 1639–1649). Singapore: World Scientific.

Han, J., Kamber, M., & Pei, J. (2012). *Data mining-concepts and techniques* (3rd ed.). Burlington: Morgan Kaufmann.

Han, X., Sun, L., & Zhao, J. (2011). Collective entity linking in web text: A graph-based method. In *Proceedings of SIGIR*.

Han, X., & Zhao, J. (2009a). Named entity disambiguation by leveraging wikipedia semantic knowledge. In *Proceedings of the 18th ACM Conference on Information and Knowledge Management* (pp. 215–224).

Han, X., & Zhao, J. (2009b). NLPR_KBP in TAC 2009 KBP track: A two-stage method to entity linking. In *Proceedings of TAC 2009 Workshop*.

Harris, Z. S. (1954). Distributional structure. *Word, 10*(2–3), 146–162.

Hashimoto, K., & Tsuruoka, Y. (2016). Adaptive joint learning of compositional and non-compositional phrase embeddings. In *Proceedings of ACL* (pp. 205–215).

Hatzivassiloglou, V., & McKeown, K. R. (1997). Predicting the semantic orientation of adjectives. In *Proceedings of EACL* (pp. 174–181). Stroudsburg: Association for Computational Linguistics.

He, Q., Chang, K., & Lim, E.-P. (2007a). Analyzing feature trajectories for event detection. In *Proceedings of SIGIR* (pp. 207–214).

He, Q., Chang, K., Lim, E.-P., & Zhang, J. (2007b). Bursty feature representation for clustering text streams. In *Proceedings of the 2007 SIAM International Conference on Data Mining* (pp. 491–496). Philadelphia: SIAM.

He, Z., Liu, S., Li, M., Zhou, M., Zhang, L., & Wang, H. (2013). Learning entity representation for entity disambiguation. In *Proceedings of the 51st Annual Meeting of the Association for Computational Linguistics (Volume 2: Short Papers)* (pp. 30–34).

Heinrich, G. (2005). Parameter estimation for text analysis. Technical report.

Hochreiter, S., & Schmidhuber, J. (1997). Long short-term memory. *Neural Computation, 9*(8), 1735–1780.

Hoffmann, R., Zhang, C., Ling, X., Zettlemoyer, L., & Weld, D. S. (2011). Knowledge-based weak supervision for information extraction of overlapping relations. In *Proceedings of the 49th Annual Meeting of the Association for Computational Linguistics: Human Language Technologies* (Vol. 1, pp. 541–550).

Hofmann, T. (1999). Probabilistic latent semantic indexing. In *Proceedings of SIGIR* (pp. 50–57).

Hu, M., & Liu, B. (2004). Mining and summarizing customer reviews. In *Proceedings of ACM SIGKDD* (pp. 168–177).

Hu, W., Zhang, J., & Zheng, N. (2016). Different contexts lead to different word embeddings. In *Proceedings of COLING 2016, the 26th International Conference on Computational Linguistics: Technical Papers* (pp. 762–771).

Huang, J., Gretton, A., Borgwardt, K., Schölkopf, B., & Smola, A. J. (2007). Correcting sample selection bias by unlabeled data. In *Advances in Neural Information Processing Systems* (pp. 601–608).

Huang, L., Fayong, S., & Guo, Y. (2012). Structured perceptron with inexact search. In *Proceedings of the 2012 Conference of the North American Chapter of the Association for Computational Linguistics: Human Language Technologies* (pp. 142–151).

Huang, L., & Huang, L. (2013). Optimized event storyline generation based on mixture-event-aspect model. In *Proceedings of the 2013 Conference on Empirical Methods in Natural Language Processing* (pp. 726–735).

Huang, X., Wan, X., & Xiao, J. (2011). Comparative news summarization using linear programming. In *Proceedings of the 49th Annual Meeting of the Association for Computational Linguistics: Human Language Technologies: Short Papers* (Vol. 2, pp. 648–653). Stroudsburg: Association for Computational Linguistics.

Huang, Z., Xu, W., & Yu, K. (2015). Bidirectional LSTM-CRF models for sequence tagging. Preprint, arXiv:1508.01991.

Ikeda, D., Takamura, H., Ratinov, L., & Okumura, M. (2008). Learning to shift the polarity of words for sentiment classification. In *Proceedings of IJCNLP* (pp. 296–303).

Inderjeet, M. (2001). *Automatic summarization*. Amsterdam: John Benjamins Publishing Co.

Irsoy, O., & Cardie, C. (2014). Deep recursive neural networks for compositionality in language. In *Advances in Neural Information Processing Systems* (pp. 2096–2104).

Jain, A., & Pennacchiotti, M. (2010). Open entity extraction from web search query logs. In *Proceedings of the 23rd International Conference on Computational Linguistics, COLING '10* (pp. 510–518). Cambridge: Association for Computational Linguistics.

Jakob, N., & Gurevych, I. (2010). Extracting opinion targets in a single and cross-domain setting with conditional random fields. In *Proceedings of EMNLP* (pp. 1035–1045). Cambridge, MA: Association for Computational Linguistics.

Jiang, J., & Zhai, C. (2007). Instance weighting for domain adaptation in NLP. In *Proceedings of ACL* (pp. 264–271).

Jiang, L., Yu, M., Zhou, M., Liu, X., & Zhao, T. (2011). Target-dependent twitter sentiment classification. In *Proceedings of NAACL* (pp. 151–160). Stroudsburg: Association for Computational Linguistics.

Jiao, X., Yin, Y., Shang, L., Jiang, X., Chen, X., Li, L., et al. (2019). Tinybert: Distilling bert for natural language understanding. Preprint, arXiv:1909.10351.

Jin, W., Ho, H. H., & Srihari, R. K. (2009). A novel lexicalized HMM-based learning framework for web opinion mining. In *Proceedings of the 26th Annual International Conference on Machine Learning* (Vol. 10). Citeseer.

Jo, Y., & Oh, A. H. (2011). Aspect and sentiment unification model for online review analysis. In *Proceedings of WSDM* (pp. 815–824).

Jurafsky, D., & Martin, J. H. (2008). *Speech and language processing: An introduction to natural language processing, computational linguistics, and speech recognition* (2nd ed.). Upper Saddle River: Prentice Hall.

Kalchbrenner, N., Grefenstette, E., & Blunsom, P. (2014). A convolutional neural network for modelling sentences. In *Proceedings of ACL* (pp. 655–665).

Kamps, J., Marx, M., Mokken, R. J., & De Rijke, M. (2004). Using wordnet to measure semantic orientations of adjectives. In *Proceedings of LREC* (Vol. 4, pp. 1115–1118). Citeseer.

Kanayama, H., & Nasukawa, T. (2006). Fully automatic lexicon expansion for domain-oriented sentiment analysis. In *Proceedings of EMNLP* (pp. 355–363).

Kennedy, A., & Inkpen, D. (2006). Sentiment classification of movie reviews using contextual valence shifters. *Computational Intelligence, 22*(2), 110–125.

Kim, S.-M., & Hovy, E. (2004). Determining the sentiment of opinions. In *Proceedings of COLING* (pp. 1367–1373). Stroudsburg: Association for Computational Linguistics.

Kim, Y. (2014). Convolutional neural networks for sentence classification. In *Proceedings of EMNLP* (pp. 1746–1751).

Kiritchenko, S., Zhu, X., Cherry, C., & Mohammad, S. (2014). NRC-Canada-2014: Detecting aspects and sentiment in customer reviews. In *Proceedings of SemEval 2014* (pp. 437–442).

Kiros, R., Zhu, Y., Salakhutdinov, R. R., Zemel, R., Urtasun, R., Torralba, A., et al. (2015). Skip-thought vectors. In *Advances in Neural Information Processing Systems* (pp. 3294–3302).

Kleinberg, J. (2003). Bursty and hierarchical structure in streams. *Data Mining and Knowledge Discovery, 7*(4), 373–397.

Knight, K., & Marcu, D. (2002). Summarization beyond sentence extraction: A probabilistic approach to sentence compression. *Artificial Intelligence, 139*(1), 91–107.

Kobayashi, N., Inui, K., & Matsumoto, Y. (2007). Extracting aspect-evaluation and aspect-of relations in opinion mining. In *Proceedings of EMNLP and CoNLL* (pp. 1065–1074).

Ku, L.-W., Liang, Y.-T., & Chen, H.-H. (2006). Opinion extraction, summarization and tracking in news and blog corpora. In *Proceedings of AAAI* (pp. 100–107).

Kumaran, G., & Allan, J. (2004). Text classification and named entities for new event detection. In *Proceedings of SIGIR* (pp. 297–304).

Kumaran, G., & Allan, J. (2005). Using names and topics for new event detection. In *Proceedings of HLT-EMNLP* (pp. 121–128). Vancouver, BC: Association for Computational Linguistics.

Kurisinkel, L. J., Mishra, P., Muralidaran, V., Varma, V., & Misra Sharma, D. (2016). Non-decreasing sub-modular function for comprehensible summarization. In *Proceedings of the NAACL Student Research Workshop* (pp. 94–101).

Lafferty, J., McCallum, A., & Pereira, F. C. (2001). Conditional random fields: Probabilistic models for segmenting and labeling sequence data. In *Proceedings of ICML*.

Lai, S., Liu, K., He, S., & Zhao, J. (2016). How to generate a good word embedding. *IEEE Intelligent Systems, 31*(6), 5–14.

Lan, Z., Chen, M., Goodman, S., Gimpel, K., Sharma, P., & Soricut, R. (2019). Albert: A lite bert for self-supervised learning of language representations. Preprint, arXiv:1909.11942.

Landauer, T. K. (2006). *Latent semantic analysis*. New York: Wiley.

Larkey, L. S., & Croft, W. B. (1996). Combining classifiers in text categorization. In *Proceedings of SIGIR* (pp. 289–297).

Lavrenko, V., & Croft, W. B. (2001). Relevance based language models. In *Proceedings of SIGIR, SIGIR '01* (pp. 120–127). New York, NY: Association for Computing Machinery.

LDC. (2005). ACE (automatic content extraction) English annotation guidelines for entities (version 5.5.1). https://www.ldc.upenn.edu/sites/www.ldc.upenn.edu/files/chinese-events-guidelines-v5.5.1.pdf.

Le, Q., & Mikolov, T. (2014). Distributed representations of sentences and documents. In *Proeceedings of ICML* (pp. 1188–1196).

Lee, R., & Sumiya, K. (2010). Measuring geographical regularities of crowd behaviors for twitter-based geo-social event detection. In *Proceedings of ACM SIGSPATIAL* (pp. 1–10).

Leek, T., Schwartz, R., & Sista, S. (2002). Probabilistic approaches to topic detection and tracking. In *Topic detection and tracking* (pp. 67–83). Berlin: Springer.

Li, B., Liu, T., Zhao, Z., Tang, B., Drozd, A., Rogers, A., et al. (2017a). Investigating different syntactic context types and context representations for learning word embeddings. In *Proceedings of the 2017 Conference on Empirical Methods in Natural Language Processing* (pp. 2421–2431).

Li, C., Liu, F., Weng, F., & Liu, Y. (2013a). Document summarization via guided sentence compression. In *Proceedings of the 2013 Conference on Empirical Methods in Natural Language Processing* (pp. 490–500).

Li, H. (2019). *Statistical machine learning* (2nd ed.). Beijing: Tsinghua University Press (in Chinese).

Li, H., Zhang, J., Zhou, Y., & Zong, C. (2016a). Guiderank: A guided ranking graph model for multilingual multi-document summarization. In *Proceedings of NLPCC* (pp. 608–620).

Li, H., Zhu, J., Ma, C., Zhang, J., & Zong, C. (2017b). Multi-modal summarization for asynchronous collection of text, image, audio and video. In *Proceedings of the 2017 Conference on Empirical Methods in Natural Language Processing* (pp. 1092–1102).

Li, J., Luong, M.-T., & Jurafsky, D. (2015). A hierarchical neural autoencoder for paragraphs and documents. In *Proceedings of ACL* (pp. 1106–1115).

Li, Q., & Ji, H. (2014). Incremental joint extraction of entity mentions and relations. In *Proceedings of the 52nd Annual Meeting of the Association for Computational Linguistics (Volume 1: Long Papers)* (pp. 402–412).

Li, Q., Ji, H., & Huang, L. (2013b). Joint event extraction via structured prediction with global features. In *Proceedings of the 51st Annual Meeting of the Association for Computational Linguistics (Volume 1: Long Papers)* (pp. 73–82).

Li, S., Chua, T.-S., Zhu, J., & Miao, C. (2016b). Generative topic embedding: A continuous representation of documents. In *Proceedings of the 54th Annual Meeting of the Association for Computational Linguistics* (pp. 666–675).

Li, S., & Huang, C.-R. (2009). Sentiment classification considering negation and contrast transition. In *Proceedings of PACLIC* (pp. 307–316).

Li, S., Lee, S. Y., Chen, Y., Huang, C.-R., & Zhou, G. (2010a). Sentiment classification and polarity shifting. In *Proceedings of COLING* (pp. 635–643).

Li, S., Xia, R., Zong, C., & Huang, C.-R. (2009a). A framework of feature selection methods for text categorization. In *Proceedings of ACL-IJCNLP* (pp. 692–700).

Li, T., Zhang, Y., & Sindhwani, V. (2009b). A non-negative matrix tri-factorization approach to sentiment classification with lexical prior knowledge. In *Proceedings of ACL-IJCNLP* (pp. 244–252). Stroudsburg: Association for Computational Linguistics.

Li, W., & McCallum, A. (2006). Pachinko allocation: DAG-structured mixture models of topic correlations. In *Proceedings of ICML* (pp. 577–584).

Li, X., Dong, Y., & Li, J. (2010b). *Data mining and knowledge discovering*. Beijing: High Education Press (in Chinese).
Li, X., & Lam, W. (2017). Deep multi-task learning for aspect term extraction with memory interaction. In *Proceedings of EMNLP* (pp. 616–626).
Li, Z., Wei, Y., Zhang, Y., & Yang, Q. (2018). Hierarchical attention transfer network for cross-domain sentiment classification. In *Proceedings of AAAI* (pp. 5852–5859).
Li, Z., Zhang, Y., Wei, Y., Wu, Y., & Yang, Q. (2017c). End-to-end adversarial memory network for cross-domain sentiment classification. In *Proceedings of IJCAI* (pp. 2237–2243).
Liao, W., & Veeramachaneni, S. (2009). A simple semi-supervised algorithm for named entity recognition. In *Proceedings of the NAACL HLT 2009 Workshop on Semi-Supervised Learning for Natural Language Processing* (pp. 58–65).
Lin, C., & He, Y. (2009). Joint sentiment/topic model for sentiment analysis. In *Proceedings of CIKM* (pp. 375–384).
Lin, C.-Y. (2004). ROUGE: A package for automatic evaluation of summaries. In *Text summarization branches out* (pp. 74–81).
Lin, J., Snow, R., & Morgan, W. (2011). Smoothing techniques for adaptive online language models: Topic tracking in tweet streams. In *Proceedings of ACM SIGKDD* (pp. 422–429).
Lin, Y., Shen, S., Liu, Z., Luan, H., & Sun, M. (2016). Neural relation extraction with selective attention over instances. In *Proceedings of the 54th Annual Meeting of the Association for Computational Linguistics (Volume 1: Long Papers)* (pp. 2124–2133).
Ling, W., Tsvetkov, Y., Amir, S., Fermandez, R., Dyer, C., Black, A. W., et al. (2015). Not all contexts are created equal: Better word representations with variable attention. In *Proceedings of the 2015 Conference on Empirical Methods in Natural Language Processing* (pp. 1367–1372).
Liu, B. (2011). *Web data mining: Exploring hyperlinks, contents, and usage data*. Berlin: Springer.
Liu, B. (2012). Sentiment analysis and opinion mining. *Synthesis Lectures on Human Language Technologies, 5*(1), 1–167.
Liu, B. (2015). *Sentiment analysis: Mining opinions, sentiments, and emotions*. Cambridge: Cambridge University Press.
Liu, F., Flanigan, J., Thomson, S., Sadeh, N., & Smith, N. A. (2015a). Toward abstractive summarization using semantic representations. In *Proceedings of the 2015 Conference of the North American Chapter of the Association for Computational Linguistics: Human Language Technologies* (pp. 1077–1086).
Liu, J., & Zhang, Y. (2017). Attention modeling for targeted sentiment. In *Proceedings of EACL* (pp. 572–577).
Liu, K. (2000). *Chinese text word segmentation and annotation*. Beijing: Commercial Press (in Chinese).
Liu, P., Joty, S., & Meng, H. (2015b). Fine-grained opinion mining with recurrent neural networks and word embeddings. In *Proceedings of EMNLP* (pp. 1433–1443).
Liu, Y., Li, Z., Xiong, H., Gao, X., & Wu, J. (2010). Understanding of internal clustering validation measures. In *2010 IEEE International Conference on Data Mining* (pp. 911–916). New York: IEEE.
Liu, Y., Ott, M., Goyal, N., Du, J., Joshi, M., Chen, D., et al. (2019). Roberta: A robustly optimized BERT pretraining approach. CoRR, abs/1907.11692.
Lovins, J. B. (1968). Development of a stemming algorithm. *Translation and Computational Linguistics, 11*(1), 22–31.
Luhn, H. P. (1958). The automatic creation of literature abstracts. *IBM Journal of Research and Development, 2*(2), 159–165.
Luo, B., Feng, Y., Wang, Z., Zhu, Z., Huang, S., Yan, R., et al. (2017). Learning with noise: Enhance distantly supervised relation extraction with dynamic transition matrix. In *Proceedings of ACL*.
Ma, D., Li, S., Zhang, X., & Wang, H. (2017). Interactive attention networks for aspect-level sentiment classification. In *Proceedings of IJCAI* (pp. 4068–4074).

Malin, W., Airoldi, E., & Carley, K. (2005). A network analysis model for disambiguation of names in lists. *Computational & Mathematical Organization Theory, 11*, 119–139.

Mann, G., & Yarowsky, D. (2003). Unsupervised personal name disambiguation. In *Proceedings of the Seventh Conference on Natural Language Learning at HLT-NAACL 2003* (pp. 33–40).

Manning, C. D., & Schütze, H. (1999). *Foundations of statistical natural language processing*. Cambridge: MIT Press.

Mao, G., Duan, L., & Wang, S. (2007). *Principles and algorithms on data mining*. Beijing: Tsinghua University Press (in Chinese).

Mao, Y., & Lebanon, G. (2007). Isotonic conditional random fields and local sentiment flow. In *Advances in Neural Information Processing Systems* (pp. 961–968).

Marcu, D. (2000). *The theory and practice of discorse parsing and summarization*. Cambridge: MIT Press.

Massoudi, K., Tsagkias, M., De Rijke, M., & Weerkamp, W. (2011). Incorporating query expansion and quality indicators in searching microblog posts. In *European Conference on Information Retrieval* (pp. 362–367). Berlin: Springer.

Mausam, Schmitz, M., Soderland, S., Bart, R., & Etzioni, O. (2012). Open language learning for information extraction. In *Proceedings of the 2012 Joint Conference on Empirical Methods in Natural Language Processing and Computational Natural Language Learning* (pp. 523–534).

Mcauliffe, J. D., & Blei, D. M. (2008). Supervised topic models. In *Advances in Neural Information Processing Systems* (pp. 121–128).

McCallum, A., Corrada-Emmanuel, A., & Wang, X. (2005). Topic and role discovery in social networks. https://scholarworks.umass.edu/cs_faculty_pubs

McCallum, A., & Li, W. (2003). Early results for named entity recognition with conditional random fields, feature induction and web-enhanced lexicons. In *Proceedings of the Seventh Conference on Natural Language Learning at HLT-NAACL 2003* (pp. 188–191).

McCallum, A., & Nigam, K. (1998). A comparison of event models for naive bayes text classification. In *AAAI-98 Workshop on Learning for Text Categorization* (Vol. 752, pp. 41–48). Citeseer.

McDonald, R., Hannan, K., Neylon, T., Wells, M., & Reynar, J. (2007). Structured models for fine-to-coarse sentiment analysis. In *Proceedings of the 45th Annual Meeting of the Association of Computational Linguistics* (pp. 432–439). Prague: Association for Computational Linguistics.

Mei, Q., Ling, X., Wondra, M., Su, H., & Zhai, C. (2007). Topic sentiment mixture: Modeling facets and opinions in weblogs. In *Proceedings of WWW* (pp. 171–180).

Mei, Q., & Zhai, C. (2001). A note on EM algorithm for probabilistic latent semantic analysis. In *Proceedings of CIKM*.

Mihalcea, R., & Tarau, P. (2004). Textrank: Bringing order into text. In *Proceedings of the 2004 Conference on Empirical Methods in Natural Language Processing* (pp. 404–411).

Mikolov, T., Chen, K., Corrado, G., & Dean, J. (2013a). Efficient estimation of word representations in vector space. In *Proceedings of ICLR Workshop Track*.

Mikolov, T., Karafiát, M., Burget, L., Černockỳ, J., & Khudanpur, S. (2010). Recurrent neural network based language model. In *Eleventh Annual Conference of the International Speech Communication Association*.

Mikolov, T., Sutskever, I., Chen, K., Corrado, G. S., & Dean, J. (2013b). Distributed representations of words and phrases and their compositionality. In *Advances in Neural Information Processing Systems* (pp. 3111–3119).

Minkov, E., Cohen, W. W., & Ng, A. Y. (2006). Contextual search and name disambiguation in email using graphs. In *Proceedings of SIGIR* (pp. 27–34).

Mintz, M., Bills, S., Snow, R., & Jurafsky, D. (2009). Distant supervision for relation extraction without labeled data. In *Proceedings of the Joint Conference of the 47th Annual Meeting of the ACL and the 4th International Joint Conference on Natural Language Processing of the AFNLP* (pp. 1003–1011).

Miwa, M., & Bansal, M. (2016). End-to-end relation extraction using LSTMs on sequences and tree structures. In *Proceedings of the 54th Annual Meeting of the Association for Computational Linguistics (Volume 1: Long Papers)* (pp. 1105–1116).

Mohammad, S., Kiritchenko, S., Sobhani, P., Zhu, X., & Cherry, C. (2016). SemEval-2016 task 6: Detecting stance in tweets. In *Proceedings of the 10th International Workshop on Semantic Evaluation (SemEval-2016)* (pp. 31–41). San Diego, CA: Association for Computational Linguistics.

Mohammad, S., Kiritchenko, S., & Zhu, X. (2013). NRC-Canada: Building the state-of-the-art in sentiment analysis of tweets. In *Proceedings of SemEval* (pp. 321–327). Atlanta, GA: Association for Computational Linguistics.

Mukherjee, A., & Liu, B. (2012). Aspect extraction through semi-supervised modeling. In *Proceedings of ACL* (pp. 339–348). Stroudsburg: Association for Computational Linguistics.

Mullen, T., & Collier, N. (2004). Sentiment analysis using support vector machines with diverse information sources. In *Proceedings of EMNLP* (pp. 412–418).

Na, J.-C., Sui, H., Khoo, C. S., Chan, S., & Zhou, Y. (2004). Effectiveness of simple linguistic processing in automatic sentiment classification of product reviews. In *Knowledge Organization and the Global Information Society: Proceedings of the Eighth International ISKO Conference* (pp. 49–54).

Nakagawa, T., Inui, K., & Kurohashi, S. (2010). Dependency tree-based sentiment classification using CRFs with hidden variables. In *Proceedings of NAACL* (pp. 786–794).

Nallapati, R., Zhai, F., & Zhou, B. (2017). Summarunner: A recurrent neural network based sequence model for extractive summarization of documents. In *Thirty-First AAAI Conference on Artificial Intelligence*.

Nallapati, R., Zhou, B., dos santos, C. N., Gulcehre, C., Xiang, B. (2016). Abstractive text summarization using sequence-to-sequence rnns and beyond. Preprint, arXiv:1602.06023.

Narasimhan, K., Yala, A., & Barzilay, R. (2016). Improving information extraction by acquiring external evidence with reinforcement learning. In *Proceedings of the 2016 Conference on Empirical Methods in Natural Language Processing* (pp. 2355–2365).

Nema, P., Khapra, M. M., Laha, A., & Ravindran, B. (2017). Diversity driven attention model for query-based abstractive summarization. In *Proceedings of the 55th Annual Meeting of the Association for Computational Linguistics (Volume 1: Long Papers)* (pp. 1063–1072).

Nenkova, A., & Passonneau, R. J. (2004). Evaluating content selection in summarization: The pyramid method. In *Proceedings of the Human Language Technology Conference of the North American Chapter of the Association for Computational Linguistics: HLT-NAACL 2004* (pp. 145–152).

Ng, A. Y., & Jordan, M. I. (2002). On discriminative vs. generative classifiers: A comparison of logistic regression and naive bayes. In *Advances in Neural Information Processing Systems* (pp. 841–848).

Ng, V., Dasgupta, S., & Arifin, S. N. (2006). Examining the role of linguistic knowledge sources in the automatic identification and classification of reviews. In *Proceedings of COLING/ACL* (pp. 611–618).

Nigam, K., McCallum, A. K., Thrun, S., & Mitchell, T. (2000). Text classification from labeled and unlabeled documents using EM. *Machine Learning, 39*(2–3), 103–134.

Orimaye, S. O., Alhashmi, S. M., & Siew, E.-G. (2012). Buy it-don't buy it: Sentiment classification on amazon reviews using sentence polarity shift. In *Proceedings of PRICAI* (pp. 386–399). Berlin: Springer.

Osborne, M. (2002). Using maximum entropy for sentence extraction. In *Proceedings of the ACL-02 Workshop on Automatic Summarization-Volume 4* (pp. 1–8). Stroudsburg: Association for Computational Linguistics.

Page, L., & Brin, S. (1998). The anatomy of a large-scale hypertextual web search engine. *Computer Networks and ISDN Systems, 30*(17), 107–117.

Paice, C. D. (1990). Another stemmer. *ACM SIGIR Forum, 24*(3), 56–61.

Pan, S. J., Ni, X., Sun, J.-T., Yang, Q., & Chen, Z. (2010a). Cross-domain sentiment classification via spectral feature alignment. In *Proceedings of WWW* (pp. 751–760).

Pan, S. J., Tsang, I. W., Kwok, J. T., & Yang, Q. (2010b). Domain adaptation via transfer component analysis. *IEEE Transactions on Neural Networks, 22*(2), 199–210.

Pan, S. J., & Yang, Q. (2009). A survey on transfer learning. *IEEE Transactions on Knowledge and Data Engineering, 22*(10), 1345–1359.
Pang, B., & Lee, L. (2004). A sentimental education: Sentiment analysis using subjectivity summarization based on minimum cuts. In *Proceedings of ACL* (pp. 271–278). Stroudsburg: Association for Computational Linguistics.
Pang, B., & Lee, L. (2008). Opinion mining and sentiment analysis. *Foundations and Trends in Information Retrieval, 2*(1–2), 1–135.
Pang, B., Lee, L., & Vaithyanathan, S. (2002). Thumbs up? Sentiment classification using machine learning techniques. In *Proceedings of EMNLP* (pp. 79–86). Stroudsburg: Association for Computational Linguistics.
Papineni, K., Roukos, S., Ward, T., & Zhu, W.-J. (2002). Bleu: A method for automatic evaluation of machine translation. In *Proceedings of the 40th Annual Meeting of the Association for Computational Linguistics* (pp. 311–318).
Pedersen, T., Purandare, A., & Kulkarni, A. (2005). Name discrimination by clustering similar contexts. In *International Conference on Intelligent Text Processing and Computational Linguistics* (pp. 226–237).
Peng, N., Poon, H., Quirk, C., Toutanova, K., & Yih, W.-T. (2017). Cross-sentence N-ary relation extraction with graph LSTMs. *Transactions of the Association for Computational Linguistics, 5*, 101–115.
Pennacchiotti, M., & Pantel, P. (2009). Entity extraction via ensemble semantics. In *Proceedings of the 2009 Conference on Empirical Methods in Natural Language Processing* (Vol. 1, pp. 238–247). Stroudsburg: Association for Computational Linguistics.
Peters, M., Neumann, M., Iyyer, M., Gardner, M., Clark, C., Lee, K., et al. (2018). Deep contextualized word representations. In *Proceedings of the 2018 Conference of the North American Chapter of the Association for Computational Linguistics: Human Language Technologies, Volume 1 (Long Papers)* (pp. 2227–2237).
Petrov, S., & McDonald, R. (2012). Overview of the 2012 shared task on parsing the web. In *Notes of the First Workshop on Syntactic Analysis of Non-Canonical Language (SANCL)*.
Petrović, S., Osborne, M., & Lavrenko, V. (2010). Streaming first story detection with application to twitter. In *Proceedings of NAACL-HLT, HLT '10* (pp. 181–189). Stroudsburg: Association for Computational Linguistics.
Peyrard, M., & Eckle-Kohler, J. (2017). Supervised learning of automatic pyramid for optimization-based multi-document summarization. In *Proceedings of the 55th Annual Meeting of the Association for Computational Linguistics (Volume 1: Long Papers)* (pp. 1084–1094).
Phuvipadawat, S., & Murata, T. (2010). Breaking news detection and tracking in twitter. In *Proceedings of IEEE/WIC/ACM WI-IAT, WI-IAT '10* (pp. 120–123). New York: IEEE Computer Society.
Pinter, Y., Guthrie, R., & Eisenstein, J. (2017). Mimicking word embeddings using subword RNNs. In *Proceedings of EMNLP* (pp. 102–112).
Platt, J. (1998). Sequential minimal optimization: A fast algorithm for training support vector machines (pp. 212–223).
Plutchik, R., & Kellerman, H. (1986). *Emotion: Theory, research and experience. Volume 3 in biological foundations of emotions*. Oxford: Pergamon.
Polanyi, L., & Zaenen, A. (2006). Contextual valence shifters. In *Computing attitude and affect in text: Theory and applications* (pp. 1–10). Berlin: Springer.
Popescu, A.-M., & Etzioni, O. (2007). Extracting product features and opinions from reviews. In *Natural language processing and text mining* (pp. 9–28). Berlin: Springer.
Popescu, A.-M., & Pennacchiotti, M. (2010). Detecting controversial events from twitter. In *Proceedings of CIKM, CIKM '10* (pp. 1873–1876). New York, NY: Association for Computing Machinery.
Popescu, A.-M., Pennacchiotti, M., & Paranjpe, D. (2011). Extracting events and event descriptions from twitter. In *Proceedings of WWW* (pp. 105–106).
Porter, M. F. (1980). An algorithm for suffix stripping. *Program, 14*(3), 130–137.

Qian, Q., Huang, M., Lei, J., & Zhu, X. (2017). Linguistically regularized LSTM for sentiment classification. In *Proceedings of ACL* (pp. 1679–1689).

Qiu, G., Liu, B., Bu, J., & Chen, C. (2011). Opinion word expansion and target extraction through double propagation. *Computational Linguistics, 37*(1), 9–27.

Qiu, X., Sun, T., Xu, Y., Shao, Y., Dai, N., & Huang, X. (2020). Pre-trained models for natural language processing: A survey. Preprint, arXiv:2003.08271.

Rabiner, L., & Juang, B. (1986). An introduction to hidden markov models. *IEEE ASSP Magazine, 3*(1), 4–16.

Radford, A., Narasimhan, K., Salimans, T., & Sutskever, I. (2018). Improving language understanding by generative pre-training. https://s3-us-west-2.amazonaws.com/openai-assets/research-covers/language-unsupervised/language_understanding_paper.pdf

Radford, A., Wu, J., Child, R., Luan, D., Amodei, D., & Sutskever, I. (2019). Language models are unsupervised multitask learners. Technical report, OpenAI.

Ramage, D., Hall, D., Nallapati, R., & Manning, C. D. (2009). Labeled LDA: A supervised topic model for credit attribution in multi-labeled corpora. In *Proceedings of EMNLP* (pp. 248–256). Stroudsburg: Association for Computational Linguistics.

Ratinov, L., & Roth, D. (2009). Design challenges and misconceptions in named entity recognition. In *Proceedings of the Thirteenth Conference on Computational Natural Language Learning (CoNLL-2009)* (pp. 147–155).

Riedel, S., Yao, L., & McCallum, A. (2010). Modeling relations and their mentions without labeled text. In *Proceedings of ECML* (pp. 148–163).

Rush, A. M., Chopra, S., & Weston, J. (2015). A neural attention model for abstractive sentence summarization. In *Proceedings of EMNLP*.

Salton, G., Wong, A., & Yang, C.-S. (1975). A vector space model for automatic indexing. *Communications of the ACM, 18*(11), 613–620.

Sanh, V., Debut, L., Chaumond, J., & Wolf, T. (2019). Distilbert, a distilled version of bert: Smaller, faster, cheaper and lighter. Preprint, arXiv:1910.01108.

Sarawagi, S. (2008). Information extraction. *Foundations and Trends in Databases, 1*(3), 261–377.

Schapire, R., & Singer, Y. (2000). Boostexter: A boosting-based system for text categorization. *Machine Learning, 39*(2), 135–168.

Schuster, M., & Paliwal, K. K. (1997). Bidirectional recurrent neural networks. *IEEE transactions on Signal Processing, 45*(11), 2673–2681.

Sebastiani, F. (2002). Machine learning in automated text categorization. *ACM Computing Surveys (CSUR), 34*(1), 1–47.

See, A., Liu, P., & Manning, C. (2017). Get to the point: Summarization with pointer-generator networks. In *Proceedings of ACL*.

Shen, W., Wang, J., & Han, J. (2015). Entity linking with a knowledge base: Issues, techniques, and solutions. *IEEE Transactions on Knowledge and Data Engineering, 27*(2), 443–460.

Shimodaira, H. (2000). Improving predictive inference under covariate shift by weighting the log-likelihood function. *Journal of Statistical Planning and Inference, 90*(2), 227–244.

Snyder, B., & Barzilay, R. (2007). Multiple aspect ranking using the good grief algorithm. In *Proceedings of NAACL* (pp. 300–307). Rochester, NY: Association for Computational Linguistics.

Socher, R., Huval, B., Manning, C. D., & Ng, A. Y. (2012). Semantic compositionality through recursive matrix-vector spaces. In *Proceedings of EMNLP* (pp. 1201–1211). Stroudsburg: Association for Computational Linguistics.

Socher, R., Lin, C. C., Manning, C., & Ng, A. Y. (2011a). Parsing natural scenes and natural language with recursive neural networks. In *Proceedings of ICML* (pp. 129–136).

Socher, R., Pennington, J., Huang, E. H., Ng, A. Y., & Manning, C. D. (2011b). Semi-supervised recursive autoencoders for predicting sentiment distributions. In *Proceedings of the Conference on Empirical Methods in Natural Language Processing* (pp. 151–161). Stroudsburg: Association for Computational Linguistics.

Socher, R., Perelygin, A., Wu, J., Chuang, J., Manning, C. D., Ng, A. Y., et al.(2013). Recursive deep models for semantic compositionality over a sentiment treebank. In *Proceedings of the 2013 Conference on Empirical Methods in Natural Language Processing* (pp. 1631–1642).

Song, K., Tan, X., Qin, T., Lu, J., & Liu, T.-Y. (2019). Mass: Masked sequence to sequence pre-training for language generation. In *Proceedings of ICML*.

Stanovsky, G., & Dagan, I. (2016). Creating a large benchmark for open information extraction. In *Proceedings of the 2016 Conference on Empirical Methods in Natural Language Processing* (pp. 2300–2305).

Steyvers, M., Smyth, P., Rosen-Zvi, M., & Griffiths, T. (2004). Probabilistic author-topic models for information discovery. In *Proceedings of ACM SIGKDD* (pp. 306–315).

Strapparava, C., & Valitutti, A. (2004). Wordnet affect: An affective extension of wordnet. In *Proceedings of LREC* (Vol. 4, p. 40). Citeseer.

Sugiyama, M., Nakajima, S., Kashima, H., Buenau, P. V., & Kawanabe, M. (2008). Direct importance estimation with model selection and its application to covariate shift adaptation. In *Advances in Neural Information Processing Systems* (pp. 1433–1440).

Sun, Y., Lin, L., Tang, D., Yang, N., Ji, Z., & Wang, X. (2015). Modeling mention, context and entity with neural networks for entity disambiguation. In *Twenty-Fourth International Joint Conference on Artificial Intelligence*.

Sun, Y., Wang, S., Li, Y., Feng, S., Chen, X., Zhang, H., et al. (2019). ERNIE: Enhanced representation through knowledge integration. Preprint, arXiv:1904.09223.

Sun, Y., Wang, S., Li, Y., Feng, S., Tian, H., Wu, H., et al. (2020). ERNIE 2.0: A continual pre-training framework for language understanding. In *Proceedings of AAAI*.

Surdeanu, M., Tibshirani, J., Nallapati, R., & Manning, C. D. (2012). Multi-instance multi-label learning for relation extraction. In *Proceedings of the 2012 Joint Conference on Empirical Methods in Natural Language Processing and Computational Natural Language Learning* (pp. 455–465).

Sutton, C., & McCallum, A. (2012). An introduction to conditional random fields. *Foundations and Trends® in Machine Learning, 4*(4), 267–373.

Suzuki, J., & Isozaki, H. (2008). Semi-supervised sequential labeling and segmentation using giga-word scale unlabeled data. In *Proceedings of ACL-08: HLT* (pp. 665–673).

Taboada, M., Brooke, J., Tofiloski, M., Voll, K., & Stede, M. (2011). Lexicon-based methods for sentiment analysis. *Computational Linguistics, 37*(2), 267–307.

Tai, K. S., Socher, R., & Manning, C. D. (2015). Improved semantic representations from tree-structured long short-term memory networks. *Proceedings of ACL and IJCNLP* (pp. 1556–1565).

Tan, J., Wan, X., & Xiao, J. (2017). Abstractive document summarization with a graph-based attentional neural model. In *Proceedings of the 55th Annual Meeting of the Association for Computational Linguistics (Volume 1: Long Papers)* (pp. 1171–1181).

Tang, D., Qin, B., Feng, X., & Liu, T. (2015a). Effective LSTMs for target-dependent sentiment classification. *Proceedings of COLING* (pp. 3298–3307).

Tang, D., Qin, B., & Liu, T. (2015b). Document modeling with gated recurrent neural network for sentiment classification. In *Proceedings of the 2015 Conference on Empirical Methods in Natural Language Processing* (pp. 1422–1432).

Tang, D., Qin, B., & Liu, T. (2016). Aspect level sentiment classification with deep memory network. *Proceedings of EMNLP* (pp. 3298–3307).

Tang, D., Wei, F., Qin, B., Zhou, M., & Liu, T. (2014a). Building large-scale twitter-specific sentiment lexicon: A representation learning approach. In *Proceedings of COLING* (pp. 172–182).

Tang, D., Wei, F., Yang, N., Zhou, M., Liu, T., & Qin, B. (2014b). Learning sentiment-specific word embedding for twitter sentiment classification. In *Proceedings of ACL* (pp. 1555–1565).

Tang, R., Lu, Y., Liu, L., Mou, L., Vechtomova, O., & Lin, J. (2019). Distilling task-specific knowledge from bert into simple neural networks. Preprint, arXiv:1903.12136.

Thet, T. T., Na, J.-C., & Khoo, C. S. G. (2010). Aspect-based sentiment analysis of movie reviews on discussion boards. *Journal of Information Science, 36*, 823–848.

Tissier, J., Gravier, C., & Habrard, A. (2017). Dict2vec: Learning word embeddings using lexical dictionaries. In *Proceedings of EMNLP* (pp. 254–263).

Titov, I., & McDonald, R. (2008). A joint model of text and aspect ratings for sentiment summarization. In *Proceedings of ACL* (pp. 308–316). Columbus, OH: Association for Computational Linguistics.

Toh, Z., & Wang, W. (2014). DLIREC: Aspect term extraction and term polarity classification system. In *Association for Computational Linguistics and Dublin City University* (pp. 235–240). Citeseer.

Turner, J., & Charniak, E. (2005). Supervised and unsupervised learning for sentence compression. In *Proceedings of the 43rd Annual Meeting on Association for Computational Linguistics* (pp. 290–297). Stroudsburg: Association for Computational Linguistics.

Turney, P. D. (2002). Thumbs up or thumbs down? Semantic orientation applied to unsupervised classification of reviews. In *Proceedings of ACL* (pp. 417–424). Stroudsburg: Association for Computational Linguistics.

Turney, P. D., & Littman, M. L. (2003). Measuring praise and criticism: Inference of semantic orientation from association. *ACM Transactions on Information Systems (TOIS)*, 21(4):315–346.

Vawani, A., Shazeer, N., Parmar, N., Uszkoreit, J., Jones, L., Gomez, A. N., et al. (2017). Attention is all you need. In *Proceedings of NeurIPS*.

Vo, D.-T., & Zhang, Y. (2015). Target-dependent twitter sentiment classification with rich automatic features. In *Proceedings of IJCAI* (pp. 1347–1353).

Vo, D. T., & Zhang, Y. (2016). Don't count, predict! an automatic approach to learning sentiment lexicons for short text. In *Proceedings of ACL* (pp. 219–224). Berlin: Association for Computational Linguistics.

Wan, X. (2011). Using bilingual information for cross-language document summarization. In *Proceedings of the 49th Annual Meeting of the Association for Computational Linguistics: Human Language Technologies-Volume 1* (pp. 1546–1555). Stroudsburg: Association for Computational Linguistics.

Wan, X., Li, H., & Xiao, J. (2010). Cross-language document summarization based on machine translation quality prediction. In *Proceedings of the 48th Annual Meeting of the Association for Computational Linguistics* (pp. 917–926). Stroudsburg: Association for Computational Linguistics.

Wang, K., & Xia, R. (2015). An approach to Chinese sentiment lexicon construction based on conjunction relation. In *Proceedings of CCL*.

Wang, K., & Xia, R. (2016). A survey on automatical construction methods of sentiment lexicons. *Acta Automatica Sinica, 42*(4), 495–511.

Wang, K., Zong, C., & Su, K.-Y. (2012). Integrating generative and discriminative character-based models for chinese word segmentation. *ACM Transactions on Asian Language Information Processing, 11*(2), 1–41.

Wang, L., & Xia, R. (2017). Sentiment lexicon construction with representation learning based on hierarchical sentiment supervision. In *Proceedings of EMNLP* (pp. 502–510).

Wang, S., Zhang, J., Lin, N., & Zong, C. (2018). Investigating inner properties of multimodal representation and semantic compositionality with brain-based componential semantics. In *Proceedings of AAAI* (pp. 5964–5972).

Wang, S., Zhang, J., & Zong, C. (2017a). Exploiting word internal structures for generic Chinese sentence representation. In *Proceedings of EMNLP* (pp. 298–303).

Wang, S., Zhang, J., & Zong, C. (2017b). Learning sentence representation with guidance of human attention. In *Proceedings of IJCAI* (pp. 4137–4143).

Wang, S., & Zong, C. (2017). Comparison study on critical components in composition model for phrase representation. *ACM Transactions on Asian and Low-Resource Language Information Processing (TALLIP), 16*(3), 1–25.

Wang, W. Y., Mehdad, Y., Radev, D. R., & Stent, A. (2016a). A low-rank approximation approach to learning joint embeddings of news stories and images for timeline summarization. In *Proceedings of ACL* (pp. 58–68).

Wang, X., & McCallum, A. (2006). Topics over time: A non-markov continuous-time model of topical trends. In *Proceedings of ACM SIGKDD* (pp. 424–433).

Wang, Y., Huang, H.-Y., Feng, C., Zhou, Q., Gu, J., & Gao, X. (2016b). CSE: Conceptual sentence embeddings based on attention model. In *Proceedings of the 54th Annual Meeting of the Association for Computational Linguistics* (pp. 505–515).

Wang, Y., Huang, M., Zhu, X., & Zhao, L. (2016c). Attention-based LSTM for aspect-level sentiment classification. In *Proceedings of EMNLP* (pp. 606–615).

Wang, Z., Zhang, J., Feng, J., & Chen, Z. (2014). Knowledge graph and text jointly embedding. In *Proceedings of the 2014 Conference on Empirical Methods in Natural Language Processing (EMNLP)* (pp. 1591–1601).

Whitehead, M., & Yaeger, L. (2010). Sentiment mining using ensemble classification models. In *Innovations and advances in computer sciences and engineering* (pp. 509–514). Berlin: Springer.

Whitelaw, C., Garg, N., & Argamon, S. (2005). Using appraisal groups for sentiment analysis. In *Proceedings of CIKM* (pp. 625–631).

Wiebe, J., Wilson, T., Bruce, R., Bell, M., & Martin, M. (2004). Learning subjective language. *Computational Linguistics, 30*(3), 277–308.

Wiebe, J. M., Bruce, R. F., & O'Hara, T. P. (1999). Development and use of a gold-standard data set for subjectivity classifications. In *Proceedings of the 37th Annual Meeting of the Association for Computational Linguistics* (pp. 246–253). College Park, MD: Association for Computational Linguistics.

Wieting, J., & Gimpel, K. (2017). Revisiting recurrent networks for paraphrastic sentence embeddings. In *Proceedings of ACL* (pp. 2078–2088).

Wilson, T., Wiebe, J., & Hoffmann, P. (2005). Recognizing contextual polarity in phrase-level sentiment analysis. In *Proceedings of HLT-EMNLP* (pp. 347–354).

Wu, X., Kumar, V., Ross, J., Joydeep, Q., Yang, G. Q., Motoda, H., et al. (2008). Top 10 algorithms in data mining. *Knowledge and Information Systems, 14*, 1–37.

Wu, Y., Bamman, D., & Russell, S. (2017). Adversarial training for relation extraction. In *Proceedings of the 2017 Conference on Empirical Methods in Natural Language Processing* (pp. 1778–1783).

Xia, R., & Ding, Z. (2019). Emotion-cause pair extraction: A new task to emotion analysis in texts. In *Proceedings of ACL* (pp. 1003–1012).

Xia, R., Hu, X., Lu, J., Yang, J., & Zong, C. (2013a). Instance selection and instance weighting for cross-domain sentiment classification via PU learning. In *Proceedings of IJCAI* (pp. 2176–2182).

Xia, R., Pan, Z., & Xu, F. (2018). Instance weighting for domain adaptation via trading off sample selection bias and variance. In *Proceedings of IJCAI* (pp. 4489–4495). Palo Alto, CA: AAAI Press.

Xia, R., Wang, C., Dai, X., & Li, T. (2015a). Co-training for semi-supervised sentiment classification based on dual-view bags-of-words representation. In *Proceedings of ACL-IJCNLP* (pp. 1054–1063).

Xia, R., Wang, T., Hu, X., Li, S., & Zong, C. (2013b). Dual training and dual prediction for polarity classification. In *Proceedings of ACL* (pp. 521–525).

Xia, R., Xu, F., Yu, J., Qi, Y., & Cambria, E. (2016). Polarity shift detection, elimination and ensemble: A three-stage model for document-level sentiment analysis. *Information Processing & Management, 52*(1), 36–45.

Xia, R., Xu, F., Zong, C., Li, Q., Qi, Y., & Li, T. (2015b). Dual sentiment analysis: Considering two sides of one review. *IEEE Transactions on Knowledge and Data Engineering, 27*(8), 2120–2133.

Xia, R., Yu, J., Xu, F., & Wang, S. (2014). Instance-based domain adaptation in NLP via in-target-domain logistic approximation. In *Proceedings of AAAI* (pp. 1600–1606).

Xia, R., Zhang, M., & Ding, Z. (2019). RTHN: A RNN-transformer hierarchical network for emotion cause extraction. In *Proceedings of IJCAI* (pp. 5285–5291). Palo Alto, CA: AAAI Press.

Xia, R., & Zong, C. (2011). A POS-based ensemble model for cross-domain sentiment classification. In *Proceedings of IJCNLP* (pp. 614–622).
Xia, R., Zong, C., & Li, S. (2011). Ensemble of feature sets and classification algorithms for sentiment classification. *Information Sciences, 181*(6), 1138–1152.
Xu, J., Liu, J., Zhang, L., Li, Z., & Chen, H. (2016). Improve Chinese word embeddings by exploiting internal structure. In *Proceedings of the 2016 Conference of the North American Chapter of the Association for Computational Linguistics: Human Language Technologies* (pp. 1041–1050).
Xue, G.-R., Dai, W., Yang, Q., & Yu, Y. (2008). Topic-bridged PLSA for cross-domain text classification. In *Proceedings of SIGIR* (pp. 627–634).
Yaghoobzadeh, Y., & Schütze, H. (2016). Intrinsic subspace evaluation of word embedding representations. In *Proceedings of ACL* (pp. 236–246).
Yamron, J. P., Knecht, S., & Mulbregt, P. V. (2000). Dragon's tracking and detection systems for the TDT2000 evaluation. In *Proceedings of the Broadcast News Transcription and Understanding Workshop* (pp. 75–79).
Yan, R., Wan, X., Otterbacher, J., Kong, L., Li, X., & Zhang, Y. (2011). Evolutionary timeline summarization: A balanced optimization framework via iterative substitution. In *Proceedings of the 34th International ACM SIGIR Conference on Research and Development in Information Retrieval* (pp. 745–754).
Yang, Y., & Liu, X. (1999). A re-examination of text categorization methods. In *Proceedings of SIGIR* (pp. 42–49).
Yang, Y., & Pedersen, J. O. (1997). A comparative study on feature selection in text categorization. In *Proceedings of ICML, Nashville, TN, USA* (Vol. 97, pp. 35).
Yang, Y., Pierce, T., & Carbonell, J. (1998). A study of retrospective and on-line event detection. In *Proceedings of SIGIR* (pp. 28–36).
Yang, Z., Dai, Z., Yang, Y., Carbonell, J., Salakhutdinov, R. R., & Le, Q. V. (2019). XLNet: Generalized autoregressive pretraining for language understanding. In *Advances in Neural Information Processing Systems* (Vol. 32, pp. 5753–5763).
Yang, Z., Yang, D., Dyer, C., He, X., Smola, A., & Hovy, E. (2016). Hierarchical attention networks for document classification. In *Proceedings of NAACL* (pp. 1480–1489).
Yao, J., Wan, X., & Xiao, J. (2017). Recent advances in document summarization. *Knowledge and Information Systems, 53*(2), 297–336.
Yu, H., Zhang, Y., Ting, L., & Sheng, L. (2007). Topic detection and tracking review. *Journal of Chinese Information Processing, 6*(21), 77–79.
Yu, J. (2017). *Machine learning: From axiom to algorithm*. Beijing: Tsinghua University Press (in Chinese).
Yu, J., & Jiang, J. (2016). Learning sentence embeddings with auxiliary tasks for cross-domain sentiment classification. In *Proceedings of EMNLP* (pp. 236–246).
Yu, J., Zha, Z.-J., Wang, M., & Chua, T.-S. (2011). Aspect ranking: Identifying important product aspects from online consumer reviews. In *Proceedings of NAACL* (pp. 1496–1505). Stroudsburg: Association for Computational Linguistics.
Yu, M., & Dredze, M. (2015). Learning composition models for phrase embeddings. *Transactions of the Association for Computational Linguistics, 3*, 227–242.
Zadrozny, B. (2004). Learning and evaluating classifiers under sample selection bias. In *Proceedings of ICML* (pp. 114–121).
Zelenko, D., Aone, C., & Richardella, A. (2003). Kernel methods for relation extraction. *Journal of Machine Learning Research, 3*, 1083–1106.
Zeng, D., Liu, K., Chen, Y., & Zhao, J. (2015). Distant supervision for relation extraction via piecewise convolutional neural networks. In *Proceedings of the 2015 Conference on Empirical Methods in Natural Language Processing* (pp. 1753–1762).
Zeng, D., Liu, K., Lai, S., Zhou, G., & Zhao, J. (2014). Relation classification via convolutional deep neural network. In *Proceedings of COLING*.
Zhang, J., Liu, S., Li, M., Zhou, M., & Zong, C. (2014). Bilingually-constrained phrase embeddings for machine translation. In *Proceedings of ACL*.

Zhang, J., Zhou, Y., & Zong, C. (2016a). Abstractive cross-language summarization via translation model enhanced predicate argument structure fusing. *IEEE/ACM Transactions on Audio, Speech, and Language Processing, 24*(10), 1842–1853.

Zhang, M., Zhang, Y., & Fu, G. (2017). End-to-end neural relation extraction with global optimization. In *Proceedings of the 2017 Conference on Empirical Methods in Natural Language Processing* (pp. 1730–1740).

Zhang, M., Zhang, Y., & Vo, D.-T. (2016b). Gated neural networks for targeted sentiment analysis. In *Proceedings of AAAI* (pp. 3087–3093).

Zhang, M., Zhou, G., & Aw, A. (2008). Exploring syntactic structured features over parse trees for relation extraction using kernel methods. *Information Processing & Management, 44*(2), 687–701.

Zhang, X. (2016). *Pattern recognition* (3rd ed.). Beijing: Tsinghua University Press (in Chinese).

Zhang, X., Zhao, J., & LeCun, Y. (2015). Character-level convolutional networks for text classification. In *Advances in Neural Information Processing Systems* (pp. 649–657).

Zhang, Z. (2014). Research and implementation of sentiment analysis methods on Chinese weibo. Master Thesis (in Chinese).

Zhang, Z., Han, X., Liu, Z., Jiang, X., Sun, M., & Liu, Q. (2019). ERNIE: Enhanced language representation with informative entities. Preprint, arXiv:1905.07129.

Zhao, W. X., Jiang, J., Weng, J., He, J., Lim, E.-P., Yan, H., et al. (2011). Comparing twitter and traditional media using topic models. In *European Conference on Information Retrieval* (pp. 338–349). Berlin: Springer.

Zhao, W. X., Jiang, J., Yan, H., & Li, X. (2010). Jointly modeling aspects and opinions with a MaxEnt-LDA hybrid. In *Proceedings of EMNLP* (pp. 56–65). Stroudsburg: Association for Computational Linguistics.

Zheng, S., Wang, F., Bao, H., Hao, Y., Zhou, P., & Xu, B. (2017). Joint extraction of entities and relations based on a novel tagging scheme. In *Proceedings of ACL*.

Zhou, G., & Su, J. (2002). Named entity recognition using an HMM-based chunk tagger. In *Proceedings of the 40th Annual Meeting on Association for Computational Linguistics* (pp. 473–480).

Zhou, G., Su, J., Zhang, J., & Zhang, M. (2005). Exploring various knowledge in relation extraction. In *Proceedings of the 43rd Annual Meeting on Association for Computational Linguistics* (pp. 427–434).

Zhou, L., Zhang, J., & Zong, C. (2019). Synchronous bidirectional neural machine translation. *Transactions of the Association for Computational Linguistics, 7*, 91–105.

Zhou, Q., Yang, N., Wei, F., & Zhou, M. (2017). Selective encoding for abstractive sentence summarization. In *Proceedings of ACL*.

Zhou, Z. (2016). *Machine learning*. Beijing: Tsinghua University Press (in Chinese).

Zhu, J., Li, H., Liu, T., Zhou, Y., Zhang, J., & Zong, C. (2018). MSMO: Multimodal summarization with multimodal output. In *Proceedings of the 2018 Conference on Empirical Methods in Natural Language Processing* (pp. 4154–4164).

Zhu, J., Zhou, Y., Zhang, J., Li, H., Zong, C., & Li, C. (2020). Multimodal summarization with guidance of multimodal reference. In *Proceedings of AAAI*.

Zhuang, L., Jing, F., & Zhu, X.-Y. (2006). Movie review mining and summarization. In *Proceedings of CIKM* (pp. 43–50). New York, NY: Association for Computing Machinery.

Zong, C. (2013). *Statistical natural language processing* (2nd ed.). Beijing: Tsinghua University Press (in Chinese).

Printed in the United States
by Baker & Taylor Publisher Services